5th International Conference on Advances in Materials, Machinery, Electronics (AMME 2022)

MATEC Web of Conferences Volume 363 (2022)

Dali, China
13-15 May 2022

Editors:

Lin Liu
Jiaru Wu

ISBN: 978-1-7138-5943-7

Printed from e-media with permission by:

Curran Associates, Inc.
57 Morehouse Lane
Red Hook, NY 12571

Some format issues inherent in the e-media version may also appear in this print version.

No changes have been made to the content of these proceedings. There may be changes to pagination and minor adjustments for aesthetics.

Printed with permission by Curran Associates, Inc. (2023)

For additional information, please contact EDP Sciences – Web of Conferences at the address below.

EDP Sciences – Web of Conferences
17, Avenue du Hoggar
Parc d'Activité de Courtabœuf
BP 112
F-91944 Les Ulis Cedex A
France

Phone: +33 (0) 1 69 18 75 75
Fax: +33 (0) 1 69 28 84 91

contact-edps@webofconferences.org

Additional copies of this publication are available from:

Curran Associates, Inc.
57 Morehouse Lane
Red Hook, NY 12571 USA
Phone: 845-758-0400
Fax: 845-758-2633
Email: curran@proceedings.com
Web: www.proceedings.com

TABLE OF CONTENTS

Unveiling the HER and ORR activity origin of isolated Co sites supported on N-doped carbon

Aijian Huang and Zhiguo Wang*

School of Electronics Science and Engineering, University of Electronic Science and Technology of China, Chengdu, 610054, P.R. China.

Abstract. Electrocatalysis plays very important role in clean energy conversion. In which, developing high active and robust electrocatalysts for the hydrogen evolution reaction (HER) and oxygen reduction reaction (ORR) are highly crucial and challenging. Using density functional theory (DFT) calculations, the Gibbs free energy diagrams and electronic structure of N-graphene, Ir-N4 and Co-N4 are investigated in this work. The results show that Co-N4 have the appropriate H adsorption and lower ORR overpotentials, which gives expectation of high multifunctional electrocatalytic performance. This theoretical study provides vital insights into the enhanced electrocatalytic mechanism of Co sites supported on N-doped carbon.

1. Introduction

Electrochemical reactions, including hydrogen evolution reaction (HER) and oxygen reduction reaction (ORR) are significantly essential in energy-related applications and future renewable energy technologies. Rational developing the highly active and robust electrocatalysis are crucial and desirable. [1]. Among various promising alternatives, single-atom catalysts (SACs) with the isolated active sites have attracted tremendous interest, owing to its unique coordination geometry, strong metal–support interaction and maximum atomic efficiency [2]. The above characteristics also offer ideal configurations for exploring the origin of the activity. Though great efforts have been devoted, the electronic structure of SACs has remained elusive and the electrocatalytic activity origin of SACs are still not clear.
Recently, single-atom iridium catalysts [3] and single-atom cobalt on N-doped carbon [4] catalysts has been synthesized and exhibited excellent electrochemical ability for ORR. Meanwhile, single-atom catalysts also show good performance in application of HER [5]. Hence, gain deep insights into the electronic structure and catalytic mechanism of these catalysts are vitally important. In this work, we investigate the Gibbs free energy diagrams and electronic structure of pure N-doped graphene (N-graphene), isolated iridium on N-doped carbon (Ir-N4) and isolated cobalt on N-doped carbon (Co-N4). The results show that Co-N4 have the appropriate H adsorption and lower overpotentials for ORR. Meanwhile, unique electronic interactions between dopants of Co and N form near the Fermi level, ensuring fast charge transfer kinetics and guaranteeing excellent electrocatalytic performance.

2. Computational methods

All calculations here are performed by spin polarized density functional theory (DFT) methods using Vienna Ab initio Simulation Package (VASP) [6]. The Projector augmented wave (PAW) method is adopted to describe electron-ion interaction [7]. And Perdew-Burke-Ernzerhof (PBE) exchange-correlation functional within a generalized gradient approximation (GGA) is employed, while a 520 eV cut off energy for the plane-wave basis set is used for the valence electrons. The self-consistent filed (SCF) calculations are performed with an energy and force convergence of 10-5 eV and 0.02 eV Å-1, respectively. To avoid the interactions between two adjacent periodic images, the vacuum thickness was set to be 15 Å. All atomic configurations are analyzed using the VESTA code [8]. In order to model the Ir-N4 and Co-N4 monolayers, the 6×6 supercell consisting of 72 carbon atoms is used firstly as the substrate. Then, the Ir/Co atoms are anchored via displacing two neighboring carbon atoms, and four carbon atoms bond with metal atom are replaced by N atom, forming the Ir-N4 and Co-N4 catalyst model.
For HER, the integral adsorption energy is defined by:

$$\Delta E_H = E(\text{surface+H}) - E(\text{surface}) - \frac{1}{2}E(H_2) \quad (1)$$

where the E(surface+H) and E(surface) are the total energy of metal surface with H atoms adsorption and without, respectively. E(H2) is the total energy of hydrogen molecule in the gaseous phase. The Gibbs free energy of H atoms (ΔGH) is an important parameter for the descriptor of HER activity:

$$\Delta G_H = \Delta E_H + \Delta ZPE - T\Delta S \quad (2)$$

where ZPE is the zero-point vibrational energy using the harmonic approximation, T is the kelvin temperature, and

* Corresponding author: zgwang@uestc.edu.cn

https://doi.org/10.1051/matecconf/202236301001

S is the entropy. According to thermodynamics, large positive values of ΔGH indicate that the hydrogen atom is difficult to bind to the catalytic sites, whereas large negative values of ΔGH indicate that molecular hydrogen is difficult to be released from the catalytic sites [9].

The reaction steps considered for the electrochemical reduction of ORR under base condition are generally reported to proceed as follow:

O2(g) + * → *O2

*O2 + H2O(l) + e- → *OOH + OH-

*OOH + e- → *O + OH-

*O + H2O(l) + e- → *OH + OH-

OH*+ e- → * + OH-

where * represents either the catalytic active sites of vacant surface, or intermediate species adsorbed on the active sites. By considering the zero-point energy (ZPE) and entropy corrections, the Gibbs free energy for OER can be calculated with the following equation [10]:

$$\Delta G=\Delta E+\Delta ZPE-T\Delta S \quad (3)$$

where ΔE indicates the adsorption energy difference for each species adsorbed on the catalyst, ΔZPE and ΔS are the zone point energy and entropy difference between the adsorbed state and corresponding free-standing state, respectively.

3. Results and discussion.

We perform the density functional theory (DFT) calculation to understand the HER intrinsic activity of N-graphene, Ir-N4 and Co-N4. Various active sites are considered and we calculate their Gibbs free energy for H (ΔGH), which is widely accepted as the HER activity descriptor that has an optimal value of zero when the HER activity reaches the maximum [11]. For N-graphene (Figure 1a), H atoms thermodynamically preferred to adsorb on N sites owing to their lower ΔGH (-1.41 eV), the lower value of ΔGH also suggest the molecular hydrogen is difficult to be released on N-graphene and not active for hydrogen evolution. As shown in Figure 1b, the values of ΔGH on Ir-N4 increased to -0.37 eV getting much closer to optimum value (0 eV), which indicates that doping of Ir greatly assisted the HER performance. Remarkably, the ΔGH on Co-N4 is 0.15 eV (Figure 1c), which is better than N-graphene and Ir-N4. The catalytic activity is also related to the positions of i0 and ΔGH, the closer the position of these values to the peak, the better the catalyst is. As can be seen from Fig. 1d, the Co site on Co-N4 locate near the peak of the volcanic curve indicating Co-N4 is more catalytically active for HER.

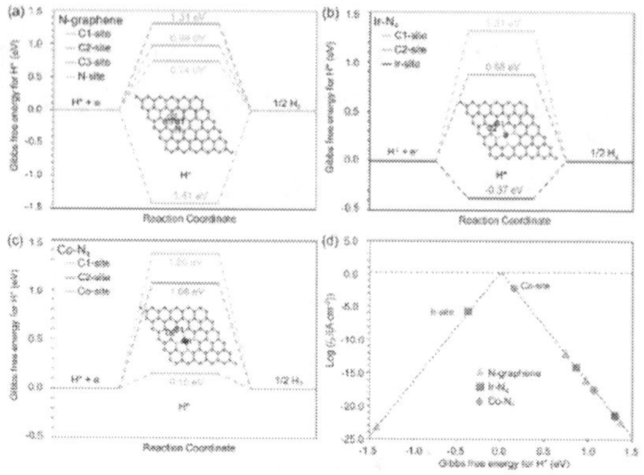

Figure 1. Gibbs free energy diagrams of hydrogen adsorption on different active sites of (a) N-graphene, (b) Ir-N_4 and (c) Co-N_4. (d) The calculated exchange current (log (i_0)) for hydrogen evolution over different configurations as a function of Gibbs free energy for H*.

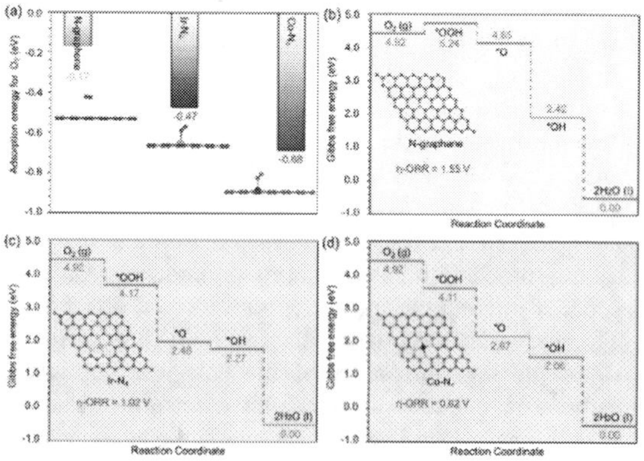

Figure 2. (a) The adsorption energies and corresponding configurations of O_2 on N-graphene, Ir-N_4 and Co-N_4. The calculated Gibbs free energy evolution diagrams for through a $4e^-$ ORR pathway on different active sites of (b) N-graphene, (c) Ir-N_4 and (d) Co-N_4.

For oxygen reduction reaction (ORR), the metal active sites will adsorb oxygen molecules in the first step and the negative adsorption energies of O_2 reveal that the above structures can spontaneously activate O_2 molecules (Figure 2a). The Co-N_4 show lower O_2 adsorption than that of N-graphene and Ir-N_4, demonstrating the better activation ability of O_2 molecules. However, due to the weak adsorption of OOH species, the N-graphene site shows a high overpotential of 1.55 V (Figure 2b). After the introducing of Ir and Co, the overpotential on Ir-N_4 and Co-N_4 structures decline to 1.02 V, and 0.62 V, respectively (Figure 2c and 2d), suggesting that the doping of Ir and Co are beneficial to optimize the adsorption energies of ORR intermediates. Here, the Co-N_4 shows lowest overpotential, confirming its better ORR electrocatalytic performance.

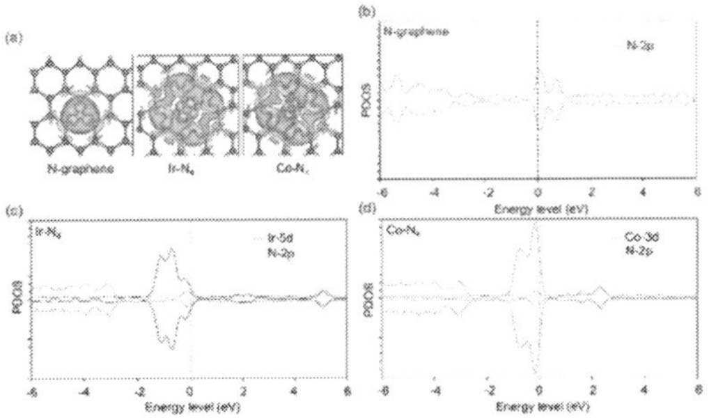

Figure 3. (a) Charge density differences of N-graphene, Ir-N_4 and Co-N_4 configurations. The isosurface levels are set to 0.003 e $Å^{-3}$, where charge depletion and accumulation were depicted by cyan and yellow, respectively. Projected density of state (PDOS) analysis of (b) N-graphene, (c) Ir-N_4 and (d) Co-N_4. The Fermi level is set to be zero and the spin-polarization is considered.

In order to gain deep insights into the catalytic mechanism and electronic structure. we further investigate the projected density of states (PDOS) and charge density of different catalyst configurations. Interestingly, the interaction between Co and N result in unique electronic structure for Co-N_4 (Figure 3a), more charge depletion region exists. And as can be observed from the PDOS (Figure 3b,3c and 3d), for Co-N_4, there are more hybridization effect between Co 3d and N 2p orbitals compared with that of pure N-graphene and Ir-N_4, which leads to more electron overlaps in the vicinity of Fermi level, demonstrating fast charge transfer kinetics and guaranteeing excellent electrocatalytic performance. Based on the above theoretical analysis, we give expectation of high HER and ORR electrocatalytic performance of Co-N_4.

4. Conclusion

In summary, DFT calculations have been carried out to investigate the Gibbs free energy diagrams and electronic structure of N-graphene, Ir-N_4 and Co-N_4. The results here show that Co-N_4 show appropriate H adsorption (0.15 eV) and lower ORR overpotentials (0.62 V) than that of N-graphene and Ir-N_4, which shows expectation of high multifunctional electrocatalytic performance. This theoretical study gives vital insights into the enhanced catalytic mechanism of Co-N_4 and also helps to institute the principles for designing high performance SACs.

References

1. Z. W. Seh, J. Kibsgaard, C. F. Dickens, I. Chorkendorff, J. K. Norskov,T. F. Jaramillo, Combining theory and experiment in electrocatalysis: Insights into materials design, Science 355 (2017)
2. S. K. Kaiser, Z. Chen, D. Faust Akl, S. Mitchell ,J. Perez-Ramirez, Single-Atom Catalysts across the Periodic Table, Chem Rev 120 (2020) 11703-809.
3. J. Z. Meiling Xiao+, Gaoran Li, Na Li, Shuang Li, Zachary Paul Cano, Lu Ma,, P. X. Peixin Cui, Gaopeng Jiang, Huile Jin, Shun Wang,* Tianpin Wu,* Jun Lu, Aiping Yu, ,a. Z. C. Dong Su, A Single-Atom Iridium Heterogeneous Catalyst in Oxygen Reduction Reaction, Angew. Chem. Int. Ed. (2019) 9640 –5.
4. X. Xie, C. He, B. Li, Y. He, D. A. Cullen, E. C. Wegener, A. J. Kropf, U. Martinez, Y. Cheng, M. H. Engelhard, M. E. Bowden, M. Song, T. Lemmon, X. S. Li, Z. Nie, J. Liu, D. J. Myers, P. Zelenay, G. Wang, G. Wu, V. Ramani ,Y. Shao, Performance enhancement and degradation mechanism identification of a single-atom Co–N–C catalyst for proton exchange membrane fuel cells, Nature Catalysis 3 (2020) 1044-54.
5. Y. Pan, S. Liu, K. Sun, X. Chen, B. Wang, K. Wu, X. Cao, W. C. Cheong, R. Shen, A. Han, Z. Chen, L. Zheng, J. Luo, Y. Lin, Y. Liu, D. Wang, Q. Peng, Q. Zhang, C. Chen ,Y. Li, A Bimetallic Zn/Fe Polyphthalocyanine-Derived Single-Atom Fe-N4 Catalytic Site:A Superior Trifunctional Catalyst for Overall Water Splitting and Zn-Air Batteries, Angew Chem Int Ed Engl 57 (2018) 8614-8.
6. G. Kresse, Efficient iterative schemes for ab initio total-energy calculations using a plane-wave basis set, PHYSICAL REVIEW B 54 (1996) 11169e86.
7. G. K. J. Furthmu¨ller, Efficient iterative schemes for ab initio total-energy calculations using a plane-wave basis set, PHYSICAL REVIEW B 54 (1999) 11169-86.
8. K. Momma,F. Izumi, VESTA: a three-dimensional visualization system for electronic and structural analysis, Journal of Applied Crystallography 41 (2008) 653-8.
9. J. K. Nørskov, T. Bligaard, A. Logadottir, J. R. Kitchin, J. G. Chen, S. Pandelov ,U. Stimming,

Trends in the Exchange Current for Hydrogen Evolution, J Electrochem Soc 152 (2005)

10. M. Li, L. Zhang, Q. Xu, J. Niu ,Z. Xia, N-doped graphene as catalysts for oxygen reduction and oxygen evolution reactions: Theoretical considerations, Journal of Catalysis 314 (2014) 66-72.
11. J. K. Nørskov, T. Bligaard, A. Logadottir, J. R. Kitchin, J. G. Chen, S. Pandelov ,U. Stimming, Trends in the Exchange Current for Hydrogen Evolution, Journal of The Electrochemical Society 152 (2005) J23.

Study of AC - 20 asphalt mixture ratio on the runway of of Chang Sha Airport

Shuaituan Tian [1, 2], Zhiyan Li[1,2], Lei Guo [1, 2, *], Yapei Yan [1, 2]

[1] China Airport Construction Group Co., Ltd, Beijing, 100101, China
[2] Beijing Super-Creative Technology Co., Ltd, Beijing, 100621, China

Abstract. After years of use, there's oil loss on the asphalt pavement. the bond strength of aggregate decreases. Loose threshing is serious. In order to improve the pavement performance, Extend the service life of pavement, Improve the safety of aircraft operation. We test the AC-20 modified asphalt mixture on the taxiway and runway of Chang Sha Airport from many factors , including raw materials, admixtures, gradation design, optimum ratio of oil to stone, then we find the best ways to resolve it .

Keyword: AC-20; raw materials; admixtures, gradation design

1. Raw materials

1.1 Aggregate

There are limestone aggregates, We teste the aggregate indexes in accordance with the relevant regulations, the test results are shown in the table below.

Table 1. The Basalt Coarse aggregate technical indicators

Test items		Test results	Technical requirements	Test method
Bulk specific gravity	15~20mm	2.773	≥2.50	T 0304
	10~15mm	2.729		
	5~10mm	2.764		
	3~5mm	2.740		
Apparent specific gravity	10~20mm	2.810	≥2.50	T 0304
	10~15mm	2.775		
	5~10mm	2.813		
	3~5mm	2.780		

From the chart, we can see all the technical indicators of the limestone coarse aggregate meet the specification requirements, and they can be used in the design and engineering.

Table 2. The Basalt Coarse aggregate Particle gradation

mesh size (mm)	P (%)							
	10 ~ 20mm		10 ~ 15mm		5 ~ 10mm		3 ~ 5mm	
	Technical requirements	Test results	Technical requirements	Test results	Technical requirements	Test results	Technical requirements	Test results
31.5	100		-		-		-	
26.5	100	100	100	100	-		-	
19	90 ~ 100	82.1	100	100	-		-	
16	-	36.2	100	100	-	100	-	
13.2	-	6.2	90 ~ 100	84.3	100	100	-	
9.5	0 ~ 15	1.4	0 ~ 15	8.4	90 ~ 100	90.8	100	100
4.75	0 ~ 5	0.1	0 ~ 5	0.7	0 ~ 15	5.4	90 ~ 100	94.8
2.36	-	0.6	-	0.5	0 ~ 5	1.4	0 ~ 15	52.8
1.18	-	0.6	-	0.5	-	1.1	-	32.9
0.6	-	0.6	-	0.5	-	1.1	0 ~ 3	22.1
0.3	-	0.6	-	0.5	-	1.1	-	16.3
0.15	-	0.6	-	0.5	-	1.1	-	10.9
0.075	-	0.6	-	0.5	-	1.1	-	7.7

From the chart, we can see some passing rate of limestone coarse aggregate can not meet the requirements of specification. We should strictly control the quality of the aggregate in the construction in order to make the passing rate and meet the specification requirements.

Table 3. The Basalt FINE aggregate technical indicators

Test items	Technical requirements	Test results
		0~3mm
Apparent specific gravity (g/cm^3)	≥2.50	2.728
Sand equivalent (%)	≥60	78
Sturdiness (%)	≤12	5.0

Table 4. The Basalt FINE aggregate Particle gradation

mesh size (mm)	P (%)	
	0 ~ 3mm	
	Technical requirements	Test results
9.5		100
4.75	100	99.7
2.36	80 ~ 100	77.7
1.18	50 ~ 80	57.5
0.6	25 ~ 60	45.5
0.3	8 ~ 45	33.4
0.15	0 ~ 25	23.2
0.075	0 ~ 15	17.5

From the chart, we can see some passing rate of limestone fine aggregate can not meet the requirements of specification. We should strictly control the quality of the aggregate in the construction in order to make the passing rate and meet the specification requirements.

* Corresponding author: 420069269@qq.com

1.2 Filler

The filler is milled limestone powder, they all meet the technical requirements, the test results are shown in the table below.

Table 4. The Mineral filler technical indicators

Test items		Technical requirements	Test results
Apparent specific gravity (g/cm³)		≥2.50	2.649
Water content (%)		≤1	0.2
Particle gradation	< 0.6mm	100	100
	< 0.15mm	90 ~ 100	93.2
	< 0.075mm	75 ~ 100	80.1
Hydrophilic coefficient		≤1	0.7

From the chart, we can see filler technical indexes meets the specification requirements, it can be used in the design and engineering.

1.3 Asphalt

There are Cnooc modified asphalt, We test the asphalt indexes in accordance with the relevant regulations, they all meet the technical requirements.

Table 5. Cnooc modified asphalt performance test results

Test items		Technical requirements	Test results
Penetration (25℃, 100g, 5s) (0.1mm)		40~60	53
Softening point (℃)		≥75	93.0
Ductility (5cm/min, 10℃) (cm)		≥20	29
Flash point (COC) (℃)		> 230	357
Density (25℃) (g/cm³)		Measured	1.024
Elastic recovery (25℃)		≥75	97
Filmy heating operational test163℃/5h	Mass loss (%)	≤±0.8	-0.07
	Penetration ratio (%)	≥65	92
	Ductility (10℃) (cm)	≥15	24

Test results show that the modified asphalt technology indexes meet the requirements.

1.4 Anti-rutting agent

Add 0.5% of the asphalt mixture quality anti-rutting agent can gain higher dynamic stability of mixture. Anti rutting agent basic indexes are testsd in table.

Table 5. Anti rutting agent basic indexes test results

Test items	Technical requirements	Test results
Density (g/cm³)	0.9~1.1	0.92
Melt flow rate (190℃, 2.16kg) (g/10min)	≥0.3	1.17
Water content (%)	< 0.5	0.1

From the chart, we can see anti-rutting agent of all the indicators meet the technical requirements and it can be used in the design and engineering.

1.5 Fiber

Fiber can get oil Absorption and reinforcement. Add 0.4% of the asphalt mixture quality fiber can gain higher dynamic stability and water damage resistance of mixture. Fiber agent basic indexes are testsd in table.

Table 6. Fiber basic indexes test results

Test items	Technical requirements	Test results
Diameter (μm)	14~20	165
Length (mm)	6±1.5	6.6
Tensile strength (MPa)	≥900	1035
Elongation at break (%)	8~12	10.2
Heat resistance (℃)	≥210°C, no change in volume	no change in volume

From the chart, we can see fiber of all the indicators meet the technical requirements and it can be used in the design and engineering.

2. Mix design of AC-20 asphalt mixture

2.1 Aggregate gradation design

The mix design is adopted for the mineral aggregate gradation of skeleton dense structure. The matching is in table.

Table 7. Ac-20 ratio of mineral aggregate gradation

specifications	10 ~ 20mm	10 ~ 15mm	5 ~ 10mm	3 ~ 5mm	0 ~ 3mm	Mineral powder
The percentage (%)	20	16	24	23	12	5

Synthetic mineral aggregate gradation is shown in table and figure.

Table 8. AC-20 mineral synthesis aggregate gradation

Mesh size (mm)	19	16	13.2	9.5	4.75	2.36	1.18	0.6	0.3	0.15	0.075
upper	100	90	80	72	58	46	34	27	20	14	8
lower	90	75	62	52	38	23	18	12	7	4	3
middle	95	82.5	71	62	48	34.5	26	19.5	13.5	7	5.5
synthetic	96.4	87.2	78.7	63.4	40.4	27.0	19.9	16.0	13.2	10.4	8.3

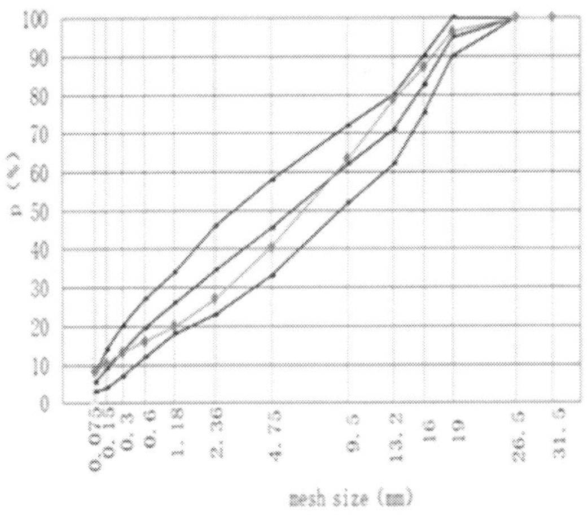

Figure 1. AC-20 grading curve

2.2 The determination of the optimum proportion

We select 5 asphalt aggregate ratio of Marshall test and calculate their physical indicators in order to determine the optimum proportion , the test results are shown in the table and figure below.

Table 9. The Marshall test results of different asphalt aggregate ratio

Asphalt-aggregate ratio (%)	Theoretical density (g/cm³)	Bulk density (g/cm³)	VV (%)	VMA (%)	VFA (%)	MS (KN)	流值 (0.1mm)
3.2	2.644	2.491	5.8	13.3	56.6	14.99	34.60
3.7	2.624	2.496	4.9	13.6	64.2	14.48	36.00
4.2	2.605	2.502	4.0	13.8	71.3	13.89	31.48
4.7	2.586	2.509	3.0	14.0	78.7	14.72	39.85
5.2	2.567	2.518	1.9	14.1	86.4	13.64	39.30
Technical requirements	—	—	3 ~ 5	≥13	65 ~ 75	> 9	20 ~ 40

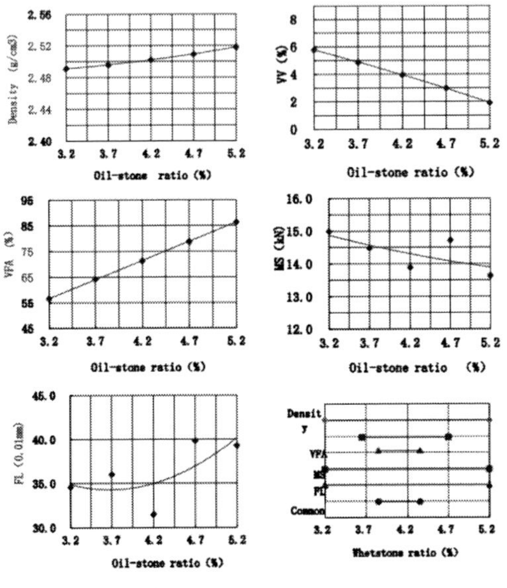

Figure 2. The determination of AC-20 optimum proportion

According to the requirements of the relevant specification, calculated the optimum proportion of 4.2%.

2.3 Road performance verification

we make road performance of verification AC-20 asphalt mixture in the optimum asphalt aggregate ratio of 4.2%, the test results are shown in the table below.

Table 10. The Road performance verification test results

Test items	Test results	Technical requirements
MS (KN)	13.9	> 9
FL (0.1mm)	31.5	20 ~ 40
MSo (%)	98.1	≥90
TSR (%)	88.3	≥85
DS (time/mm)	14540	≥10000
Cw (mL/min)	No seepage	–

MS is strength index, MSo and TSR are water damage resistance index, DS is high temperature performance index, Cw is water permeability index.

From the above test results, all the indexes can meet the requirements of related technologies in the optimum proportion 4.2%.

Adding anti-rutting agent quality of 0.5% asphalt mixture, dynamic stability of asphalt mixture arrives at 14540 times/mm, MS arrives at 13.9 (KN) , FL arrives at 31.5 (0.1mm) , MSo arrives at 98.1 %, TSR arrives at 88.3%, Cw arrives at no seepage. They all meet the design requirements. we can see that the mixture has formed the skeleton dense structure from the profile of the specimen, it meets with the skeleton dense type AC asphalt mixture design intent.

3. Conclusion

For the analysis of AC - 20 mix proportion design of asphalt mixture, in order to improve the high temperature performance of asphalt mixture, we can adopt the following measures.

I) We should strictly control each index of the aggregate, all indicators are within the specification limits, including aggregate, filler and asphalt. The important indicators of aggregate are bulk specific gravity, apparent specific gravity, sand equivalent, sturdiness and p%; The important indicators of filler are apparent specific gravity, water content, Particle gradation and hydrophilic coefficient; The important indicators of asphalt are penetration, softening point, ductility, equivalent softening point, equivalent brittle pointT1.2, flash point, density, elastic recovery and filmy heating operational test163℃/5h.

II) We should be reasonable to add additives, such as anti-rutting agent and fiber. The important indicators of anti-rutting agent are density, melt flow rate (190°C, 2.16kg) , water content and softening point; The important indicators of fiber are diameter, length, tensile strength, elongation at break and heat resistance.

Ⅲ) it is better to choose skeleton dense type of grading, such as 10 ~ 20mm: 10 ~ 15mm: 5 ~ 10mm: 3 ~ 5mm: 0 ~ 3mm:fiber=20:16:24:23:12:5.

References

1. The civil aviation administration of China. JTG F40-2004. Specification for asphalt concrete pavement construction of civil airports [S].
2. The ministry of communications highway engineering science institute. JTG F40-2004. Technical specification for construction of highway asphalt pavement [S].
3. The ministry of communications highway engineering science institute. JTJ E20-2011.Standard test methods of bitumen and bituminous mixtures for highway engineering [S].
4. The ministry of communications highway engineering science institute. JTG E42-2005. Test methods of aggregate for Highway Engineering [S].
5. Xiaoming,Huang. Asphalt and asphalt mixture [M].

Research of SMA - 13 asphalt mixture ratio on the taxiway of HuangHua Airport

Shuaituan Tian [1,2,*], Yapei Yan [1,2], Peng Wang [1,2]

[1] China Airport Construction Group Co., Ltd, Beijing, 100101, China
[2] Beijing Super-Creative Technology Co., Ltd, Beijing, 100621, China

Abstract. Because of the large load and large impact force of airport pavement under, it often appears the phenomenon of serious rut. In order to improve this kind of phenomenon, combined with the taxiway of HuangHua Airport area reconstruction project, we analyze the upper layer type of SMA-13 modified asphalt mixture ratio design. we analysis from raw material selection, mixture gradation design and admixture added, and finally choose the excellent performance of asphalt mixture.

Keywords: SMA-13; Oil loss; Mix ratio; Performance test.

1. Introduction

The main raw materials required for asphalt mixture mix ratio test include asphalt, gravel, stone chips, mineral powder, fiber and anti-rut agent. All materials are tested according to relevant specifications. Raw materials are determined by the construction unit and the design unit after investigation, and provide sufficient number of samples for the test. In order to ensure the accuracy of test data, all raw materials are sampled according to the standard sampling method stipulated in the specification to meet the needs of various test projects. Of asphalt mixture pavement involved in a variety of materials, quality must meet the design requirements, at the same time should be in strict accordance with the design requirements of dosage and proportion of added, otherwise will largely reduce the performance of asphalt mixture material, in order to ensure project quality to meet the design intent, specific should meet the following requirements.

2. Material composition

2.1 Coarse aggregate and fine aggregate

The reason why SMA has a good high temperature stability is mainly based on the embedding of a lot of coarse aggregates, so its crushing resistance of coarse aggregates is very high. Coarse aggregate must use tough, rough, angular high quality stone; The needle and flake particle content of coarse aggregate must be strictly limited. The gravel used should not be broken by jaw plate stone crusher, but by hammer or counterattack (rotary) stone crusher. At the same time, because SMA is used for the surface layer of the road surface, aircraft takeoff and landing on the road surface anti-skid performance requirements are very high, coarse aggregate must meet the requirements of anti-skid performance, which requires that it must have a good polishing value.

Through the investigation and comparison of the airport construction party, the technicians of our company selected basalt and limestone coarse aggregate for the target mix ratio design test in the upper layer. Perform raw material test on selected coarse aggregate,

The fine aggregate of asphalt pavement should be machine-made sand, and the fine aggregate should be clean, dry, without impurities, and have appropriate particle gradation. According to the relevant technical requirements, limestone machine-made sand is mainly used in the fine aggregate of asphalt mixture mix ratio test.

2.2 Asphalt

The asphalt binder in asphalt mixture must have high viscosity and good adhesion with aggregate to ensure sufficient high temperature stability and low temperature toughness.

When sampling asphalt test, according to the requirements of the specification, the quantity required by the test will be collected and stored separately, and the samples will be tested according to the specifications. Only the asphalt that meets the requirements of the "Specification for Asphalt Concrete Pavement Design of Civil Airport" (MH5010-2017) can be used.

According to the climate characteristics of Huanghua area and the operation status of Huanghua Airport, it is necessary to enhance the ability of the middle and lower layer of asphalt pavement to resist permanent deformation

* Corresponding author: tianst@cacc.com.cn

(anti-rutting), so SBS modified asphalt is adopted in this mix design.

2.3 Filler

In this overhaul project, mineral powder is used as the filler of asphalt mixture. Its role is very important, and its quality will directly affect the performance of the mixture. Ore powder must be ground from limestone alkaline rock. Mineral powder should be dry, loose particles, no agglomeration, rain or damp agglomeration of mineral powder shall not be used

2.4 Fiber

Fiber stabilizer for SMA mainly includes lignin fiber, mineral fiber, polymer chemical fiber, etc. Its function is to improve the performance of asphalt mixture, adsorption asphalt, reduce leakage. Fiber should be able to withstand the minimum environmental temperature of 250 ℃ without deterioration, and do not cause harm to the environment, do not harm to health.

According to the experience of similar airport projects in the past, polyester fiber is used in the test. Polyester fiber can play the role of oil absorption and reinforcement in asphalt mixture, and can improve the performance of rutting resistance and water damage resistance of the mixture.

2.5 High adhesive

High viscosity asphalt modifier is a new kind of additive specially developed for preparing high viscosity asphalt and high viscosity asphalt mixture. High viscosity asphalt mixture with excellent performance is prepared by directly putting it into mixing pot. The dosage of asphalt mixture is 0.5%. The temperature sensitivity of the prepared asphalt mixture decreases, the high temperature stability, aging resistance and low temperature performance are significantly improved

3. Mix design of SMA-13 asphalt mixture

SMA is a kind of asphalt mixture composed of asphalt, fiber stabilizer, mineral powder and a small amount of fine aggregate. SMA material structure characteristics are: three more and one less, more coarse aggregate, more ore powder, more asphalt, less fine aggregate.

Due to the amount of SMA ore powder, the aggregate specific surface area is large, and more asphalt is needed to cover. At the same time, the skeleton void formed by a large number of coarse aggregate needs more asphalt cement to fill, so the asphalt content of SMA gradation mixture is larger, which forms another characteristic of SMA, the amount of asphalt. Because SMA belongs to discontinuous grading, so from the traditional point of view, THE SMA mixture is easy to segregate, but another characteristic of SMA is rich in oil mud, and the cohesion of mud because of ore powder and larger, so that SMA overcomes the shortcomings of discontinuous grading easy segregation. At the same time, oil-rich cement enhanced the low temperature cracking resistance and aging resistance of SMA. The high content of coarse aggregate in SMA mixture also forms a skeleton structure that can transfer vehicle load efficiently, so SMA has a very high temperature resistance to rutting ability. The high content of coarse aggregate also increases the roughness of the surface of the mixture, so that the SMA surface has a larger structural depth, enhance the anti-skid ability of the road surface and the safety of high-speed driving in rainy and snowy days. In addition, intermittent gradation makes it easy to fill the cement aggregate skeleton, sufficient amount of cement to make the final compacting SMA mixture has a low level of voidage, so that the WATERPROOF performance of SMA is very strong, anti-aging, anti-fatigue advantages outstanding, has a high durability, long service life.

3.1 Selected raw materials

The cementing material is SBS modified asphalt. Coarse aggregate basalt and limestone stone, material specifications of basalt 10 ~ 15mm, 5 ~ 10mm and 3 ~ 5mm; Fine aggregate is limestone, the specification is 0 ~ 3mm; The filling is limestone powder.

The fiber used in this mixing ratio test is polyester fiber, and the dosage is 0.4% of the weight of asphalt mixture, and the anti-rutting agent and high viscosity agent are 0.5% of the weight of asphalt mixture.

3.2 The determination of aggregate gradation ratio

In accordance with the "Civil Airport Asphalt Pavement Construction Technical Specifications" (MH/T 5011-2019), "Highway Asphalt Pavement Construction Technical Specifications" (JTG F40-2004) and "Highway Asphalt Mhoa resin gravel Pavement Technical Guide" (SHC F40-01-2002), the first SMA ore gradation design, To prepare asphalt mixture with coarse, medium and fine grading, it is required that the passing rate of 4.75mm sieve (key sieve) is about ±4% (that is, the passing rate is about 22%, 26% and 30%) in the guideline SM-13 grading range, and the passing rate of 0.075mm is about 10%.

The aggregate gradation ratio of SMA-13 asphalt mixture is formed with five different raw materials. we choose three mix designs .All the mix designs and raw materials are put on the below table.

Table 1. SMA-13 ratio of mineral aggregate gradation (%)

Specifications	10 ~ 15mm	5 ~ 10mm	3 ~ 5mm	0 ~ 3mm	filler
Coarse gradation	49	26	7	8	10
Middle gradation	48	28	8	9	10
Fine gradation	47	24	9	10	10

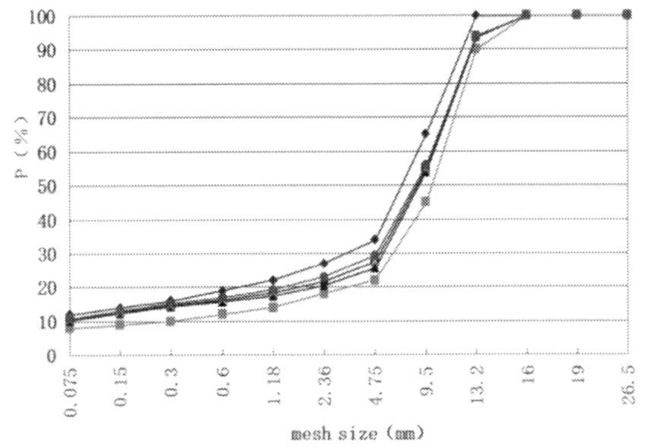

Figure I. SMA-13 grading curve

Content of coarse aggregate P_{CA} and Clearance rate VCA_{DRC} of more than 4.75mm in three mixtures are tested in table3.

Table 2. The Marshall test results of different asphalt aggregate ratio

items	loose unit weight (g/cm³)	the passing rate of 4.75 mm (%)	bulk specific gravity of above4.75mm (g/cm³)	P_{CA}	VCA_{DRC}
coarse	1.654	25.5	2.899	0.703	42.95
middle	1.646	27.4	2.889	0.685	43.03
fine	1.638	29.4	2.866	0.666	43.24

Based on the experience of previous similar airport engineering, we use ratio of 6.0% as a first try oil-stone ratio in Marshall design method, and then mold specimens in accordance with the specification requirements, measure the physical indexes of the specimens.

According to the recommendation of "Technical Specifications for Construction of Asphalt Pavement in Civil Airports" (MH/T 5011-2019), the volume index of selected gradation VCAmix must be less than VCADRC, and THE VMA should be greater than 16.5%. Under the condition that the void ratio meets the requirements of the specification, the high temperature stability (anti-rutting performance) of asphalt mixture can be improved by using coarser aggregate gradation. The coarse gradation can be determined as the design gradation of mineral material by analyzing the above table.

3.3 The determination of the optimum proportion

Marshall design method is adopted to determine the optimum asphalt-stone ratio by rough grading according to the results of ore material grading design.

According to the previous experience of asphalt mixture mixture ratio test in similar engineering, three asphalt-stone ratios were selected for Marshall test, which were 5.7%, 6.0% and 6.3% respectively.

According to the "Civil Airport Asphalt Pavement Construction Technical Specifications" (MH/T 5011-2019), "Highway Asphalt Mma Foot resin gravel Pavement Construction Technical Requirements" (SHC F40-01-2002) and "Highway Asphalt Pavement Construction Technical Specifications" (JTG F40-2004), for SMA asphalt mixture, The density and voidage of the Marshall test should be calculated using the relative density of gross volume measured by the surface dry method, and the maximum theoretical density of mineral materials should be calculated using the gross volume density of aggregate.

We select 3asphalt aggregate ratio of Marshall test and calculate their physical indicators in order to determine the optimum proportion , the test results are shown in the table4.

Table 3. The Marshall test results of different asphalt aggregate ratio

Test items	oil-stone ratio (%)			Specification requirements
	5.7	6.0	6.3	
Bulk specific gravity (g/cm³)	2.502	2.507	2.510	-
Theoretical density (g/cm³)	2.617	2.605	2.594	-
VV (%)	4.4	3.8	3.2	3～4
VMA (%)	17.2	17.2	17.4	≥16.5
VFA (%)	74	78	81	-
VCA_{mix} (%)	39.17	39.22	39.32	≤VCA_{DRC}
VCA_{DRC} (%)	42.95	42.95	42.95	-
MS (KN)	7.92	7.64	8.13	≥6
FL (0.1mm)	27.6	30.3	34.0	-

Based on the data analysis in the above table, considering that when the design void ratio is 3-4.5%, the corresponding indexes VMA, VCAmix, VFA and Marshall stability meet the requirements, 6.0% is decided as the best oil/stone ratio in combination with the corresponding oil/stone ratio of the design void ratio of 3-4.5% and the actual experience.

It can be seen from the test piece that the coarse aggregate in the asphalt mixture has formed a good skeleton. The asphalt mace has filled the gap of mineral material fully and distributed evenly, and the whole mixture has become a good skeleton dense structure. To sum up, the target mixture ratio of SM-13 asphalt mixture is reasonable and feasible.

3.4 The optimum proportion of validation

When the optimum asphalto-stone ratio is 6.0%, all the Marshall test indexes meet the requirements. Further performance test of SMA is carried out. The method and corresponding performance test are as follows.

1. Schellenberg asphalt leakage test: to check whether the maximum amount of SMA asphalt is appropriate, which is related to the success or failure of SMA structure. Therefore, after the ratio is determined, it is necessary to check whether too much asphalt will cause leakage or too little asphalt will cause dispersion. There are beaker method, enamel plate method, net basket method and so on in the international leakage test. The highway department of Our country has decided to adopt beaker

method as the standard method and formulated the corresponding test regulations. Therefore, the beaker method is used in the mix ratio test to evaluate whether the maximum amount of asphalt is appropriate.
2. Kent fort dispersion test: used to test whether the amount of asphalt is too little or insufficient asphalt adhesion. Asphalt mixture under the action of traffic load, aggregate may fall off from the road surface and disperse, so it must be tested by flying test under certain conditions.
3. Rutting test: SMA is usually used as the surface layer of the road surface, which needs to have good high temperature performance. Whether the asphalt mixture has good high temperature performance can be characterized by the high temperature anti-rut ability of the asphalt mixture. Rutting tests in many projects show that the dynamic stability of modified asphalt can reach 5000 times /mm or more.
4. Water stability test: In China, immersion Marshall test and freeze-thaw splitting test are generally used to test the water stability of asphalt mixture.
5. Water seepage test: an important feature of SMA is that the void rate between the coarse aggregate is fully filled by asphalt ma Foot grease part, the void rate is very small, good SMA structure after rolling molding should be impervious to water or water permeability is very slow, the guidelines require indoor test in the wheel forming plate specimen water seepage coefficient detection.
6. Structural depth test: structural depth represents the sliding resistance of asphalt mixture. In Our country, sand paving method is generally used to test the structural depth on the plate specimen formed by wheel grinding.
In the optimum asphalt aggregate ratio of 6.0%, we make Marshall specimen and the dynamic stability of specimen, then test their Physical and mechanical performance. we put the results on the below table.

Table 4. The Road performance verification test results

Test items	Test results
△M (%)	0.09
△S (%)	2.2
DS (time/mm)	10862
MSo (%)	97.8
TSR (%)	90.9
Cw (mL/min)	No seepage
TD (mm)	1.08

From the results, it appears that all the test results can meet the requirements of related technologies.

4. Conclusion

In combination with the taxiway and runway of HuangHua Airport, in order to slove the serious problems of SMA - 13 asphalt mixture, we achieve these important ways.
First; Starting from the raw material, we should choose the qualified raw materials. including aggregate, filler and asphalt. The important indicators of aggregate are bulk specific gravity, apparent specific gravity, sand equivalent, sturdiness and p%; The important indicators of filler are apparent specific gravity, water content, Particle gradation and hydrophilic coefficient; The important indicators of asphalt are penetration, softening point, ductility, equivalent softening point, equivalent brittle pointT1.2, flash point, density, elastic recovery and filmy heating operational test163℃/5h.
Second: we should choose good aggregate gradation.such as 10 ~ 15mm: 5 ~ 10mm: 3 ~ 5mm: 0 ~ 3mm:fiber=49:26:7:8:10.
Third: In order to obviously improve the dynamic stability of mixture, it is reasonable to adding anti-rutting agent and high adhesive.

References

1. The ministry of communications highway engineering science institute. JTJ E20-2011.Standard test methods of bitumen and bituminous mixtures for highway engineering [S].
2. Xiaoming,Huang. Asphalt and asphalt mixture [M].
3. The ministry of communications highway engineering science institute. JTG F40-2004. Technical specification for construction of highway asphalt pavement [S].
4. The civil aviation administration of China. JTG F40-2004. Specification for asphalt concrete pavement construction of civil airports [S].
5. The ministry of communications highway engineering science institute. JTG E42-2005. Test methods of aggregate for Highway Engineering [S].

Study on Influence of Tin Plating Layer of Solder Strip on Electrical Properties of Photovoltaic Modules

Xiankun Ren [1,4], Yuying Ma [2,*], Chuan Jiang [3] and Xiaojun Yan [3]

[1] Shandong Linuo Solar Power Holdings Co., Ltd, Jinan 250103, China
[2] Shandong Vocational and Technical University of Engineering, Jinan 250200, China
[3] Shandong Linuo Solar Power Technology Co., Ltd, Jinan 250103, China
[4] Shandong Linuo Photovoltaic High Tech Co., Ltd, Jinan 250103, China

Abstract. With the continuous development of social economy, the cost reduction and efficiency increase of enterprises are very important for the development of enterprises. Starting from the packaging materials of photovoltaic modules, this paper mainly studies the influence of welding ribbon with different tin layer thicknesses on the performance of photovoltaic modules, and compares the resistivity of welding ribbon with different thickness of tin layers. The experimental results show that with the reduction of the thickness of the tin-plated layer on the non-soldering surface of the welding ribbon, the resistivity of the welding ribbon decreases, and the output power of the photovoltaic module is effectively improved.

Keywords: Photovoltaic; Modules; Tin layer; Welding ribbon; Resistivity.

1. Introduction

Solar energy is the cleanest, safe and reliable energy source in the future, and the photovoltaic industry is increasingly becoming another explosive industry [1-3]. The solar cell module is the core part of the solar power generation system and the most important part of the solar power generation system. The quality and cost of solar cell modules will directly determine the quality and cost of the entire system.

The photovoltaic cells inside the solar module are connected in series to form an array, which is then encapsulated by glass, backplane, ethylene-vinyl acetate copolymer (EVA), silica gel and other protective materials. Photovoltaic modules are installed and constructed to form a module array system to convert light energy into electrical energy [4]. The structure of the photovoltaic module is shown in Figure 1. The efficiency of photoelectric conversion is the most important point for converting light energy into electrical energy. The higher the photoelectric conversion efficiency, the higher the utilization rate of sunlight and the greater the power generation under the same irradiation area.

The power generation principle of photovoltaic modules is the process of converting the absorbed solar energy into electrical energy by the solar cells in the module. In photovoltaic modules, the photovoltaic ribbon mainly plays the role of electrical connection, and transmits the current collected by the main grid of the solar cell through the photovoltaic ribbon. The power loss of photovoltaic modules mainly includes optical loss and electrical loss [5]. The optical loss is mainly caused by the shading of photovoltaic ribbons, the transmittance and optical mismatch of packaging materials such as glass and EVA. The electrical aspect mainly refers to the power loss caused by the resistance loss in the module, including the resistance of the welding tape, contact, and junction box, and is also affected by the current mismatch of the cell and the influence of the design version of the module. For photovoltaic practitioners, it is an important task to reduce production costs and increase the power generation of modules by continuously increasing the power of modules.

Photovoltaic ribbon is an important part of photovoltaic modules. It is made of high-quality oxygen-free copper and tinned on both sides. Photovoltaic modules connect the cells through ribbon welding, and collect the current generated by the battery and output it to the outside through the junction box. Therefore, the resistance and welding performance of the ribbon itself will have a certain impact on the power of the photovoltaic module.

* Corresponding author: yyyjbd@163.com

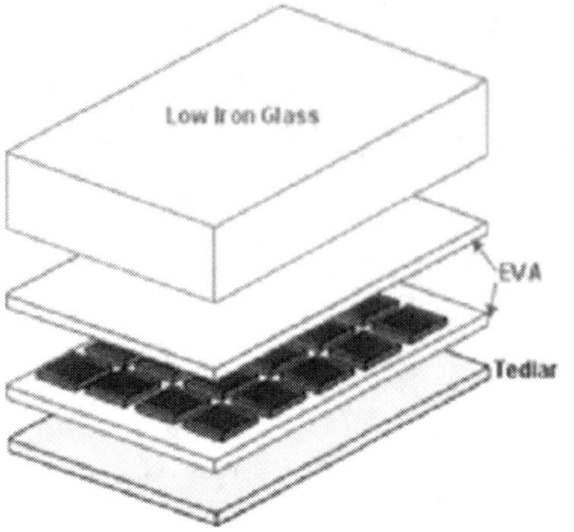

Fig. 1 The stacking sequence of packaging materials for photovoltaic modules

In this paper, an experiment is carried out on the thickness of the tin-plated layer on the non-soldering surface of the photovoltaic module welding strip, and the resistivity of the welding strip with different tin-plated layer thicknesses is tested. The PV modules were packaged with different thicknesses of tin-coated ribbons to encapsulate the modules and the encapsulation efficiency of the PV modules was collected. The influence of the thickness of the tin-coated ribbons on the output performance of PV modules was discussed.

2. Experimental section

In this experiment, the welding tape produced by Taicang Juren Photovoltaic Materials Co., Ltd. was selected, and the ZY9987 digital micro-ohmmeter was used for the resistivity test. The photovoltaic modules use P-type monocrystalline PERC cells with a size of 182 mm × 182 mm and a conversion efficiency of 22.8%. The component package layout design is shown in Figure 2.

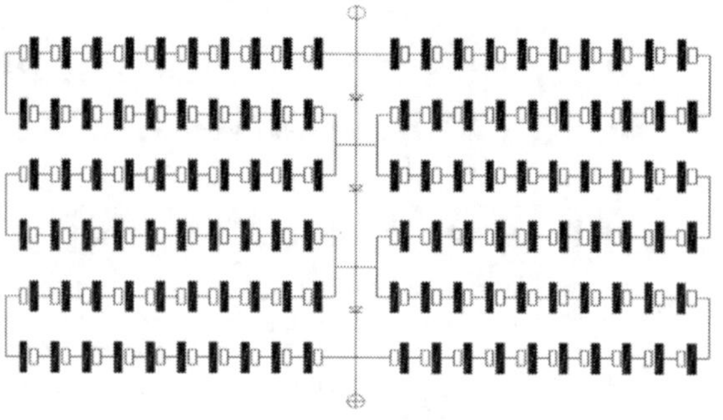

Fig. 2 Schematic diagram of electrical connection of experimental components (back)

3. Results and analysis

At present, the specifications of photovoltaic ribbons on the market and the thickness of the tinned layer are different. In order to study the influence of the thickness of the tinned layer on the resistivity of the ribbons, three types of ribbons are selected for experiments. The structures and parameters of the ribbons are shown in the table 1, and the cross section of the ribbon is shown in Figure 3.

Table 1 Specification and parameter table of welding strip for experiment

Experiment	Total thickness	Soldering surface tin layer thickness	Thickness of tin plating on non-soldering surface
A	0.3mm	20-25um	20-25um
B	0.3mm	20-25um	10-15um
C	0.3mm	20-25um	1-5um

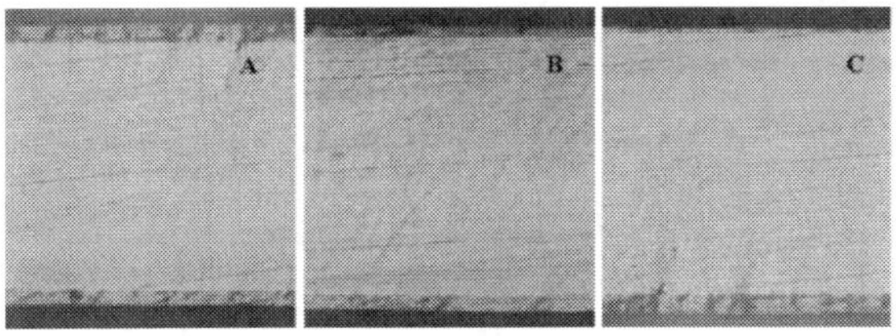

Fig. 3 Cross-sectional views of solder ribbons with different tin plating thicknesses

3.1 Influence of the thickness of the tinned layer on the resistivity of the ribbon

The resistivity test was carried out on the ribbon with different tin-plated layer structures, and the test results are shown in Figure 4. As can be seen from the figure, as the thickness of the tin-plated layer on the non-welding surface of the ribbon gradually decreases, the resistivity of the ribbon gradually decreases. This is mainly because the tin-plated layer of the ribbon is a tin-lead alloy, and the resistivity of tin and lead is much higher than that of copper, so as the tin-plated layer with high resistivity is thinned, the resistivity of the ribbon gradually decreases[6,7].

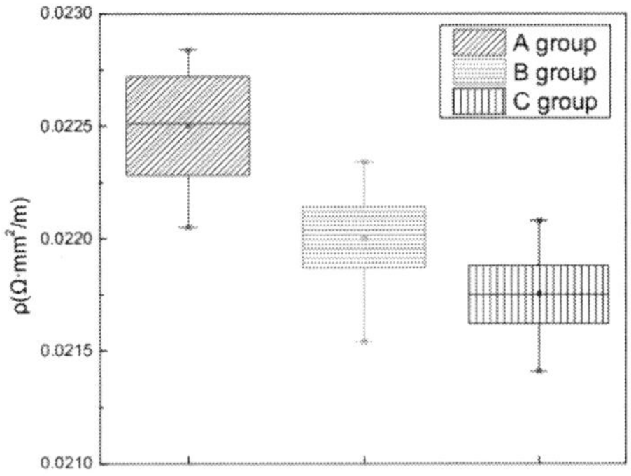

Fig. 4 Distribution of resistivity test of test ribbons in groups A, B and C

3.2 Influence of the thickness of the tin plating layer on the encapsulation power of photovoltaic modules

In order to avoid the influence of other factors on the power of photovoltaic modules, on the basis of ensuring the consistency of cells and packaging materials, single-crystal silicon photovoltaic modules were prepared by using the above three kinds of welding tapes, each type of photovoltaic module was 20 plates, and the photovoltaic modules were packaged The output power test was carried out for three groups of experiments, and the test results are shown in Figure 5. The data in the figure shows that as the thickness of the tinned layer of the solder ribbon decreases, the output power of the module shows an upward trend. This is because the thinning of the tin plating layer reduces the resistivity of the solder ribbon, which reduces the series resistance of the photovoltaic module and improves the output power of the photovoltaic module.

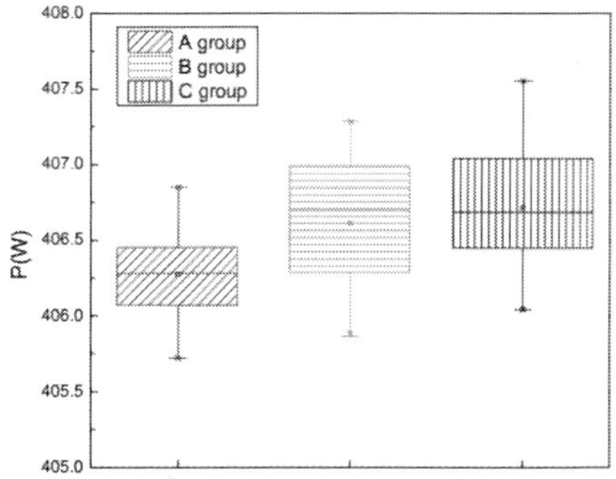

Fig. 5 Package power test distribution of test components in groups A, B and C

4. Summary

This paper takes the welding strip for photovoltaic module materials as the research object, designs welding strips with different thicknesses of tinned layers, compares the resistivity of the welding strips with different structures and the output power of photovoltaic modules, and draws the following conclusions: The thinning of the thickness of the tinned layer on the surface reduces the resistivity of the soldering strip, and the output power of the photovoltaic module gradually increases.

References

1. Karakosta C, Pappas C, Marinakis V, et al. Renewable energy and nuclear power towards sustainable development: Characteristics and prospects[J]. Renewable and sustainable energy reviews, 2013, 22:187-197.
2. Ellabban O, Abu-Rub H, Blaabjerg F. Renewable energy resources: Current status, future prospects and their enabling technology[J]. Renewable and Sustainable Energy Reviews, 2014, 39:748-764.

3. Polman A, Knight M, Garnett EC, et al. Photovoltaic materials: Present efficiencies and future challenges[J]. Science, 2016, 352.
4. Xiankun Ren , XiaojunYang and Yuying Ma. The Influence of Partial Shading on the Output Performance of PV Modules[J]. World Scientific Research Journal, 2020, 6(9) : 216-220.
5. Hong Yang, He Wang. Performance analysis of crystalline silicon solar modules with the same peak power and the different structure[J]. Clean Technologies & Environmental Policy, 2011, 13(3):527-533.
6. Brandes E A, Brook G B. Smithells Metals Reference Book[M]. Butterworth-Heinemann, 1983.
7. Watanabe M, Ishibashi A, Miura H . Basic study of dynamic recrystallization in Cu-Sn-P alloy for high strength copper[J]. R and D: Research and Development Kobe Steel Engineering Reports, 2012, 62(2):96-101.

Preparation of sheep manure biochar and its enhancement effect on wastewater treatment performance of CRI systems

Jiao Chen[1,2], Shuxin Li[1], Qin Huang[1], Haolin Liu[1], Jieyu Sun[1], Ying Zhang[1], Yixin Lu[1,2,3,*]

[1]School of Materials and Environmental Engineering, Chengdu Technological University, Chengdu 611730, China
[2]State Environmental Protection Key Laboratory of Synergetic Control and Joint Remediation for Soil & Water Pollution, Chengdu 610059, China
[3]Haitian Water Group Co., Ltd., Chengdu 610200, China

Abstract: Sheep manure biochar (SMB550) was prepared by limited oxygen pyrolysis at 550℃ and applied to constructed rapid infiltration (CRI) systems. The structural characteristics of SMB500 were tested and its effect on the wastewater treatment performance of CRI systems was investigated. The results showed that SMB550 had a large specific surface area, abundant pore structure and rich functional groups, showing a good adsorption potential. Using SMB550 as one of the filter material of CRI system, the average removal rates of COD, NH_4^+-N and TP were respectively 9.3%, 8.6% and 34.9% higher than those of the blank group, which enhanced the treatment performance of CRI systems for wastewater. As a new type of filter material, SMB550 had a good application prospect in the field of wastewater treatment.

Key words: CRI systems; sheep manure; biochar; wastewater treatment.

1. Introduction

As a kind of eco-friendly wastewater treatment technology, the constructed rapid infiltration (CRI) system has attracted more and more attention in recent years, and has broad application prospects in the treatment of urban domestic sewage and polluted surface water [1,2]. This technology is different from the traditional soil infiltration system. It uses natural or artificial filter materials with better infiltration performance to replace the soil for filling, so as to obtain higher hydraulic load [3]. This technology mainly relies on the adsorption, interception and microbial degradation of the filter material to remove pollutants from wastewater, so the filter material is the key functional component of the technology, and the appropriate filter material structure is essential to improve the wastewater treatment performance of CRI systems.
Biochar is a carbonaceous substance produced by pyrolysis of biomass residue under the condition of anoxic and high temperature. In general, it has the advantages of large specific surface area, simple preparation process, low cost and good adsorption potential [4]. Recently, the application of biochar in the treatment of organic matter, nitrogen, phosphorus and other pollutants in water is increasing [5-7]. At present, the traditional CRI systems mostly used river sand, zeolite sand and marble sand as the filter material, while the research of using biochar as the filter material was rarely reported.
Therefore, this study used waste sheep manure as raw material to prepare biochar, and applied it to CRI systems to investigate the variations of wastewater treatment performance before and after the addition of biochar, so as to provide a new filter material for CRI systems and explore a new way for resource utilization of wasted sheep manure.

2. Materials and Methods

2.1 Preparation of sheep manure biochar

Wasted sheep manure was taken from a sheep farm. After being dried, crushed and sieved for 60 mesh, put a proper amount of sheep manure powder into the crucible, covered it and put it into the muffle furnace, heated it to 550℃ at a heating rate of 20℃/min, and then kept it at a constant temperature for 3 hours.
After cooling to room temperature, added 1.0 mol/L hydrochloric acid with the solid-liquid ratio (m:V) of 1:20, shaken at room temperature for 3 times, 30 minutes each time, and the frequency of vibration was 150 r/min. After cleaning to neutral and drying in an oven at 105℃, sheep manure biochar (SMB550) was successfully prepared [8].

* Corresponding author: yxlu61@163.com

2.2 Experimental reactors and operation conditions

Two CRI reactors (CRI1 and CRI2) were used in this study, all of which were made of polyvinyl chloride. Fig. 1 shows the device diagram of CRI reactors. The column height and inner diameter were 120 cm and 7 cm respectively, and the filter material was 100 cm high. CRI1 reactor adopted river sand, zeolite sand and marble sand with the particle size of 0.5~1.0 mm uniformly mixed by volume ratio of 5:3:2 as filter material, and CRI2 reactor adopted river sand, zeolite sand, marble sand and SMB550 evenly mixed by volume ratio of 5:3:1:1 as filter material. A layer of gravel with a particle size of 5~10 mm was laid on the top and bottom of the filter material respectively as a buffer layer and a supporting layer. The wastewater was evenly distributed via drip infiltration, the wastewater inflow was controlled by metering pump, and the inflow time was controlled by relay. During the experiment, the controlled temperature was (25±5)℃, running for 2 cycles every day, each cycle running for 12 hours, in which the flooding and drying time were 3 hours and 9 hours respectively, and the hydraulic load was 1.0 m/d.

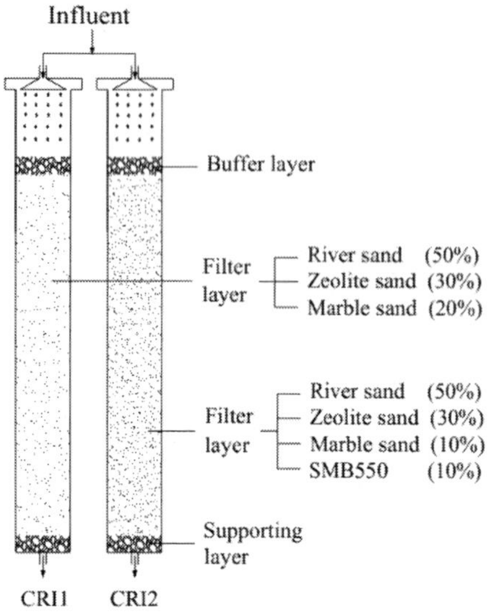

Fig. 1. Device diagram of CRI reactors.

2.3 Influent quality and inoculated sludge

The wastewater used in the experiment was taken from the domestic sewage of a college student apartment in Chengdu. The influent COD, NH_4^+-N and TP concentrations were 212.6~285.8 mg/L, 38.5~50.6 mg/L and 3.6~4.9 mg/L respectively, and the pH was 6.5~7.9. The inoculated sludge was taken from the secondary sedimentation tank of a wastewater treatment plant in Chengdu.

2.4 Analysis items and methods

The ash content of SMB550 was determined by muffle furnace (SG-XL1200, China). The content of C, H, N and O was determined by element analyzer (VARIO EL cube, Germany). The specific surface area, total pore volume and pore size distribution were determined by specific surface area analyzer (NOVA4000e, USA). The surface morphology of SMB550 was characterized by field emission scanning electron microscope (ZEISS SUPRA40, Germany). The surface functional groups were determined by FT-IR spectroscopy (FT-IR Spectrum100, USA). The analysis method of COD, NH_4^+-N and TP concentration referred to *Water and Wastewater Monitoring and Analysis Methods* (4th Edition, China).

3. Results and Discussion

3.1 Structural characteristics of SMB550

The productivity of SMB550 was 48.71%, indicating that the mass loss of the preparation process of SMB550 was appropriate. The component contents of SMB550 were shown in Fig.2. In terms of element composition and atomic ratio, SMB550 contained a large number of C, and H/C<0.6, O/C<2.0, reflecting that SMB550 had a high biochemical stability [9].

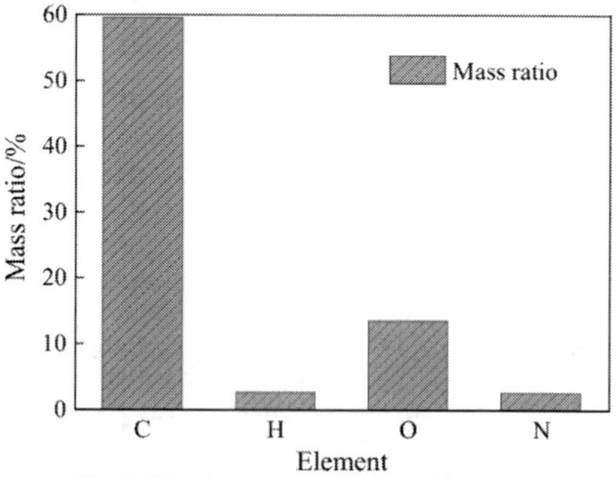

Fig. 2. The element composition of SMB550.

According to BET analysis results, the specific surface area, total pore volume and average pore diameter of SMB550 were 105.61 m²/g, 0.107 cm³/g and 8.64 nm respectively. It indicated that SMB550 had a large specific surface area, total pore volume and average pore diameter. According to the SEM results as shown in Fig.3, the surface of SMB550 was rough and it also had a large number of micro pores, which was conducive to improving its adsorption capacity [8].

(a)SEM of SMB550(×500)

(b)SEM of SMB550(×3000)

Fig. 3. Scanning electron microscope of SMB550.

Fig.4 shows the FT-IR spectra of SMB500. The peak at about 3400 cm^{-1} belongs to the stretching vibration of —OH [10]. The absorption peak at 2919 cm^{-1} belongs to the stretching vibration of C—H. The absorption peak at 1600 cm^{-1} is attributed to the stretching vibration of C=C and C=O. The absorption peak at about 1100 cm^{-1} is attributed to the stretching vibration of C—O. The absorption peak of 790 cm^{-1} is attributed to the bending vibration of C—H on the aromatic ring [11]. These absorption peaks showed that SMB500 was rich in functional groups and had high aromaticity and stability [12]. These will be conducive to its adsorption of pollutants.

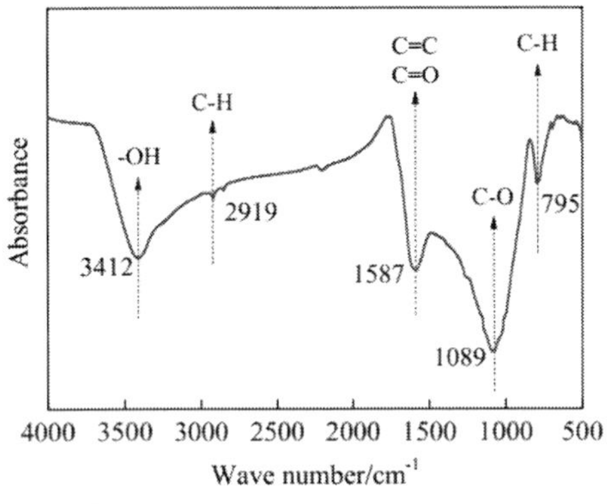

Fig. 4. The FT-IR spectra of SMB550.

3.2 Comparison of pollutant removal effect

Fig.5 shows the effect of CRI systems on COD, NH_4^+-N and TP removal before and after the addition of SMB550.

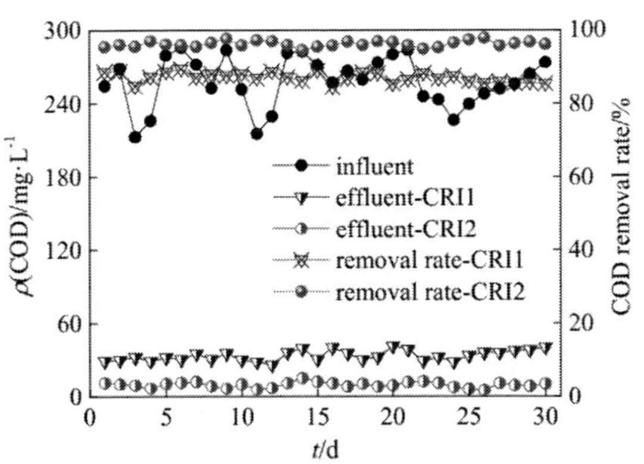

(a)COD

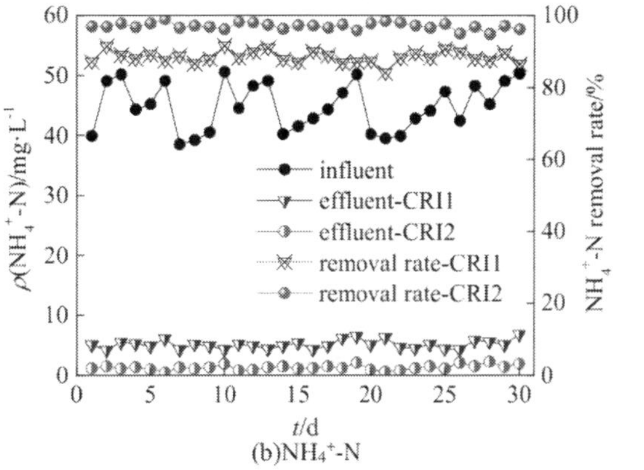

(b)NH_4^+-N

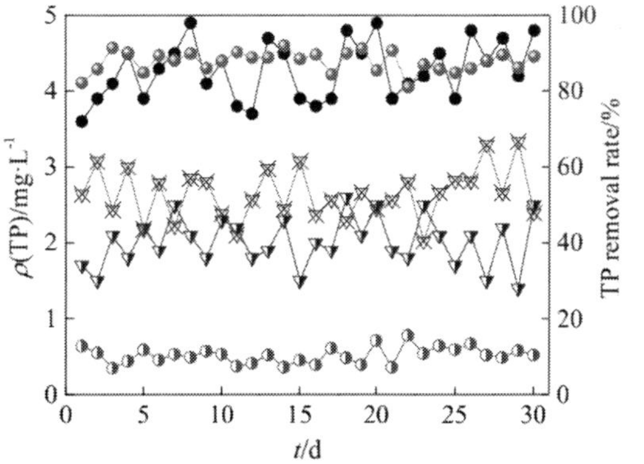

(c)TP

Fig. 5. Variations of pollutant removal effect before and after adding SMB550.

It can be seen from Fig.5 that the effluent COD, NH_4^+-N and TP concentrations of CRI1 were 26.1~41.3, 4.2~6.8 and 1.4~2.6 mg/L respectively. Compared with CRI1, the effluent COD, NH_4^+-N and TP concentrations of CRI2 were significantly reduced, which were lower than 15.2, 2.3 and 0.78 mg/L respectively. The average removal rates of CRI1 for COD, NH_4^+-N and TP were 87.0%, 88.5% and 52.9% respectively, and the average removal rates of CRI2 for COD, NH_4^+-N and TP were 96.3%, 97.1% and 87.8% respectively. Compared with CRI1, the average removal rates of COD, NH_4^+-N and TP by CRI2 increased by 9.3%, 8.6% and 34.9% respectively. It can be seen that the removal efficiency of COD, NH_4^+-N and TP by CRI system was significantly improved after adding SMB550. The addition of SMB550 had the strongest enhancement effect on the removal effect of TP in sewage.

3.3 Analysis of strengthening principle

The CRI system could realize natural reoxygenation and form aerobic area in the period of drying, while in the period of flooding, it was in the state of anoxic. Therefore, wastewater could experience aerobic and anoxic environment successively in the process of infiltration, which provided suitable living conditions for different types of microorganisms [13]. COD, N and P in the wastewater could be absorbed or retained by the filter material, and then transformed or decomposed through the microbial action on the surface of the filter material, so that the wastewater could be purified [3,13]. Therefore, the performance of filter material would directly affect the performance of CRI systems on wastewater treatment.

In this study, the CRI1 reactor used river sand, zeolite sand and marble sand as filter material, which had good treatment effect on COD and NH_4^+-N, but poor removal effect on TP. Because of the addition of SMB550 in the CRI2 reactor, the removal efficiency of COD and NH_4^+-N was further improved, and the removal efficiency of TP was greatly improved. There are three reasons for this result. Firstly, sheep manure biochar had a huge specific surface area, abundant pore structure and rich functional groups, which had good adsorption effect on COD, NH_4^+-N and TP, and could effectively adsorb or intercept pollutants from wastewater [14]. Secondly, sheep manure biochar could be used as an excellent carrier of microorganisms, and its rough structure was more conducive to the adhesion and growth of microorganisms, accelerating the formation and stability of biofilm on the surface of filter material [15]. Thirdly, the particle size of sheep manure biochar was relatively small, which enriched the particle size distribution of filter materials in CRI systems. Reasonable particle size distribution could effectively improve the capacity of holding pollutants and the utilization rate of filter materials, which was conducive to improving the treatment efficiency of wastewater [16].

4. Conclusion

By adding sheep manure biochar to CRI systems, the average removal rates of COD, NH_4^+-N and TP were 9.3%, 8.6% and 34.9% respectively higher than those without adding sheep manure biochar. The huge specific surface area, abundant pore structure and rich functional groups of sheep manure biochar provided the basis for better adsorption of COD, N and P pollutants in wastewter, and greatly promoted their removal effect. Using sheep manure biochar as the filter material of CRI system could enhance the wastewater treatment performance efficiently.

Acknowledgments

This work was financially supported by Sichuan Science and Technology Program (No.2022YFG0307, 2022NSFSC0393), Open Fund of State Environmental Protection Key Laboratory of Synergetic Control and Joint Remediation for Soil & Water Pollution (No.GHBK-2021-004), Sichuan Higher Education Talent Training Quality and Teaching Reform Project from 2021 to 2023 (No.JG2021-1409), Young Seedling Program of Chengdu Technological University (No.QM2021003), and Laboratory Open Fund Project of Chengdu Technological University (No.2022CHZH04).

References

1. L. Yang, F. Kong, M. Xi, et al, Environmental economic value calculation and sustainability assessment for constructed rapid infiltration system based on emergy analysis, Journal of Cleaner Production. 167 (2017) 582-588.
2. D. Wang, Z. Zhang, X. Li, et al, A full-scale treatment of freeway toll-gate domestic sewage using ecology filter integrated constructed rapid infiltration, Ecological Engineering. 36 (2010) 827-831.
3. J. Chen, Y. Lu, Z. Ouyang, et al, Regulation of partial nitritation in constructed rapid infiltration system and analysis of microbial community structure, Polish Journal of Environmental Studies. 29 (2020) 33-43.
4. X. Tan, Y. Liu, G. Zeng, et al, Application of biochar for the removal of pollutants from aqueous solutions, Chemosphere. 125 (2015) 70-85.
5. M. Wu, B. Pan, D. Zhang, et al, The sorption of organic contaminants on biochars derived from sediments with high organic carbon content, Chemosphere. 90 (2013) 782-788.
6. N. Liu, Z. Sun, Z. Wu, et al, Adsorption characteristics of ammonium nitrogen by biochar from diverse origins in water, Advanced Materials Research. 664 (2013) 305-312.
7. S. Wang, L. Kong, J. Long, et al, Adsorption of phosphorus by calcium-flour biochar: Isotherm, kinetic and transformation studies, Chemosphere. 195 (2018) 666-672.
8. J. Zhang, W. Huang, J. Chen, et al, Adsorption characteristics of norfloxacin in aqueous solution by sheep manure biochar, Acta Scientiae Circumstantiae. 37 (2017) 3398-3408.
9. K. A. Spokas, Review of the stability of biochar in soils: predictability of O:C molar ratios, Carbon Management. 1 (2010) 289-303.
10. H. Zheng, Z. Wang, X. Deng, et al, Characteristics and nutrient values of biochars produced from giant reed at different temperatures, Bioresource Technology. 130 (2013) 463-471.
11. X. Tan, Y. Liu, G. Zeng, et al, Application of biochar for the removal of pollutants from aqueous solutions, Chemosphere. 125 (2015) 70-85.
12. J. Tang, H. Lv, Y. Gong, et al, Preparation and characterization of a novel graphene/biochar composite for aqueous phenanthrene and mercury removal, Bioresource Technology. 196 (2015) 355-363.
13. J. Chen, Y. Lu, J. Cheng, et al, Effect of starvation on the nitrification performance of constructed rapid infiltration systems, Environmental Technology. 44 (2019) 1408-1417.
14. Q. Yin, B. Zhang, R. Wang, et al, Biochar as an adsorbent for inorganic nitrogen and phosphorus removal from water: a review, Environmental Science and Pollution Research. 24 (2017) 26297-26309.
15. W. Ding, X. Zeng, Y. Wang, et al, Characteristics and performances of biofilm carrier prepared from agro-based biochar, China Environmental Science. 31 (2011) 1451-1455.
16. R. Zhao, J. Li, C. Guo, et al, Filler improvement and purification effects of constructed rapid infiltration facility, Environmental Science and Pollution Research International, 26 (2019) 33654-33669.

Research on Palletizing Robot System of Multi - axis Synchronous Control Technology

Yuan Yang [1], Lei Zhang [2], Ligang Qiang [3], Jinbing Wang [4], Yong Zhang [5], Xudong Yang [6]

[1] School of mechanical engineering, Guizhou University, Guiyang 550000, China
[2] Guizhou tobacco company Guiyang company, Guiyang 550000, China
[3] Aerospace Jiangnan Group Co., Ltd, Guiyang 550000, China
[4] Guizhou tobacco company Guiyang company, Guiyang 550000, China
[5] Aerospace Jiangnan Group Co., Ltd, Guiyang 550000, China
[6] School of mechanical engineering, Guizhou University, Guiyang 550000, China.

Abstract: Due to the increasing demand for personalized objects, the palletizing packaging of personalized objects has attracted people ' s attention. To improve the efficiency of the palletizing system, a palletizing robot based on multi-axis synchronous control technology is proposed in this paper. The robot takes SIMATIC S7 - 1200 PLC as the control core, uses the master command synchronization method to achieve multi-axis synchronization control, and realizes human-computer interaction through PROFINET communication with the touch screen. The multi-sensor is connected with the input and output modules of the PLC, and the review of the grasping object is realized. After the robot is put into use, the flexibility of the palletizing system is improved, the multi-axis synchronous control technology is improved, and the palletizing efficiency of personalized objects is greatly increased.

Keywords: Robot; multi-axis synchronous control; master synchronous; PLC.

1. Introduction

With the increase of people's demand for substance and culture, people's demand for personalized objects is also gradually increasing, so the sales of personalized objects are expanding. As the only way in the production and sales process, handing and palletizing have also ushered in unprecedented challenges. As an important support device for the process, the palletizing robot determines the efficiency of the process. Therefore, a large number of researchers have done a lot of research on the palletizing robot. For example, Hu Min has greatly improved the intelligence of the palletizing robot by combining the machine vision system and the automatic palletizing algorithm into the palletizing system. However, the end effector of the palletizing system is a rigid suction cup, which can only grab at most 2 objects at a time; Yang Shu-yuan established a theoretical model of the electrical system design of the palletizing robot by using the PID control principle combined with the structural characteristics and motion interpolation functions of each articulated arm, which greatly improved the stacking accuracy and the obstacle avoidance success rate of each articulated arm.

In this paper, by using a PLC controller as the control core of the palletizing robot system, a set of multi-axis palletizing robot control systems using multi-axis synchronous control technology is designed.

2. Palletizing System

2.1 Structure of The Palletizing System

The palletizing system is mainly composed of a belt, manipulator, manipulator end actuator, touch screen, double-speed chain, etc. Its structure is shown in Figure 1. The belt is powered by a servo motor, and the start and stop of the motor are controlled by a servo drive. A pair of length measuring sensors, 5 pairs of in-position sensors, and blocking bars are installed on the belt, and their function is to judge the size of the object and its in-position condition. The model used is EPSON LS20-B804S, LS20-B is a high-performance robot that continues at high speed and is cost-effective, high capacity is achieved by increasing the allowable U-axis moment to $1.00kg \cdot m^2$. Secondly, the clock time is improved by high-speed movement, and the time by swinging the arm is shortened. The manipulator end actuator is composed of 5 sets of suction cups, each set of suction

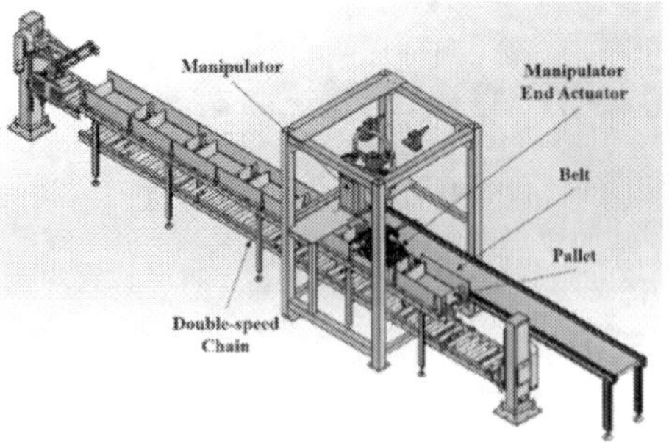

Fig.1 System of palletizing

cups has an independent groove sensor and height measuring sensor, which are used for the detection of the origin and the review of the height of the object. The height measured sensor also has the function of object drop protection. As the benchmark of the manipulator end actuator, the third group can only move in the Z-axis direction, and its fourth group can also move in the X-axis direction. Each set of suction cups consists of 1 suction cupboard and 8 suction cups, each suction cupboard is connected with a solenoid valve to control the suction and blowing of the suction cups. As a human-computer interaction interface, the touch screen can realize the parameter setting of the system, the manual control of the end actuator, and the signal display of each electrical control link. The double-speed chain is used for pallet switching, which can move the palletized pallet to the next station, move the empty pallet to the palletizing position, and send a allow palletizing signal at the same time.

2.2 Rigid palletizing system

The rigid palletizing system has high requirements for objects, and all objects must be highly consistent. In the face of the rapid development of personalized object sales, the system can only be seized by multiple times and with fewer bars. Although it has been further improved compared with manual stacking, it is far from enough to face the current market diversification. As shown in Fig. 2, the three-dimensional figure is a rigid palletizing system with four grips, which can simultaneously adsorb four objects of the same height. In this system, it can grasp multiple objects, but the orders of diversification cannot be coped with. To achieve the orders of diversification, the volume of the manipulator end actuator needs to be reduced, and the number of grasping objects at the same time also decreases. The flexible palletizing system can not only deal with the problem of orders of diversification, but also realize the simultaneous capture of multiple objects, which meets the current market demand.

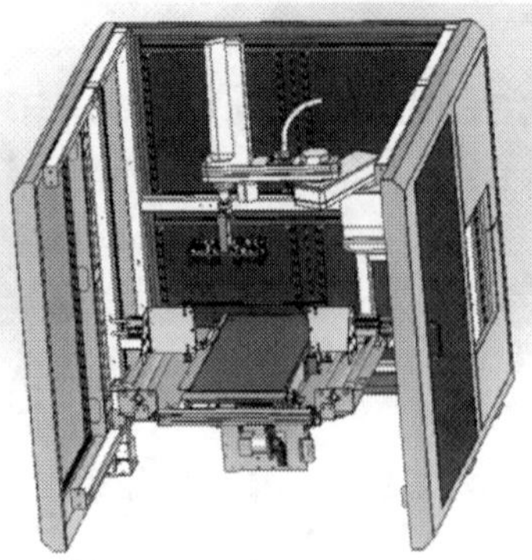

Fig.2 Rigid palletizing system

3. Multi-axis Synchronization Control Technology

Based on the control mode, multi-axis synchronization control can be divided into mechanical synchronization and electronic control synchronization. Mechanical synchronization is composed of a multi-axis and gear system, relying on the mechanical structure to achieve synchronization, this synchronization control mechanism is complex, high machining accuracy requirements, so it has been eliminated. Electronic control synchronization methods mainly include master-slave, cross-coupling, master synchronous, and virtual axis synchronization.[6]-[8].

Master-slave synchronization takes the actual output of the motor as the given parameter of the slave motor to realize the synchronization from the motor to the main motor. The synchronization performance of the control method is difficult to do well, mainly because the given signal is transmitted from the main motor to the slave motor, which makes the synchronization control command of the slave motor lag.

Master synchronization is similar to master-slave synchronization, which is driven by a separate motor for each axis. Each motor driver is controlled by the same command signal source, and then the controller assigns the signal source and sends the allocated signal to the motor drive to complete the whole synchronization control.

Virtual axis synchronization is the combination of master synchronization and cross-coupling synchronization. The control command is filtered by the virtual main motor as the given value of each actual motor. However, this synchronization method needs to complete the calculation of the virtual motor, and there is a problem that the calculated amount of parameter dependence is not small[9].

Based on the above several synchronization methods, because there is no coupling between the motors in the master synchronization method, any motor axis disturbed by the external load will not affect the normal operation of other shafts. Therefore, this paper selects the master synchronization method to control the motion of the end actuator. The system structure diagram is shown in Figure. 3.

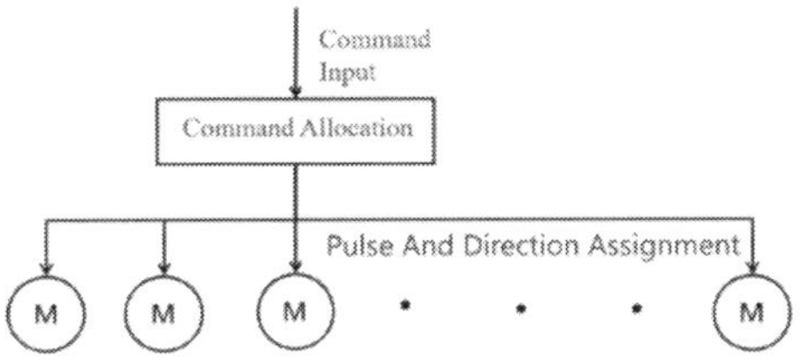

Fig.3 Schematic diagram of master control

4. Hardware Design

The control system is mainly composed of PLC (master and slave), stepper motor, driver, touch screen, and groove sensor. The system includes four X-axis origin sensors and five Z-axis origin sensors, which are used to detect the positions of the X-axis and Z-axis of the sucker. The separate drive of nine motors ensures that the motors do not

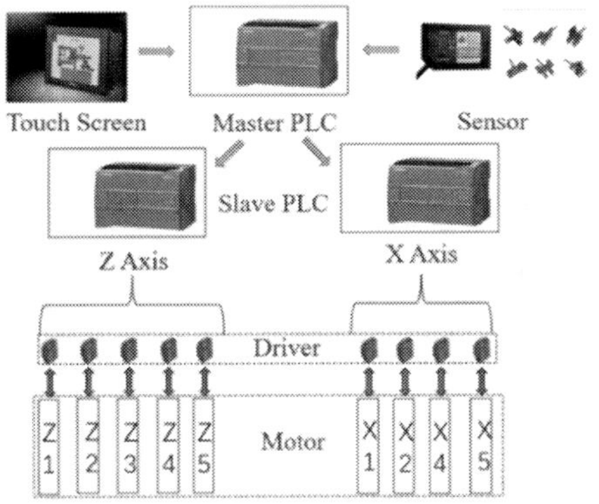

Fig.4 Hardware structure of control system

interfere with each other, and can also visually display the fault information of each motor. The hardware structure is shown in Fig. 4.

PLC is called Programmable Logic Controller, it has complex logic algorithm control ability, but also has the function of timing and counting. The system uses Siemens S7-1200 PLC and S7-1200 PLC as a compact controller that does not limit its function. The controller is mainly for simple and high-precision automation tasks, and it also has high flexibility, which can largely meet the various needs of customers. It realizes the communication with HMI and PLC through the integrated PROFINET interface and can be programmed through this interface. It can also support communication with third-party devices through the open Ethernet protocol.The addresses of the motor output signal and the origin photoelectric input signal of the multi-axis synchronization control system in the PLC controller are shown in Table 1.

Tab.1 I/O allocation

Input Signal		Output Signal	
Address	Function	Address	Function
ZKPLC I0.0-3	X1, X2, X4, X5 Groove Sensor	PLCX Q0.0, Q0.2, Q0.4, Q0.6	X1, X2, X4, X5 Pulse
ZKPLC I0.4-8	Z1, Z2, Z3, Z4, Z5 Groove Sensor	PLCX Q0.1, Q0.3, Q0.5, Q0.7	X1, X2, X4, X5 Direction
		PLCZ Q0.0, Q0.2, Q0.4, Q0.6, Q0.8	Z1, Z2, Z3, Z4, Z5 Pulse
		PLCZ Q0.1, Q0.3, Q0.5, Q0.7, Q0.9	Z1, Z2, Z3, Z4, Z5 Direction

5. Software Design

The Hardware system is the carrier of the control system, and the software system is the key to measuring the stability and reliability of the whole system. By setting the PLC program and the human-machine interface, the multi-axis synchronous control is realized, and the motion control of the human-machine interface on a single axis is realized so that the manipulator end actuator of the palletizing system can achieve synchronous operation and independent control.

5.1 PLC Program Setting

This process is based on the master synchronization mode. The system is connected in series with the enable, reset, human-machine interface reset, X-axis pitch, Z-axis pitch, X-axis, and Z-axis speed, grasping and placing coils, and controlled by a contact point to realize the multi-motor control of a signal source. The X-axis variable distance and Z-axis variable distance are distributed through the object's data sent by the host computer, and then uniformly called to enable each axis to achieve synchronous variable distance. The poor position of the sucker will appear in the process of synchronizing the varying distance of the sucker, resulting in a large load on the sucker, so the synchronization of the system is destroyed. The corresponding alarm program is designed in the system. By comparing the target position and the return target position, the unreturned target position coil is triggered to achieve the alarm effect. The alarm can be removed by reset and man-machine interface reset, which greatly reduces the time of manual troubleshooting.

5.2 Man-machine Interface

The man-machine interface of the palletizing system mainly includes the main interface of palletizing, sensor parameter interface, data interface, palletizing data interface, sucker motor interface, and manipulator interface. The sucker motor interface is the main interface of the control robot, as shown in Fig. 5. The bit state switch in the interface is directly connected to the relevant contacts in the electrical control program through the connection port Net. The corresponding address type is selected and then the corresponding contact address is input for communication, to realize the motion control of the X and Z axes. When the sucker arrives at the slot sensor, the electric opportunity automatically stops. This method is to obtain the signal of the suspension shaft by connecting a zero-point switch in series with a constant contact point of a positive and negative point of touch.

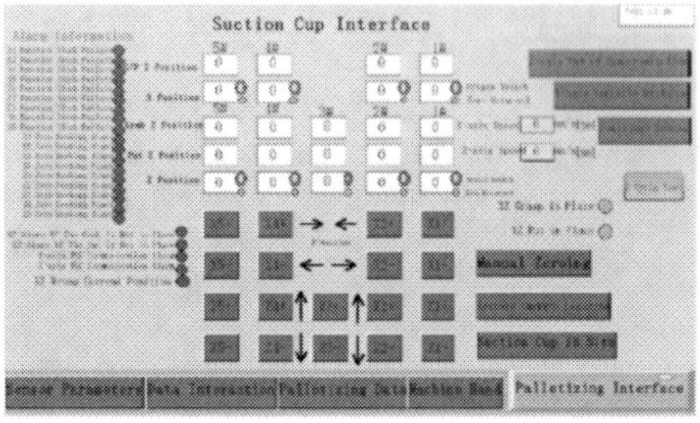

Fig.5 Human-computer interface

6. Conclusion

To improve the palletizing efficiency of personalized objects, this paper designs a palletizing system based on multi-axis synchronous control technology, which breaks the limitations of the original rigid palletizing system and realizes the simultaneous grasping of multiple objects. The system takes Siemens S7 - 1200 PLC as the control core and speeds up the operation process of the system by setting the master-slave PLC. The master-slave synchronous mode is used to control the movement of each motor, and the synchronization of each sucker is realized. Then, the human-computer interaction interface based on the touch screen is designed to control the drive of a single motor. Through field tests, the stacking efficiency of the system is far greater than that of the rigid sucker, which improves the stacking efficiency of personalized objects and has good stability.

Acknowledgments

This work was financially supported by Guizhou Tobacco Corporation Science and Technology Project (Research on Optimization of Sorting Scheduling and Delivery Service Based on Business Flow Data Driven (No. 2022)) and National Key Research and Development Plan (Feeder Receiver Disassembly Robot and Wireless Interference Intelligent Detection System [2019YFB1312704]).

References

1. Chunsheng Hu, Hongxing Wei, Xiaopeng Yan, etc. Research and Application of Palletizing Robot[J]. Computer Engineering and Application, 2022, 58(02): 57-77.
2. DOU Lei. Research on the Status and Development Palletizing Robots[J]. Times Agricultural, 2018, 45(7): 87.
3. Zhenpeng Feng, Haibo Lin, Pengcheng Liu, etc. Research on Application of Domestic Palletizing Robot[J]. Mechanical Engineer, 2017(12): 63-65.
4. Min Hu. Research on the Application of Visual System in Profiled Smoke Palletizing Robot System[D]. Chongqing University, 2017.
5. Shuyuan Yang. Design on the Electrical Control System of the Palletizing Robot[J]. Journal of Agricultural Mechanization Research, 2023, 45(03): 138-142.
6. Shigang Yang, Junyan Duan, Shedang Xin. Application of Siemens motion controller multi axis synchronous control in switch rail forging system[J]. Journal of Machine Design, 2021, 38(S1): 256-259.
7. Jing Cui. Design of Control System of Multi-Function Envelope Machine Based on Multi-axis Synchronous Control Technology[D]. Jiangsu University, 2010.
8. Yudong Zhang. The Research on Multi-axis Synchronous Control and its implementation based on Parallel Bus[D]. Huazhong University of Science and Technology, 2011.
9. Ruochen Liu. Research on Synchronous Control of Multi-motor System[D]. Zhejiang University, 2020

A novel geometric error identification and prediction approach

Jinwei Fan, Peitong Wang *, Zhuang Li

College of Mechanical Engineering & Applied Electronics Technology, Beijing University of Technology, Beijing 100124, China

Abstract. This paper proposed an integrated geometric error identification and prediction method to solve the uncertainty problem of the PDGEs of rotary axis. First, based on homogeneous transform matrix (HTM) and multi-body system (MBS) theory, The transfer matrix only considering the C-axes rotated is derived to the position error model. Then a geometric errors identification of rotary axis is introduced by measuring the error increment in three directions. Meanwhile the geometric errors of C-axis are described as truncated Fourier polynomials caused by fitting discrete values. Thus, The geometric error identification is converted into the function coefficient. Finally, the proposed new prediction and identification model of PDGEs in the global frame are verified through simulation and experiments with double ball-bar tests.

Keywords Geometric error identification, Prediction, Rotary axis, Double ball bar, Five-axis machine tool

1. Introduction

Five-axis CNC machine tools provide greater productivity, better flexibility, and less fixture time than three axis machining centers, because the cutting tool can approach the workpiece from any direction. However, the two rotary axes would bring in additional geometric errors subsequently, such as squareness between a rotary axis and a translational axis [1]. Therefore, an accurate and efficient measurement method of geometric errors is a key prerequisite for the improvement of five-axis machine tool accuracy [2]. If there is an effective method to improve or compensate for those deviations, the machining performance of multi-axis machines will be improved drastically.

In the past decades, some researches have been taken to measure and identify the errors that are inherent to a rotary axis of a multi-axis machine tool. Tsutsumi and Saito proposed a calibration method using the simultaneous four-axis control technique for five-axis machining centers with a tilting rotary table. The eight deviations were estimated by the observation equations from the ball bar measured results [3]. Zargarbashi identified eight link errors of a five-axis machine through measuring the tool center position deviations at a number of five-axis poses by using a ball-bar sensor [4]. Zargarbashi and Mayer presented a method consisting of five double ball bar tests to evaluate five trunnion axis motion errors including axial, radial and tilt errors.

However, some others may have such a capacity, require complex mathematical formulations and especially involve a complicated measurement process to identify the error parameters at present. Therefore, in this paper, a novel prediction approach for geometric errors of rotary axis is proposed, which is effective and uncomplicated to improve the accuracy of identification.

2. Description of geometric errors

Geometric errors caused mainly by manufacturing and assembly defects are considered to be systenmatic errors. The geometric errors are usually categorized as position-independent geometric errors (PIGs) and position-dependent geometric errors (PDGEs). PDGEs have a certain relationship to the machine positions, while PIGs are constant values which are usually produced during the assembly process[10]. $\delta_x(C)$, $\delta_y(C)$ $\delta_z(C)$, $\varepsilon_x(C)$, $\varepsilon_y(C)$, $\varepsilon_z(C)$ are position-dependent geometric errors of C-axis. The position -dependent geometric errors is fitted as the truncated Fourier polynomial as shown in Eq.(1). The fitting method includes Fourier polynomial, simple polynomial, Legendre polynomial, Chebyshev polynomial and so on. Compared to truncated Fourier polynomial, The random unpredictability of the error can be reflected, and the lower and upper limits of the position-dependent geometric satisfy the Dirichlet condition. Hence, the truncated Fourier polynomials is introduced to fit the PDGs. In this paper, error identification of rotary axis is the focus of research and the position-dependent geometric errors of rotary axis can be fully expressed in Eq.(1)

$$f(\theta)=\sum_{i=1}^{n} A_i \sin\frac{\theta}{2i-1} \qquad 1$$

* Corresponding author: 504789981@qq.com

Where f(θ) and θ denote geometric errors function and angles of rotation respectively. Ai represent a set of the coefficients for function.

3. Experiment and identify geometric errors of rotary axis

The identification and experiment of rotating axis geometric error are introduced in this paper. In previous studies, the identification of translational axis geometric error has been clearly specified, but the identification of rotating axis error has not been clearly specified. The geometric errors of rotary axis can be detected using renishaw XR20W double ball bar device. The remaining 10 PDGs can be derived from the multi-body theory and the Houston transformation matrix which contains six the linear errors and four angle errors. In order to express more briefly, the five PDGs of C-axis are identified by using DBB, which is taken as an example of rotary axis.

3.1 Measurement

The detection of the C axis is divided into three models: C-X, C-Y and C-Z . The detection mode of the cue meter is detected in the X, Y and Z directions respectively. As shown in Fig.1, assuming P_c is the ideal origin of C axis coordinates, P'_c is the projection of the actual position of the far point of C axis coordinate system in the XZ plane, delta x is the projection of the positioning error of C axis in the x direction, delta is the projection of the degree error of C axis in the XZ plane. The PDGs of the C-axis will inevitably lead to the error of the position of different height points of the C axis in the X direction. During the measurement, firstly to ensure the accuracy of the experiment, the angle of B and C of the double pendulum head should be kept at 0° in the initial stage. Then when the C axis is rotated, the length of the measuring rod will change in the X,Y and Z directions. Two measuring points (c_1,c_2) need to be selected that affects the accuracy of geometric error element parameters,as shown in Fig2 and Fig. 3. Finally the geometric error parameters of the C axis are deduced through the position relationship. It is obvious that the main objective of this experiment is to measure the values of delta L in the X,Y, and Z directions of c1 and c2.

$$\Delta L_c = [\Delta L_{cx} \quad \Delta L_{cy} \quad \Delta L_{cz}]^T \tag{2}$$

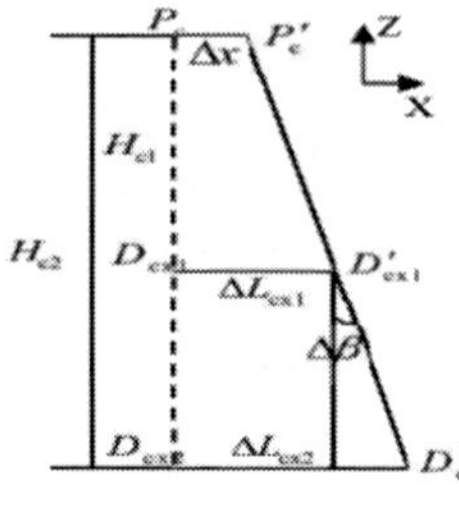

Fig.1 C-X measurement mode error generation mechanism

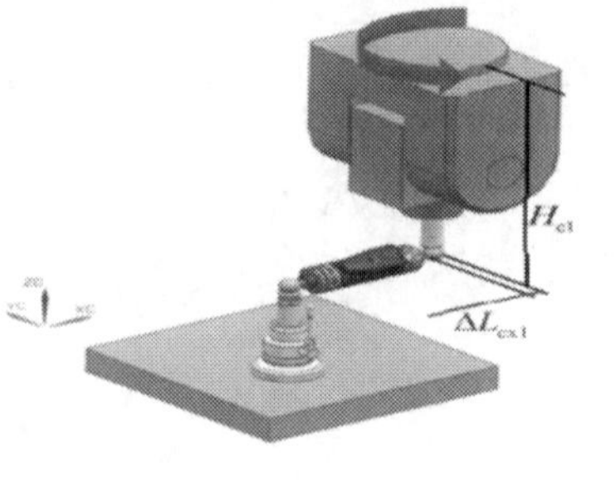

Fig.2 C-X-1model

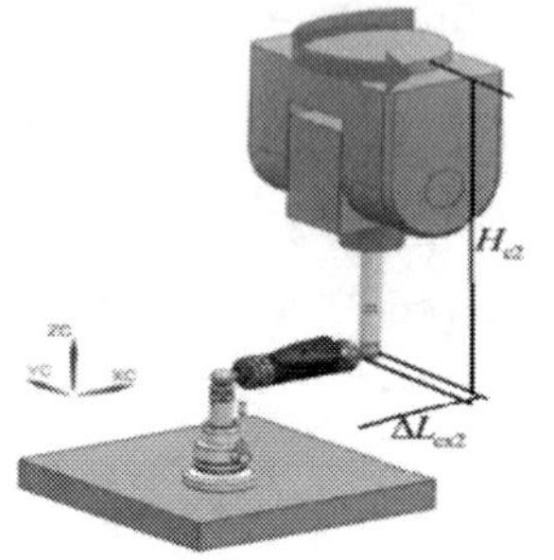

Fig.3 C-X-2 model

The ideal point D_{CX} and actual point D'_{cx} can be calculated from the HTM above Section 2. Therefore, the error increment of the length (ΔL_{cx}) of the measuring instrument in the X direction can be expressed as

$$\Delta L_{cx} = D'_{cx} - D_{cx} \tag{3}$$

$$D_{cx} = \begin{bmatrix} \cos(c) & -\sin(c) & 0 & 0 \\ \sin(c) & \cos(c) & 0 & 0 \\ 0 & 0 & 1 & 0 \\ 0 & 0 & 0 & 1 \end{bmatrix} \begin{bmatrix} 0 \\ 0 \\ -H_c \\ 1 \end{bmatrix} \tag{4}$$

$$D'_{cx} = \begin{bmatrix} 1 & 0 & S_{xc} & 0 \\ 0 & 1 & -S_{yc} & 0 \\ -S_{xc} & S_{yc} & 1 & 0 \\ 0 & 0 & 0 & 1 \end{bmatrix} \begin{bmatrix} \cos(c) & -\sin(c) & 0 & 0 \\ \sin(c) & \cos(c) & 0 & 0 \\ 0 & 0 & 1 & 0 \\ 0 & 0 & 0 & 1 \end{bmatrix} \begin{bmatrix} 1 & -\xi_z(c) & \xi_y(c) & \delta_x(c) \\ \xi_z(c) & 1 & -\xi_x(c) & \delta_y(c) \\ -\xi_y(c) & \xi_x(c) & 1 & \delta_z(c) \\ 0 & 0 & 0 & 1 \end{bmatrix}$$
$$\begin{bmatrix} 1 & -S_{bc} & 0 & 0 \\ S_{bc} & 1 & -S_{xb} & 0 \\ 0 & S_{xb} & 1 & 0 \\ 0 & 0 & 0 & 1 \end{bmatrix} \begin{bmatrix} 0 \\ 0 \\ -H_c \\ 1 \end{bmatrix} \tag{5}$$

If Eq.(11) and Eq.(12) are put into Eq(10), the result of described

simplification will be as

$$\begin{bmatrix} \cos(c) & -\sin(c) & -H_{c1}\cos(c) & -H_{c1}\sin(c) \\ \cos(c) & -\sin(c) & -H_{c2}\cos(c) & -H_{c2}\sin(c) \end{bmatrix} \begin{bmatrix} \delta_x(c) \\ \delta_y(c) \\ \xi_y(c) \\ \xi_x(c) \end{bmatrix} = \begin{bmatrix} \Delta L_{cx1} + H_{c1}S_{xc} + H_{c1}\sin(c)S_{bc} \\ \Delta L_{cx2} + H_{c2}S_{xc} + H_{c2}\sin(c)S_{bc} \end{bmatrix} \tag{6}$$

It is obvious that the matrix equation possesses four unknown parameters and the rank of coefficient matrix can be determined as 2. Considering the error source parameters unsolved in the matrix equation, another dimension matrix equation needs to be established which can identify the geometric error parameters. Therefore,

https://doi.org/10.1051/matecconf/202236301007

the error relationship needs to be confirmed in the C-Y direction.

(2) C-Y measurement

Similarly, the implementation of C-Y measurement is same to C-X measurement. During the measurement process, ΔL_{cy1} and ΔL_{cy2} need to be identified respectively. By combining Eq.10 and Eq.15, the error parameters ($\delta_x(c)$、$\delta_y(c)$、$\xi_x(c)$、$\xi_y(c)$) concerning the position of the four terms of the C-axis. C-Y measurement can be inversely solved.

$$\Delta L_{cy} = D'_{cy} - D_{cy} \qquad 7$$

$$\begin{bmatrix} \cos(c) & -\sin(c) & -H_{c1}\cos(c) & -H_{c1}\sin(c) \\ \cos(c) & -\sin(c) & -H_{c2}\cos(c) & -H_{c1}\sin(c) \\ \sin(c) & \cos(c) & -H_{c1}\sin(c) & H_{c1}\cos(c) \\ \sin(c) & \cos(c) & -H_{c1}\sin(c) & H_{c2}\cos(c) \end{bmatrix} \begin{bmatrix} \delta_x(c) \\ \delta_y(c) \\ \xi_y(c) \\ \xi_x(c) \end{bmatrix} = \begin{bmatrix} \Delta L_{cx1} + H_{c1}S_{xc} + H_{c1}\sin(c)S_{bz} \\ \Delta L_{cx2} + H_{c2}S_{xc} + H_{c2}\sin(c)S_{bz} \\ \Delta L_{cy1} - H_{c1}S_{xc} - H_{c1}\cos(c)S_{bz} \\ \Delta L_{cy2} - H_{c2}S_{xc} - H_{c2}\cos(c)S_{bz} \end{bmatrix} \qquad 8$$

It can be obtained by matrix inversion, why singular matrices have not been taken into consideration(c≠0).

$$\xi_x(c) = \frac{\cos(c)(\Delta L_{cy1} - \Delta L_{cy2}) + \sin(c)(\Delta L_{cx2} - \Delta L_{cx1})}{H_{c1} - H_{c2}} - \cos(c)S_{yc} - \sin(c)S_{xc} - S_{bz} \qquad 9$$

$$\xi_y(c) = \frac{\cos(c)(\Delta L_{cx2} - \Delta L_{cx1}) + \sin(c)(\Delta L_{cy2} - \Delta L_{cy1})}{H_{c1} - H_{c2}} - \cos(c)S_{xc} + \sin(c)S_{yc} \qquad 10$$

$$\delta_y(c) = \frac{\sin(c)(H_{c2}\Delta L_{cx1} - H_{c1}\Delta L_{cx2}) + \cos(c)(H_{c1}\Delta L_{cy2} - H_{c2}\Delta L_{cy1})}{H_{c1} - H_{c2}} \qquad 11$$

$$\delta_x(c) = \frac{\cos(c)(H_{c1}\Delta L_{cx2} - H_{c2}\Delta L_{cx1}) + \sin(c)(H_{c1}\Delta L_{cy2} - H_{c2}\Delta L_{cy1})}{H_{c1} - H_{c2}} \qquad 12$$

(3) C-Z measurement

In the C-Z detection model, double bar ball will be utilized to measure the displacement error of C-axis. When the C-axis rotates, the double bar ball will expand and contract in the axial direction. Considering that axis C is alone without the other axes involved during the motion process, the axial displacement increment ΔL_{cz} is the error source parameter $\delta_z(c)$. Thus $\delta_z(c)$ can be obtained as

$$\Delta L_{cz} = D'_{cz} - D_{cz} = \delta_z(c) \qquad 13$$

Thus, based on obtained constant, the five PDGs of C axis will be identified totally.

4. Results of calculation and simulation

In the third part, a method of geometric error identification of rotation axis is proposed, which only taken picking two points at random into consideration. However, considering the factor that the error of the guide machine varies with the movement of the machine, a comprehensive identification method for the rotary axis is proposed. For the first stage, all 16 geometric errors were described as third-orders truncated Fourier polynomials, and the identification of the polynomial coefficients were based on the selection of measurement points through an observability index. For the second stage, based upon the identification results of stage one, an integrated geometric error identification model of rotating axis is proposed to improve the identification accuracy.

In the simulations, the identification accuracy of different sets of measurement points was compared. The results of identified PDGs are illustrated in Fig. 4, which are fitted by third orders truncated Fourier series. From these, it can be seen that for both the predicted set and the measured sets of points, the motion positions of C-axis covered the entire measurement workspace and ensured a valid identification accuracy obtained from these sets of measurement points. The red lines represent the fitting curve of measured results, the blue lines represent the predicted results, and the green lines represent residual errors which are calculated by the subtraction between the measured error data and the predicted error data. It is obvious that the predicted values gets close to the measured values by comparing with the measured results. The values of R-square were about fitting curve shown in Table.1 that the maximum value of R-square gets to 0.976. As a matter of fact, the identification method used in this paper possesses less term numbers, smaller fitting residuals, and higher fitting accuracy.

Table 1 The R-square of five PDGEs

PDGEs	$\delta_x(c)$	$\delta_y(c)$	$\delta_z(c)$	$\xi_x(c)$	$\xi_y(c)$
R-square value	0.973	0.969	0.976	0.962	0.963

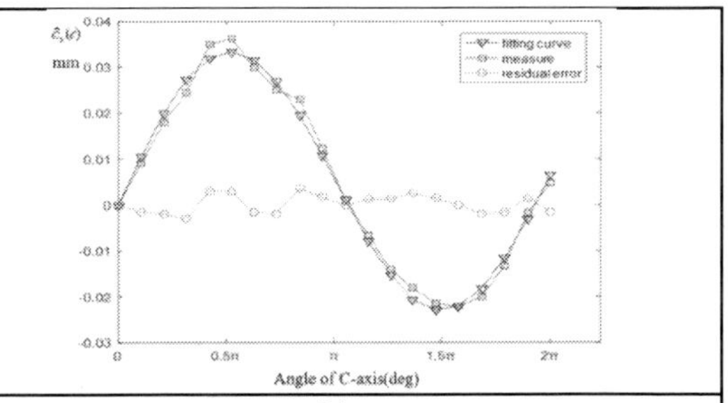

(a)

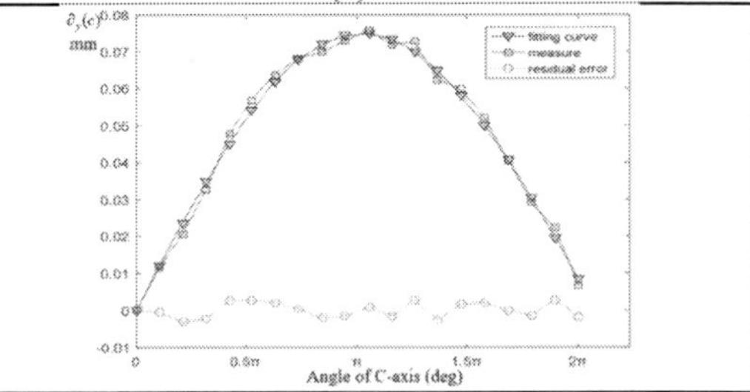

(b)

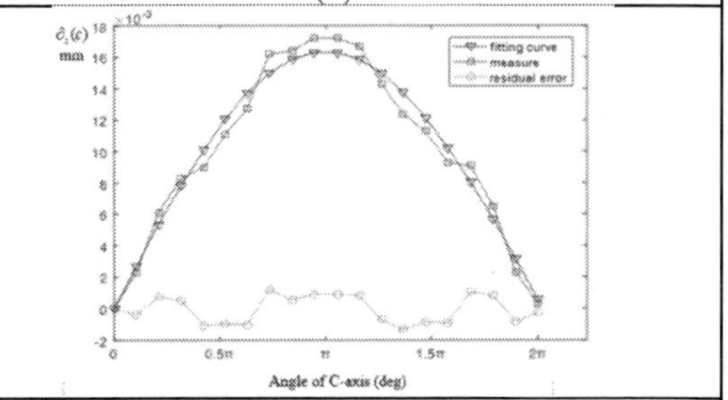

(c)

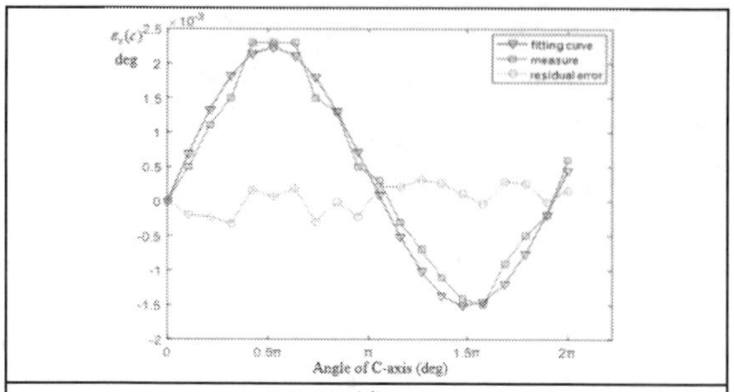

(d)

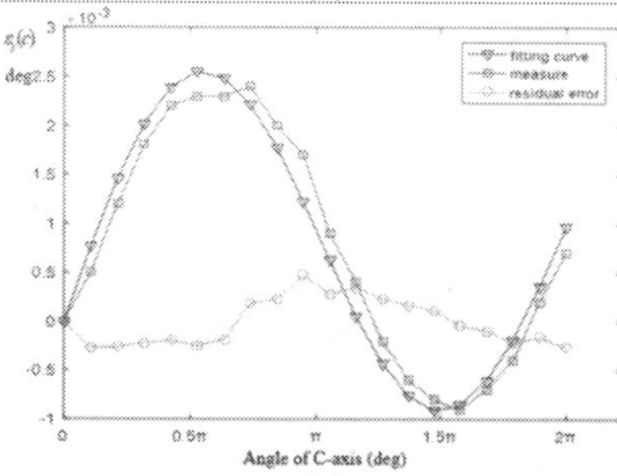

(e)

Fig.4. Result of identification

5. Conclusion

This paper has proposed a novel and efficient on-machine measurement methodology for identification and prediction of the five PDGs of the rotary axes on a five-axis machine tool by using double ball bar. Error identification model was built based on the multi-body system theory and HTM during the test by measuring length change of double ball bar in the three measure model. In this model, the PDGs of C-axis have been identified completely. Meanwhile fueled by the uncertainty of PDGs, a novel geometric error prediction approach has been devoted to support us to confirm the geometric error conveniently.

References

1. Tsutsumi M, Saito A (2003) Identification and compensation of systematic deviations particular to 5-axis machining centers. Int J Mach Tool Manuf 43(8):771–780
2. Qingzhao Li, Wei Wang Measurement method for volumetric error of fve-axis machine tool considering measurement point distribution and adaptive identifcation process International Journal of Machine Tools & Manufacture 147 (2019) 103465
3. M. Tsutsumi, A. Saito, Identification of angular and positional deviations inherent to 5-axis machinincenters with a tilting-rotary table by simultaneous four axis control movement, Int. J. Mach. Tools Manuf. 44 (12–13) (2004) 1333–1342
4. Y. Abbaszadeh-Mir, J.R.R. Mayer, G. Cloutier, C. Fortin, Theory and simulation for the identification of the link geometric errors for a five-axis machine tool using a telescoping magnetic ball-bar, Int. J. Prod. Res. 40 (18) (2002) 4781–4797
5. S.H.H. Zargarbashi, J.R.R. Mayer, Assessment of machine tool trunnion axis motion error using magnetic double ballbar, Int. J. Mach. Tools Manuf. 46 (14) (2006) 1823–1834
6. S. Weikert, A new device for accuracy measurements on five axis machine tools, CIRP Ann.-Manuf. Technol. 53 (1) (2004) 429–432.
7. S. Ibaraki, C. Oyama, H. Otsubo, Construction of an error map of rotary axes on a five-axis machining center by static r-test, Int. J. Mach. Tools Manuf. 51 (3) (2011) 190–200
8. S.H.H. Zargarbashi, J.R.R. Mayer, Single setup estimation of a five-axis machine tool eight link errors by programmed end point constraint and on the fly measurement with capball sensor, Int. J. Mach. Tools Manuf. 49 (10) (2009) 759–766
9. C.F. Hong, S. Ibaraki, C. Oyama, Graphical presentation of error motions of rotary axes on a five-axis machine tool by static r-test with separating the influence of squareness errors of linear axes, Int. J. Mach. Tools Manuf. 59 (2012) 24–33.
10. Lee KI, Yang SH (2013) Robust measurement method and uncertainty analysis for position-independent geometric errors of a rotary axis using a double ball-bar. Int J Precis Eng Manuf 14(2):231–239

Research on a new high-quality and high-efficiency SLM forming method based on interlayer strategy

Peng Wang[1, 2], Dongju Chen[1, 2] and Gang Li [1, 2]

[1] Mechanical Industry Key Laboratory of Heavy Machine Tool Digital Design and Testing, Faculty of Materials and Manufacturing, Beijing University of Technology, Beijing 100124, China.
[2] Beijing Key Laboratory of Advanced Manufacturing Technology, Faculty of Materials and Manufacturing, Beijing University of Technology, Beijing 100124, China.

Abstract. To ensure the forming quality of SLM and to improve the forming efficiency of SLM, this paper investigates the forming effect of Ti6Al4V specimens formed by SLM based on an interlayer strategy. The conditions for improving the relative density of the specimens are high laser power, small point distance, long exposure time and long hatch spacing. High laser power provides sufficient laser energy density in the formation of thicker powder layers, reduced point distance ensures wetting behavior of the melt pool, longer exposure time provides the time required to fully melt the powder and increased hatch distance facilitates dispersion of the laser energy density. Process parameters have a strong influence on the relative density of SLM formed specimens, and by adjusting the process the highest density of 99.99% can be achieved. The microstructure of the specimens consisted mainly of acicular martensite α' and columnar β grains due to the effects of repeated thermal cycling. Due to the presence of martensite α', the tensile and yield strengths of the specimens are increased, but the elongation of the specimens is reduced.

Keywords: Selective laser melting; keywords covered; microstructure and mechanical properties.

1. Introduction

Ti6Al4V is the most common nonferrous alloy and is widely used in aerospace, marine, and biomedical applications [1]. It has long-term biocompatibility, excellent strength-to-weight ratio, superior corrosion resistance, outstanding low-temperature brittleness resistance, and high fracture toughness. Despite its many advantages, Ti6Al4V has proven to be a difficult material to machine. There is significant material waste, long lead times, and high manufacturing costs using conventional casting, forging, and rolling processes [2]. Selective laser melting (SLM) technology uses a laser beam and metal powder as the fabrication tool, eliminating the need for traditional fabrication tools such as fixtures or molds, thereby significantly reducing the processing cycle time [3].

Improving the efficiency of SLM has been a hot research direction in the field of additive manufacturing. Currently, the forming efficiency can be improved by using multiple lasers, but the more important purpose of using multiple lasers is to increase the size of SLM formed parts. In previous research, smaller layer thicknesses are generally used to ensure forming accuracy and quality, which leads to a decrease in forming efficiency. Compared with increasing the scanning speed, using a high layer thickness can reduce the melting time per-layer and thus have a higher forming efficiency. Increasing the layer thickness by about 1.6 times can reduce the fabrication time by 40% while maintaining reasonable surface roughness [4]. Savalani et al. [5] investigated the effect of layer thickness on magnesium formed by SLM and found that the specimens with low layer thickness had smoother and flatter surfaces. When the thickness increased, the surface exhibited higher surface roughness values. Sufiiarov et al. [6] investigated the relationship between the layer thickness and the relative density, microstructure and mechanical properties of nickel-based superalloys formed by SLM. It was found that specimens with a layer thickness of 30μm had higher strength and lower elongation than those with a layer thickness of 50μm. Nguyen et al [7] prepared Inconel 718 specimens with different powder layer thicknesses (20, 30, 40 and 50μm). The results showed that the thinner the layer thickness, the higher the relative density and the higher the dimensional accuracy. The study by Greco [8] showed that thicker powder layers would lead to a decrease of relative density and a decrease of microhardness due to the limited depth of laser energy penetration into the powder layer. The results of GAO et al. [9] showed that the relative density increases gradually with decreasing laser volume energy density when the layer thickness is less than or equal to 40μm, while the relative density increases and then decreases with decreasing laser volume energy density when the layer thickness is greater than 40μm.

Previous studies have shown that increasing the layer thickness can improve the forming efficiency, but to a certain extent it will affect the quality of the formed specimen. Therefore, using the appropriate layer thickness at the appropriate part of the specimen is the key to ensure the high-quality and high-efficiency forming of the formed specimen. This paper focuses on a new process approach to improve forming efficiency using interlayer strategies to ensure higher forming efficiency in parts that require efficient forming. The effects of process parameters on the wettability, relative density, defects, microstructure and mechanical properties of specimens formed by the interlayer strategy are investigated in detail by single track and multi-layer SLM fabrication.

2. Experimental details

2.1 Materials and experimental equipment

The experiment uses the gas atomized Ti6Al4V powder material produced by Falcontech. The powder is nearly spherical particles (shown in Fig. 1a), with a particle size range of 15-53μm (Fig. 1b). The particle size detected by the laser particle size analyzer is D10: 16.4μm, D50: 32.13μm, D90: 57.54μm. The experiment was completed in the AM400 (Renishaw plc, London, UK) laser melting forming equipment. The entire experiment was carried out in a closed cabin filled with argon as a protective gas. AM400 uses a continuous laser mode with a maximum forming power of 400W and an Nd: YAG laser with a wavelength of 1075nm.

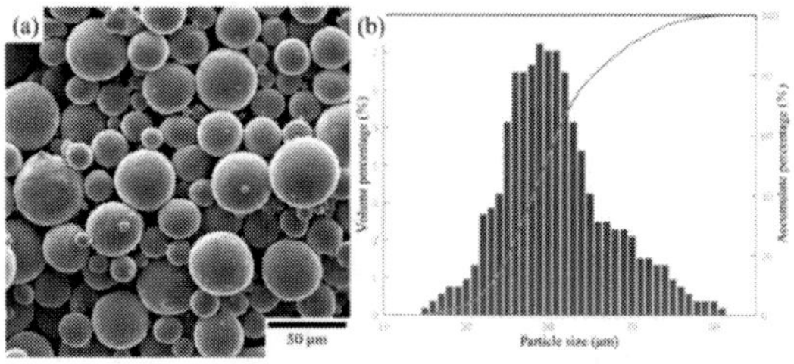

Fig. 1 Powder (a) morphology of Ti6Al4V; (b) particle size distribution of Ti6Al4V

The experiment aims to improve the forming efficiency and ensure the forming quality. The forming strategy of only spreading powder on some layers without laser forming is used to improve the forming efficiency. As shown in Fig. 2, the powder spreader does not perform laser forming after spreading a layer of powder and then spreading a layer of powder without laser forming. When spreading the third layer of powder, the laser forms the specimen according to preset parameters. In this way, the efficient strategy of the interlayer strategy is realized. Assuming that the preset layer thickness is 50μm and the actual layer thickness is 150μm, efficient forming is achieved, while the high quality forming strategy using preset layer thickness forming in critical areas can be realized. The preset experiment scanning strategy is based on the curved scanning strategy, i.e., the angle between the Nth and N+1th layers is 67°.

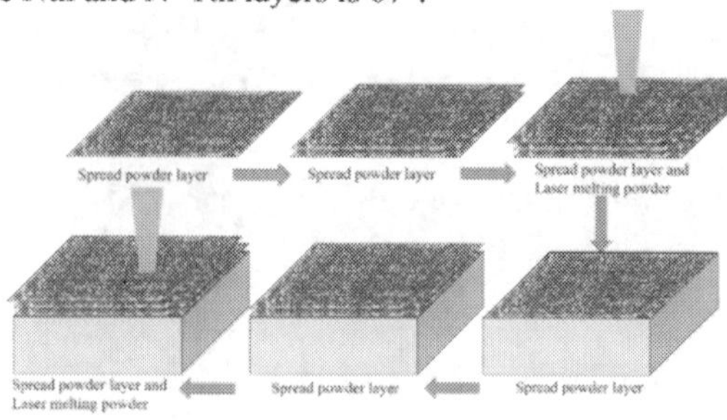

Fig. 2 Schematic diagram of the interval of the powder layer thickness

2.2 2.2 Experimental methods

The laser power, exposure time, and point distance of the single track formed on the specimen-substrate were respectively set to 300-350W (in an increment of 50W), 40-120μs (in an increment of 20μs), and 35-105μm (in an increment of 17.5μm). The preset layer thickness is 50μm, and the actual layer thickness is 150μm based on the interlayer strategy shown in Fig. 2. Table 1 shows the process parameters used in the single track. Based on the experimental results of the single track, circular fabricated parts with a diameter of 8mm were prepared according to the parameters shown in Table 2 to evaluate the quality of the process parameters. After slicing the multilayer blocks with wire cutting equipment (Cmne, Beijing, China), the metallographic specimens are grinded and polished according to standard metallographic procedures, and then processed with Kroll reagent (2ml HF, 6ml HNO3, and 90ml H2O solution) for etching. Images of specimen cross-sections were analyzed and measured using ImageJ (NIH, Bethesda, MD, USA) to calculate the relative density. The morphology of the single-track and multi-layer specimens was observed with an optical microscope (VK-X200, KEYENC, Japan), and the microstructure of the specimens was observed with a scanning electron microscope (SEM) (Phenom XL; Holland). A universal tensile testing machine (5966; Instron, Boston, MA, USA) was used to test the specimen at a tensile rate of 0.01 mm/s at room temperature.

Table 1 Process parameters of the single-track experiments

Parameter	Value	Increment
Laser Power (W)	300-350	50
Layer thickness (μm)	150 (actual layer thickness)	--
Exposure time (μs)	40-120	20
Point distance (μm)	35-105	17.5

https://doi.org/10.1051/matecconf/202236301008

Table 2 The process parameters of the different multilayer fabrication specimens

Parameter	Value	Increment
Laser Power (W)	380	--
Layer thickness (μm)	150 (actual layer thickness)	--
Exposure time (μs)	100-220	20-40
Point distance (μm)	25-65	10-15
Hatch spacing (μm)	50-110	20

3. Results and discussion

3.1 Single-track SLM experiments

Uniform, smooth and stable single-tracks are the guarantee for obtaining high relative density specimens. The morphology of the single-tracks under different process parameters is shown in Fig. 3. The horizontal coordinate indicates the point distance, with the values increasing from 35 to 105μm (in an increment of 17.5μm), and the vertical coordinate indicates the exposure time, with the values increasing from 40μs to 120μs (in an increment of 20μs). As shown in Fig. 3 (a-b), at the laser power of 300W, the single track shows an unstable or even distorted form when the exposure time is less than 100μs or the point distance is greater than 70μm. This is because the exposure time is too short or the point distance is too large, and the laser energy density of the laser input cannot completely melt the powder. When the exposure time is longer than 100μs or the point distance is less than 52.5μm, the laser energy density can reach a moderate state and the single-track is stable. As shown in Fig. 3 (c-d), a stable single track can be obtained when the laser power is 350W, at a point distance of 35μm and the exposure time higher than 60μs. When the point distance increases, it is not easy to ensure the wettability of the molten pool, and the single-track presents a distorted state. This is because the relatively high laser energy density at high laser power tends to lead to instability of the melt pool, but the high laser power can ensure sufficient melting of the thick powder layer. Compared with Fig. 3, the high laser power can ensure the stability of the single track in the relatively large floating range of exposure time.

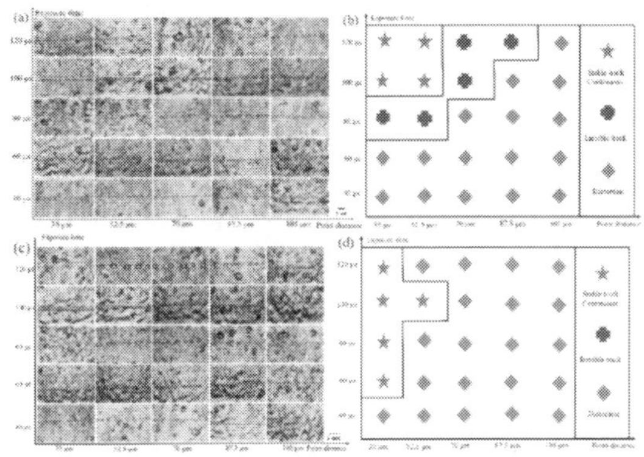

Fig. 3 Morphological characteristics of single-track (a) surface characteristics under the laser power of 300W; (b) process results of (a); (c) surface characteristics under the laser power of 350W; (d) process results of (c)

3.2 Multi-layer SLM fabrication

Based on experimental research of single-track, high laser power allows for a wide range of floating process parameters. In this section, the exposure time, point distance and hatch spacing were set to 100-220μs (in increments of 20-40μs), 25-65μm (in increments of 10-15μm) and 50-110μm (in increments of 20μm), respectively, in order to fully investigate the process required to achieve high relative density of the specimens.

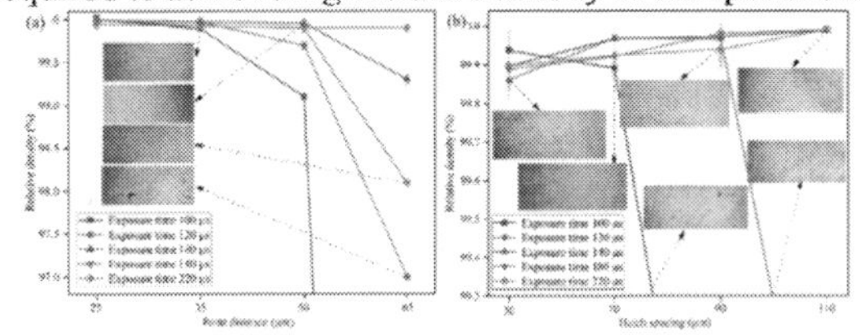

Fig. 4 Relative density curve (a) relative density curve under same hatch spacing; (b) relative density curve under different hatch spacing

As shown in Fig. 4(a), the optimal combination of exposure time and point distance was investigated at a hatch spacing of 70μm. At a fixed exposure time, the relative density showed a decreasing trend with the increase of the point distance. This is because the lap quality between the melt pools becomes poorer as the point distance increases, leading to more un-melted defects between the melt pools. Sufficient laser energy density can be ensured to melt the powder with a small point distance. In addition, a higher relative density can be maintained at the higher exposure time. This is because the actual layer thickness of the interlayer strategy is 150μm, while thicker powder layers require longer melting time to ensure complete melting of the powder. As shown in Fig. 4(b), the fixed point distance is 35μm, and the optimal combination of exposure time and hatch spacing is investigated. As the hatch spacing increases, the relative density shows an increasing trend at a fixed exposure time. This is because the energy density is higher at high laser power and small point distance, and increasing the hatch spacing can help disperse the laser energy density. When the exposure time is 100-120μs, the relative density is unstable. As the hatch spacing increases, un-melted defects appear between the melt pools. The longer exposure time allows a larger floating space for the hatch spacing. In this experiment, the relative densities of most of the formed specimens were higher than 99%, and the highest density could reach 99.99%.

The main reason affecting the relative density is defects due to the choice of process parameters. When the powder melts into a liquid and then solidifies rapidly, defects such as holes and inclusions will inevitably appear due to shrinkage of the melt, transient stresses, and the presence of gases and impurities [10]. In this section, the mechanism of defect formation is investigated by analyzing the cross-section of the specimen.

Fig. 5 shows the defects that resulted in the decrease of relative density. Fig. 5(a) shows the un-melted defects between the molten pools and un-melted powder can be observed in the un-melted region. Fig. 5(b) shows the microporous defects in the molten pool, and Fig. 5(c) shows the spheroidization inside the specimen. The defect of un-melted defects between the molten pools is caused by poor adhesion of the molten pool, as shown in Fig. 5(d). This defect is mainly caused by the improper selection of process parameters (too large setting of point distance or hatch spacing is the main cause of un-melted defects between molten pools). One is that the gas cannot escape from the bottom of the molten pool during solidification and cooling because the solubility of the gas decreases and the gas cannot be released in time, as shown in Fig. 5(e). Another reason for the micropores is that the gas in the hollow powder did not escape in time, as shown in Fig. 5(f). Spheroidization defects are mainly caused by splashing during the forming process. The spattered particles solidify during the flying process and partly fall into the molten pool and form spheroidization inside the specimen after solidification. Some of the splashes splashed onto the surface of the solidified specimen, forming surface spheroidization. Some spatters fall on the unformed powder and form inclusions. After being melted, internal spherification of the specimen may also form, as shown in Fig. 5 (g). The cause of the defect may also be the poor wetting between the single-track and the substrate, as shown in Fig. 5 (h-i). When the single-track and the substrate do not bond well, internal porosity or spherification can result.

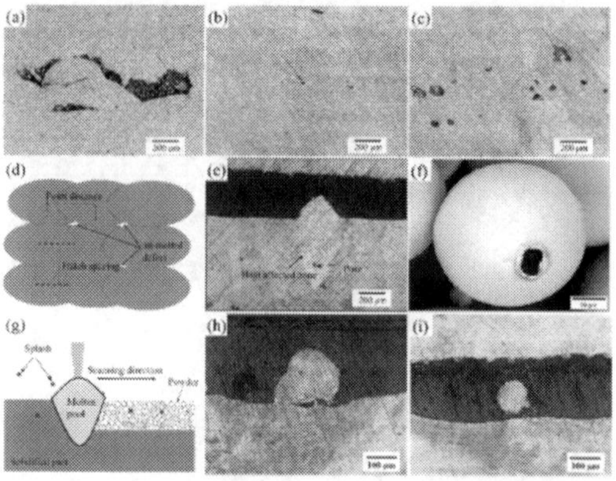

Fig. 5 Internal defects of formed specimens (a) un-melted defects between the molten pools; (b) pore defects; (c) slag inclusion defects; (d) schematic diagram of the formation of un-melted defects between the molten pools; (e) keyhole defect; (f) hollow powder; (g) schematic diagram of the formation of spheroidization defects; (h) single-track with poor wettability; (i) spheroidization

3.3 Microstructure and mechanical properties

The microstructure of the specimen based on the interlayer strategy as shown in Fig. 6. During the solidification of the specimen, mainly acicular martensite α' and columnar β grains are generated, and martensite α' fills and occupies the columnar β grains. As shown in Fig. 6(a), martensite α' can be continuously derived by iteration, and the derivatives are perpendicular to the nascent martensite α'. The martensite α' on both sides of the columnar β grain boundary shows different morphologies, as shown in Fig. 6(b). By observation, it can be found that the top martensite α' of the formed specimen is relatively coarse (Fig. 6c) and the bottom martensite α' is relatively fine. This is because, with the continuous stacking of formed layers, the heat input from the newly formed layers affects the already formed areas (as shown in Fig. 5e). The bottom part has many repetitive thermal cycles, and the fine martensitic α' structure is easily formed by successive iterations. On the other hand, the top has less repetitive thermal cycles and the coarse martensitic α' structure of the first iteration is easily formed.

The tensile strength and yield strength of the specimens formed by SLM based on the interlayer strategy are higher than that of the forged and cast specimens, but the elongation is lower than that of the forged and cast specimens. The existence of martensite α' guarantees the strength of the specimen but reduces the plasticity of the specimens. Numerous studies have shown that the SLM-formed Ti6Al4V specimens suffer from low elongation [11-13], but the elongation of SLM-formed specimens can still be improved by post-treatment methods such as heat treatment or adding reinforcement [14-16]. Fig. 7 shows the fracture topography of the tensile specimen. There are usually pores and other defects at the fracture, which reduces the strength of the specimen at this location, as shown in Fig. 7 (c-d). The fracture modes of tensile specimens include ductile fracture and brittle fracture. As shown in Fig. 7(a), the region without ductile deformation is mainly composed of the cladding formed after the solidification of the molten pool. The ductile deformation region consists of small and shallow dimples, and these dense dimples increase the strength of the formed specimen, as shown in Fig. 7(b).

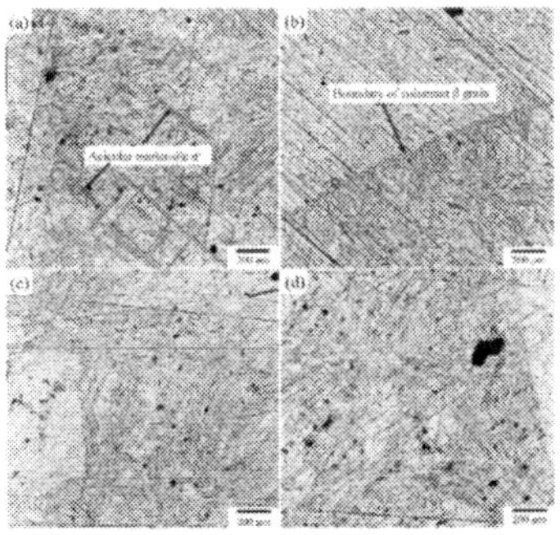

Fig. 6 Microstructure

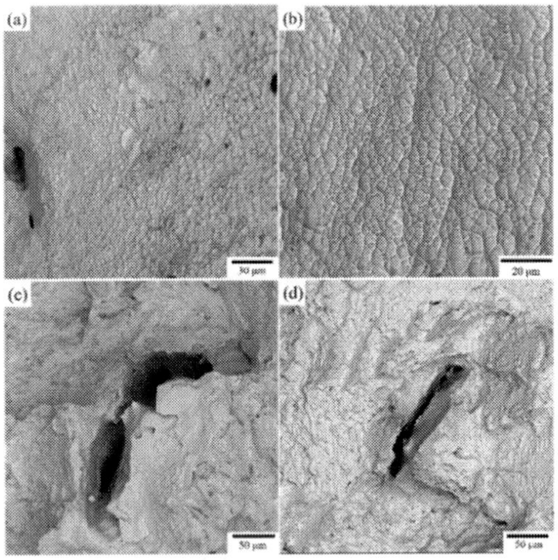

Fig.7 Morphology of tensile fracture surface (a) plastic deformation; (b) pore defects; (c-d) fracture caused by porosity

4. Summary

In this paper, the forming quality of Ti6Al4V specimens formed by SLM based on interlayer strategy was investigated. The effects of process parameters on the wettability, relative density, defects, microstructure and mechanical properties of the specimens were analyzed by single-track and multilayer fabrication. The following conclusions were drawn:

1. The single-track experiment shows that it is not easy to ensure the wettability of the molten pool by increasing the point distance in the thicker layers forming environment. High laser power can ensure adequate melting of thick powder layers, and high laser power can ensure that the single track is stable over a relatively large floating range of exposure times.

2. Multilayer fabrication experiments show that small point distance ensures sufficient laser energy density to melt the powder, while longer exposure times maintain high relative densities. Increasing the hatch spacing can help to disperse the laser energy density and increase the relative density of the specimens at high laser power and small point distance. In this experiment, the relative density of most of the formed specimens is higher than 99%, and the highest density can reach 99.99%.

3. During solidification of the specimens, mainly acicular martensite α' and columnar β grains are generated, and martensite α' can be continuously derived by iteration. Due to the effect of repeated thermal cycling, the bottom martensite α' of the formed specimens is relatively small and the top martensite α' is relatively coarse. Due to the effect of microstructure, the experimental tensile and yield strengths are higher than those of the forged and cast specimens, but the elongation is lower than that of the forged and cast specimens.

References

1. de Souza A F, Al-Rubaie K S, Marques S, et al. Effect of laser speed, layer thickness, and part position on the mechanical properties of maraging 300 parts manufactured by selective laser melting. Materials Science and Engineering: A. Vol. 767 (2019) p. 138425.
2. Zhu Y, Zou J, Yang H. Wear performance of metal parts fabricated by selective laser melting: a literature review. Journal of Zhejiang University-SCIENCE A. Vol. 19 (2018) No. 2, p. 95-110.
3. Todd I. No more tears for metal 3D printing. Nature. Vol. 549 (2017) No. 7672, p. 342-343
4. Martin J H, Yahata B D, Hundley J M, et al. 3D printing of high-strength aluminium alloys. Nature. Vol. 549 (2017) No. 7672, p. 365-369.
5. Savalani M M, Pizarro J M. Effect of preheat and layer thickness on selective laser melting (SLM) of magnesium. Rapid Prototyping Journal. Vol. 22 (2016) No. 1, p. 115-122.
6. Sufiiarov V S, Popovich A A, Borisov E V, et al. The effect of layer thickness at selective laser melting. Procedia engineering. Vol. 174 (2017) No. 7672, p. 126-134.
7. Nguyen Q B, Luu D N, Nai S M L, et al. The role of powder layer thickness on the quality of SLM printed parts. Archives of Civil and Mechanical Engineering, Vol. 18 (2018) No. 3, p. 948-955.
8. Greco S, Gutzeit K, Hotz H, et al. Selective laser melting (SLM) of AISI 316L—impact of laser power, layer thickness, and hatch spacing on roughness, density, and microhardness at constant input energy density. The International Journal of Advanced Manufacturing Technology. Vol. 108 (2020) No. 5, p. 1551-1562.
9. Gao P, Wei K, Yu H, et al. Influence of layer thickness on microstructure and mechanical properties of selective laser melted Ti-5Al-2.5 Sn alloy. Acta Metall Sin. Vol. 54 (2018) No. 7, p. 999-1009.
10. Shi W, Wang P, Liu Y, et al. Properties of 316L formed by a 400 W power laser Selective Laser

Melting with 250 μm layer thickness. Powder Technology. Vol. 360 (2020) p. 151-164.

11. Shi X, Yan C, Feng W, et al. Effect of high layer thickness on surface quality and defect behavior of Ti-6Al-4V fabricated by selective laser melting. Optics & Laser Technology. Vol. 132 (2020) p. 106471.
12. Shi W, Liu Y, Shi X, et al. Beam diameter dependence of performance in thick-layer and high-power selective laser melting of Ti-6Al-4V. Materials. Vol. 11 (2018) No. 7, p. 1237.
13. Simonelli M, Tse Y Y, Tuck C. Effect of the build orientation on the mechanical properties and fracture modes of SLM Ti–6Al–4V. Materials Science and Engineering: A. Vol. 616 (2014) p. 1-11.
14. Vrancken B, Thijs L, Kruth J P, et al. Heat treatment of Ti6Al4V produced by Selective Laser Melting: Microstructure and mechanical properties. Journal of Alloys and Compounds. Vol. 541 (2012) p. 177-185.
15. Wycisk E, Siddique S, Herzog D, et al. Fatigue performance of laser additive manufactured Ti–6Al–4V in very high cycle fatigue regime up to 109 cycles. Frontiers in Materials. Vol. 2 (2015) p. 72.
16. Yu H, Li F, Wang Z, et al. Fatigue performances of selective laser melted Ti-6Al-4V alloy: Influence of surface finishing, hot isostatic pressing and heat treatments. International Journal of Fatigue. Vol. 120 (2019) p. 175-183.

Optimization of Processing Parameters for Picosecond Laser Cutting of Carbon Fiber Reinforced Polymers

Xingfei Ren, Jinwei Fan and Peitong Wang

Faculty of Materials and Manufacturing, Beijing University of Technology, Beijing 100124, China

Abstract. With the increasingly widespread use of carbon fiber reinforced polymer in aerospace and automotive fields, the problem of difficult processing of carbon fiber reinforced polymer has become more prominent. Although laser processing can achieve more ideal processing results, the existence of the heat-affected zone has a negative impact on the use of the material. Therefore, the primary objective of this study is to reduce the width of the heat-affected zone in laser-cut carbon fiber reinforced polymer. The effect of process parameters on the heat-affected zone and the kerf width was first investigated. After that, the heat-affected zone width prediction model and the kerf width prediction model were developed based on the response surface method, and the process parameters were optimized. Finally, the excellent cutting results with almost no heat-affected zone were obtained.

Keywords: Carbon fiber reinforced polymers; heat affected zone; processing parameters; laser cutting.

1. Introduction

With the rapid development of the aerospace field, carbon fiber reinforced polymer (CFRP) has gradually replaced traditional metal materials due to their advantages of low density and high strength, and have developed into the fourth widely used aerospace structural material after aluminum alloy, titanium alloy and alloy steel [1-3]. Additionally, CFRP is becoming increasingly prevalent in high-end sporting goods and cars [4]. The advantages of CFRP, such as light weight, high strength, high hardness, good fracture resistance, corrosion resistance, fatigue resistance, vibration resistance and wear resistance, make it the ideal material for aircraft structural components. CFRP has surpassed metal as the primary structural material in many critical components of the new A350 and B787 aircraft, with a composition of more than 50% [5]. However, due to the material's heterogeneity, anisotropy, and high strength, there are issues with delamination, fiber pullout, burrs, and tool wear when processing CFRP using conventional processing techniques[6]. Numerous novel processing methods have been used to CFRP processing in recent years, including electric discharge machining (EDM), abrasive water jet machining (AWJM), ultrasonic vibration-assisted machining (UVAM) and laser processing [5, 7]. Among these, laser processing is regarded to be the most appropriate technique owing to its benefits such as avoiding tool wear and high efficiency.

However, the heat-affected zone (HAZ) produced in laser processing would cause the reduction of the mechanical characteristics of the material and lower the service life. Due to the fact that carbon fiber takes longer to vaporize than resin, a significant amount of heat is transmitted through carbon fiber and released into the matrix under the beam-material interaction, causing the matrix on both sides of the kerf to be ablated and pyrolyzed, and the fiber is exposed, forming the HAZ [8]. Thus, it is critical to minimize the breadth of the HAZ during carbon fiber reinforced composite laser processing.

In general, the HAZ may be decreased by shortening the beam-material interaction time or cooling the material from the external environment, such as via the use of ultra-short pulse lasers or water jet aided laser processing [9, 10]. In addition, the HAZ can be significantly reduced by optimizing processing parameters. Freitaga et al. pointed out that the HAZ will increase with the increase of pulse energy, pulse duration, average power and repetition frequency [11]. In the study of Hu et al., one-factor experimental design was utilized to investigate the influence of process parameters including laser power, hatch distance and cutting speed on the pulsed laser–material interaction. The process parameters were optimized by using central composite design of Response Surface Methodology [12]. The research of Xu et al. demonstrates that the carbon fiber retains a relatively complete degree of exposure when the pulse energy is low [13]. Jung et al. investigated the effect of cutting speed, time interval, defocus, and other processing parameters on cutting quality, observing that a faster cutting speed results in a smaller heat affected zone and slit width [14].

From the above considerations, it results that, it is critical to choose the right process parameters for high-quality CFRP laser cutting. Additionally, for various machines and materials, the processing parameters required to achieve the best cutting quality vary. Therefore, this article focuses on the selection of process parameters for picosecond laser cutting CFRP.

Aim of the work is to investigate the influence of process parameters (laser power, pulse frequency, scanning speed) of a high-power picosecond laser on the HAZ and to optimize the process parameters. In section II, a picosecond laser was used to carry out the cutting experiment of CFRP; in section III, the influence of process parameters on HAZ and slit width was discussed in detail, and the prediction model of HAZ and kerf width was established, and the process parameters were optimized. Conclusions are drawn in Section IV.

2. Experimental approach

A picosecond laser (Hyper Rapid 50 by Coherent), working at the fundamental wavelength of 1064 nm, was used in the experiments. The duration is less than 15 picoseconds, which can significantly reduce the HAZ. In Table 1 the main characteristics of the laser are reported. The scanning procedure is carried out using a two-dimensional scanning galvanometer. The beam waist diameter of the laser beam is about 20 μm.

Table 1 Laser system characteristics.

Characteristics	Value
Wavelength (nm)	1064
Pulse duration (ps)	<15
Average power (W)	0~50
Divergence angle (mrad)	<1
Repetitive rate (kHz)	10~1000
Beam quality parameter (M2)	≤1.2

The investigated material used in this paper is a CFRP sheet of 0.6 mm thickness. The fiberboard is composed of five layers. Two outside twill weave laminates contribute to the material's strength, while the interior twill weave laminates are unidirectional in the 0° and 90° directions. The plates were obtained by autoclave cure at 150 °C and 7 bar. The adopted reinforcing material was T300 carbon fibre (Toray, Japan), while the matrix was a HMF 934 epoxy resin. In Table 2 the main characteristics of the carbon fiber and epoxy resin are reported.

Table 2 Characteristics of the carbon fiber and epoxy resin

Characteristics	Epoxy resin	Fiber carbon
Density, ρ (kg/m3)	1300	1760
Tensile strength, Rm (MPa)	82.7	3530
Elastic modulus, E (GPa)	4.1	230
Thermal conductivity (Wm-1k-1)	0.08	50 (Axial)/5 (Radial)
Specific heat capacity (Jkg-1K-1)	700	1200
Evaporation temperature (K)	653	3573
damage temperature (K)	443	2973

Previous studies have shown that the selection of process parameters has a significant impact on cutting quality. In this article, the laser power, pulse frequency and scanning speed are taken into consideration. According to preliminary experiments, laser power (P, 10 W, 25 W and 40 W), pulse frequency (f, 200 kHz, 500 kHz and 800 kHz), and scanning speed (v, 200 mm/s, 1000 mm/s and 1800 mm/s) were taken into account in this study. In order to further study the influence of various processing parameters on the HAZ and kerf width, and establish a more accurate processing model, a full factorial experimental design was adopt. Multiple scans were performed with each set of process parameters to ensure that the material is cut through. Due to the thin thickness of the CFRP utilized in this study, the lower kerf width is almost identical to upper kerf width. Therefore, only the upper kerf width and HAZ were investigated. The transfer of laser energy in CFRP is mainly along the direction of fiber laying, and the heat transfer speed in carbon fiber is much faster than that in resin, resulting in a wider HAZ along the fiber direction [15]. As a result, only the kerf width and HAZ were assessed in this research at the region where the laser scanning direction is perpendicular to the fiber bundle (Fig. XX). The HAZ and kerf width were measured by an optical microscope (BX53MRF-S, OLYMPUS).

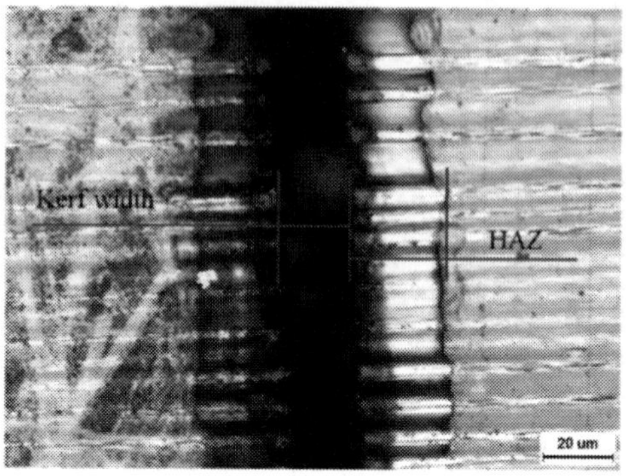

Fig. 1 HAZ and kerf width measurement

3. Results and discussion

3.1 Effect of process parameters on HAZ and kerf width

To investigate the impact of laser power on HAZ and slit width, the pulse frequency and scanning speed were set to 500kHz and 1000mm/s, respectively, and CFRP cutting tests were conducted using various laser powers. As shown in Fig.2. a), increasing the laser power from 10W to 25W results in a substantial increase in the HAZ, from 31 μm to about 67 μm. As laser power increases, the rate at which the HAZ increases lowers. As the laser power increases, the amount of energy that can be delivered to the material in a short period of time increases significantly, and the amount of heat transferred to the matrix at both sides of the kerf along the carbon fiber increases obviously, forming a more visible HAZ. However, the increasing trend of the kerf width with the laser power is not obvious, and it is stable between 20 μm and 30 μm. This may be because the energy density

https://doi.org/10.1051/matecconf/202236301009

required for carbon fiber vaporization and separation is relatively high, and the energy of the laser beam is mainly concentrated within the beam waist diameter (20 μm), and the energy transmitted through the carbon fiber is not enough to cause carbon fiber vaporization.

Compared with the laser power, the HAZ increases slightly with the pulse frequency. And the kerf width rises very little as the pulse frequency increases. An increase in pulse frequency shortens the distance between two laser pulse bombardment sites and increases the accumulation of heat, thus having an effect on the HAZ and the kerf width. However, since the laser used is a picosecond laser, the time of each laser bombardment is extremely short, so increasing only the pulse frequency will not bring about a rapid growth of the HAZ.

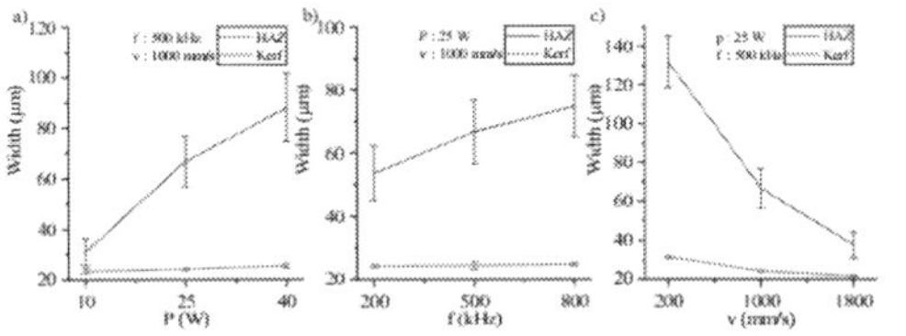

Fig. 2 Effect of process parameters on HAZ and kerf width a) laser power b) pulse frequency c) scanning speed

It is obvious from Fig. 2c) that the scanning speed has the most significant effect on HAZ and kerf width. Both HAZ and kerf width decreased significantly with the increase of the scanning speed. This is mainly due to the fact that the HAZ and kerf width of CFRP increase with increasing laser heat input (P/v) because of longer interaction period between the laser and the CFRPs at lower cutting speed [14]. And when the laser power is constant and the scanning speed increases, the laser heat input decreases, so the HAZ and kerf width decreases.

3.2 Prediction models

To explore the connection between HAZ and kerf width and process parameters in more detail, the prediction models for the HAZ and the kerf width were developed using the response surface method. The result obtained is shown in Eq. (1).

$$\begin{aligned} HAZ &= 75.382 + 2.656p - 0.004f - 0.081v + 0.001pf - 0.001pv - 0.00002fv \\ &\quad - 0.021p^2 + 0.00003f^2 + 0.00002v^2 \\ Kerf\ width &= 28.065 + 0.048p - 0.0005f - 0.004v \end{aligned} \tag{1}$$

ANOVA is one of the most commonly used statistical methods to determine the performance of the model and determine the influence of machining parameters on the surface integrity [16]. Therefore, ANOVA was used to test the performance of the two models. The ANOVAs of the HAZ at 95% confidence level are shown in Table 3.

Table 3 Analysis of variance (ANOVA) table for HAZ and Kerf width. (SS, DF and MS denote summation of squares, degrees of freedom, and mean squares, respectively.)

	Source	SS	DF	MS	F	p-value	
HAZ	Model	55674.5	9	6186.1	240.4	4.0599E-16	significant
	P	10853.6	1	10853.6	421.8	1.94E-13	
	f	2280.4	1	2280.4	88.6	3.7264E-08	
	v	39668.1	1	39668.1	1541.5	4.0112E-18	
	Error	437.5	17	25.7			
	Total	56111.9	26				
	R^2	0.982		Adjusted R^2		0.978	
	Predicted R^2	0.968		Adeq Precision		54.0221	
Kerf Width	Model	191.5	3	63.8	23.1	3.9566E-07	significant
	P	9.4	1	9.4	3.4	0.07809596	
	f	0.3	1	0.3	0.1	0.7366143	
	v	181.8	1	181.8	65.8	3.3629E-08	
	Error	63.5	23	2.8			
	Total	255.0	26				
	R^2	0.751		Adjusted R^2		0.719	
	Predicted R^2	0.659		Adeq Precision		12.6128	

In the model for HAZ, the p-value is much less than 0.05, f-value is up to 240.4, which indicates that the model for HAZ is significant. In addition, the p-values of laser power, pulse frequency and scanning speed are much less than 0.05, which indicates that these parameters are significant. The Predicted R^2 of 0.968 is in reasonable agreement with the Adjusted R^2 of 0.978, and the difference between them is less than 0.2. Adeq Precision measures the signal to noise ratio. The ratio of 54.022 indicates an adequate signal. The above analysis study demonstrates that the model developed can be used to accurately reflect the actual actual processing situation. Similarly, the ANOVA results for the kerf width prediction model also prove that the developed model can accurately predict the kerf width.

Additionally, the f-value can reflect the ability of each processing parameter to influence the heat-affected zone width and the slit width. It can be found that the scanning speed is the process parameter that has the greatest influence on the HAZ and the kerf width, followed by the laser power, while the pulse frequency has a relatively weak ability to influence the HAZ and the kerf width, which is consistent with the analysis results in Section 3.1. In addition, in the analysis of variance for the HAZ model, the f-values for each process parameter are significantly higher than those in the kerf width, which indicates a more significant effect of the process parameters on the HAZ.

On the other hand, the plotting the normal plot of residuals provides a more intuitive response to the adequacy of the model. Fig. 3 illustrates that all residuals of both the HAZ prediction model and the kerf width prediction model are near straight lines, indicating that the errors are regularly distributed. It can be considered that

 https://doi.org/10.1051/matecconf/202236301009

the HAZ prediction model and the kerf width prediction model are sufficient.

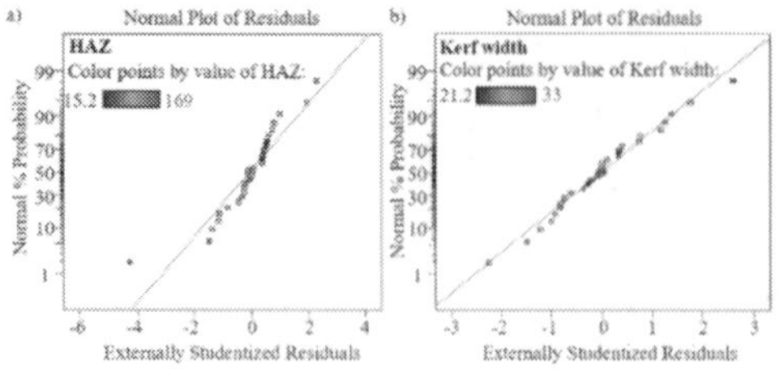

Fig.3 Normal probabilities of residuals for taper percentage a) HAZ b) kerf width

In addition, several sets of process parameters were randomly generated within the allowed range of each process parameter to verify the predictive performance of the developed model. The obtained results are shown in Fig. 4.

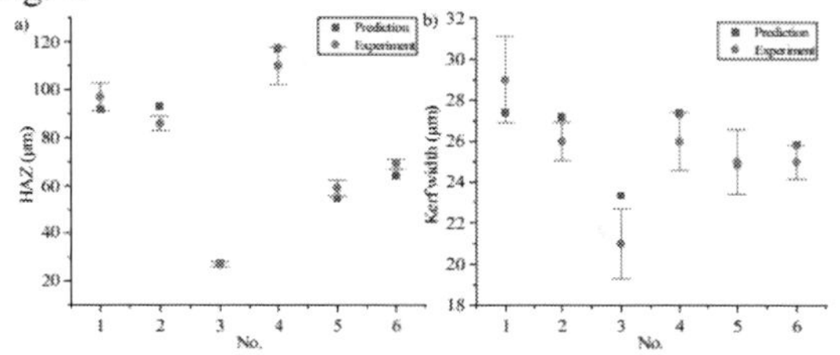

Fig. 4 Validation of model prediction results a) HAZ b) kerf width

As shown in Fig. 4, the predicted results for HAZ and kerf width using the developed model are extremely similar to the actual experiments. The established HAZ prediction model has higher prediction accuracy, and the difference between its predicted and actual values is quite small. Especially for the prediction of HAZ of the third group of machining parameters, the predicted value is almost identical to the actual value, and the maximum prediction error appeared in the prediction of HAZ of the fourth most group of machining parameters, and the maximum error is about 7 μm. It is within the acceptable range. It is expected that the model has reliable prediction accuracy, and one of the reasons for the deviation in the prediction is that there is also a small difference in the properties of the CFRPs, and thus there will be some deviation using the same process parameters. The predicted results for the kerf width are also basically consistent with the actual experimental results. In the predictions for groups 2, 3, 4 and 6 process parameters, the predictions were slightly larger than the actual experimental results. The largest prediction error occurs in the prediction of the kerf width for the third set of process parameters, and the maximum error is only 1.7 μm, which is acceptable for the prediction of the kerf width. From this, it can be concluded that the established HAZ prediction model and the kerf width prediction model both have high prediction accuracy.

3.3 Optimization of process parameters

The optimization of process parameters can effectively reduce the HAZ and the kerf width. Therefore, the optimal process parameters were obtained by using the expectation function method for multi-response optimization using Design Expert software. The optimization objective is to minimize the HAZ and the kerf width, and the constraints are that each process parameter needs to be within the appropriate range. The obtained optimization results are shown in Fig. 5.

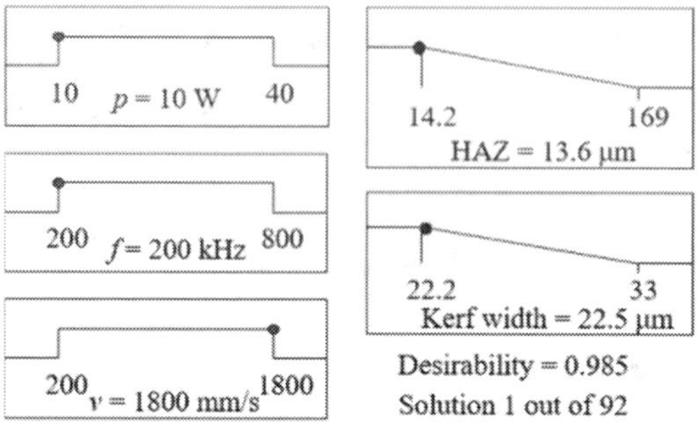

Fig. 5 Process parameter optimization results

The optimized laser power, pulse frequency and scanning speed are 10 W, 200 kHz and 1800 mm/s. The results obtained from the optimization can be explained by the laws of influence of processing parameters on HAZ and kerf width. The reduction of power, the reduction of pulse repetition frequency and the increase of scanning speed all contribute to the reduction of heat affected zone width and slit width.

The obtained kerf width is expected to be 22.5 μm and the HAZ is expected to be 13.6 μm. The machining results using the optimized machining parameters are shown in Fig. 6a). The obtained kerf width is 24 μm and the HAZ is 13.3 μm. The predicted value for the kerf width is 6.25% smaller than the experimental value, and the predicted value for the HAZ is 2.26% larger than the experimental value. The prediction errors are small and within acceptable limits. Fig. 6b) shows the results obtained by cutting CFRP using unoptimized process parameters. The comparison shows that the optimized process parameters significantly reduce the HAZ, while the reduction of the kerf width is not significant.

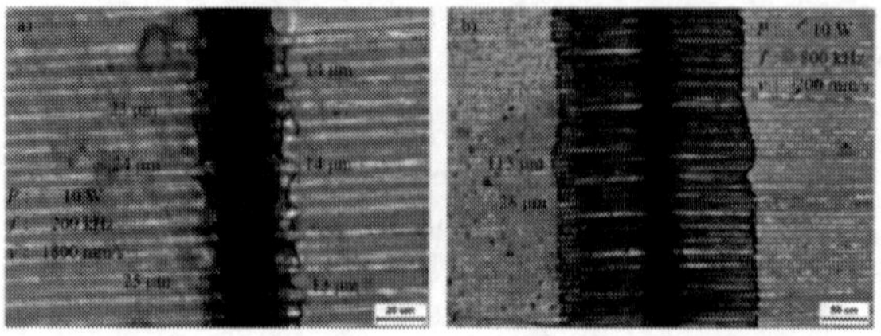

Fig. 6 Comparison of cutting quality a) with parameter optimization b) without parameter optimization

4. Summary

In this paper, the effects of laser process parameters on the HAZ and kerf width of CFRPs were investigated, and a HAZ prediction model and a kerf width prediction model were established and the process parameters were optimized. The main conclusions obtained are as follows:

- The scanning speed has the greatest effect on the HAZ and the kerf width, followed by the laser power and pulse frequency.
- The HAZ is strongly influenced by the process parameters, while the kerf width is relatively less influenced by the process parameters.
- There is a clear relationship between HAZ and kerf width and process parameters, and the cutting quality can be predicted by the established HAZ prediction model and the kerf width prediction model.
- The HAZ and the kerf width can be effectively reduced by optimizing the process parameters.

Acknowledgments

This work was supported by the construction of machine tools and equipment CNC interconnection platform and big data center and application platform (grant No. 2021-0171-1-1).

References

1. Katnam, K.B., L.F.M. Da Silva, and T.M. Young, Bonded repair of composite aircraft structures: A review of scientific challenges and opportunities. Progress in Aerospace Sciences, 2013. 61: p. 26-42.
2. Thirumalai Kumaran, S., et al., Rotary ultrasonic machining of woven CFRP composite in a cryogenic environment. Journal of Alloys and Compounds, 2017. 698: p. 984-993.
3. Jia, C., et al., Improving the Mechanical and Surface Properties of Aramid Fiber by Grafting with 1,4-Dichlorobutane under Supercritical Carbon Dioxide. Materials (Basel, Switzerland), 2019. 12(22): p. 3766.
4. Osborne, J., Automotive Composites – in Touch with Lighter and More Flexible Solutions. Metal Finishing, 2013. 111(2): p. 26-30.
5. El-Hofy, M.H. and H. El-Hofy, Laser beam machining of carbon fiber reinforced composites: a review. The International Journal of Advanced Manufacturing Technology, 2019. 101(9): p. 2965-2975.
6. Lopresto, V., A. Caggiano, and R. Teti, High Performance Cutting of Fibre Reinforced Plastic Composite Materials. Procedia CIRP, 2016. 46: p. 71-82.
7. Dutta, H., K. Debnath, and D.K. Sarma, Investigation on cutting of thin carbon fiber-reinforced polymer composite plate using sandwich electrode-assisted wire electrical-discharge machining. Proceedings of the Institution of Mechanical Engineers, Part E: Journal of Process Mechanical Engineering, 2021. 235(5): p. 1628-1638.
8. Wu, C.-W., X.-Q. Wu, and C.-G. Huang, Ablation behaviors of carbon reinforced polymer composites by laser of different operation modes. Optics & Laser Technology, 2015. 73: p. 23-28.
9. Bluemel, S., et al., Comparative Study of Achievable Quality Cutting Carbon Fibre Reinforced Thermoplastics Using Continuous Wave and Pulsed Laser Sources. Physics Procedia, 2014. 56: p. 1143-1152.
10. Wu, Y., et al., The cutting process and damage mechanism of large thickness CFRP based on water jet guided laser processing. Optics & Laser Technology, 2021. 141: p. 107140.
11. Freitag, C., et al., High-speed Observation of the Heat Flow in CFRP During Laser Processing. Physics Procedia, 2012. 39: p. 171-178.
12. Hu, J. and D. Zhu, Experimental study on the picosecond pulsed laser cutting of carbon fiber-reinforced plastics. Journal of Reinforced Plastics and Composites, 2018. 37(15): p. 993-1003.
13. Xu, L., et al., Study on the mechanism of inhomogeneous microdamage in short-pulse laser processing of carbon fiber reinforced plastic. Journal of Reinforced Plastics and Composites, 2021. 40(15-16): p. 568-576.
14. Jung, K.-W., Y. Kawahito, and S. Katayama, Ultra-high speed disk laser cutting of carbon fiber reinforced plastics. Journal of Laser Applications, 2011. 24(1): p. 012007.
15. Zhang, Y. and L. Qi. Experimental investigation on fiber laser cutting carbon fiber reinforced plastics. in Proc.SPIE. 2020.
16. Rao B, S.D., A. Sethi, and A.K. Das, Fiber laser processing of GFRP composites and multi-objective optimization of the process using response surface methodology. Journal of Composite Materials, 2018. 53(11): p. 1459-1473.

THOR 50th Dummy Neck Calibration Analysis Based on Bi-LSTM Neural Network

Gang Xu, Ruihan Yu and Bingxu Duan

CATARC Automotive Test Center (Tianjin) Co., Ltd., Tianjin, China

Abstract. In this paper, a neural network model is established based on the neck calibration data of the Thor50th crash test dummy using a bi-directional long and short-term memory (Bi-LSTM) neural network algorithm. The model input is the factors affecting the neck calibration test, and the output is the maximum value of My in the neck calibration test, and the accuracy of the model is calibrated by comparing it with the actual calibration data. The accuracy and suitability of the Bi-LSTM neural network model is further verified by comparing with the radial basis (RBF) neural network algorithm.

Keywords: Thor 50th Dummy; Neck calibration test; Bi-LSTM neural network; RBF neural network

1. Introduction

As the proportion of Thor 50M in crash test dummies has significantly increased in current state of passive safety development, which also means the increase of the requirements against dummy performances. Hence, multiple assessment protocol has been added to the calibration procedure. The dummy neck calibration is the most complicated task, which is similar to the calibration process of the Hybrid III 50th dummy. The calibration process of the Thor 50M neck can be divided into six parts, Neck Frontal Flexion, Neck Extension, Neck Lateral Left/Right Flexion, Dynamic Response Left/Right Torsion. According to calibration test data, Neck Frontal Flexion and Neck Extension appears to have minimum completeness. Peak Upper Neck My is most likely to fail among the 8 assessment criterions, thus, this particular criterion is selected as the research subject in this study.
In calibration tests, except from the peak value of each test item that is involved in the overall evaluation, the test curve should also be considered. Hence, it is more accurate to consider the neck calibration test results as a set of time-series data, however, due to the high sample frequency of calibration test, which leads to repetitive and meaningless data, so certain points can be selected instead of the entire time-series data for model fitting. In this case, Bi-LSTM neural network should be considered more suitable for time-series data processing, so Bi-LSTM neural network is selected for fitting in this study, and the conventional RBF neural network is introduced as control test to compare the fitting results of the two methods.

2. Bi-LSTM neural network

Bidirectional long-short term memory(bi-lstm) is the process of making any neural network to have the sequence information in both directions backwards (future to past) or forward (past to future).
In bidirectional, input flows in two directions, making a bi-lstm different from the regular LSTM. With the regular LSTM, the input can flow in one direction, either backwards or forward. However, in bi-directional, the input can flow in both directions to preserve the future and the past information.
Recurrent neural network (RNN) is a class of artificial neural networks where connections between nodes form a directed or undirected graph along a temporal sequence. The model structure is shown below, including input layer x, hidden layer s, and output layer o. There is a loop operation on the hidden layer s. Meanwhile, the linear relationship parameters U, W, and V are shared by the RNN at all moments, which greatly reduces the amount of parameter training.

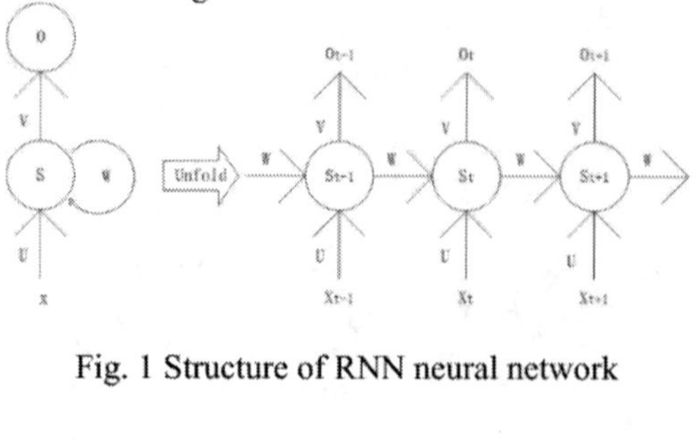

Fig. 1 Structure of RNN neural network

Long Short-Term Memory is a kind of recurrent neural network proposed to solve the problem of RNN, e.g., vanishing and exploding gradients, and unable to keep track of long-term dependencies. LSTM has a similar main structure compared to RNN. The main difference is LSTM can protect its hidden activation using gates between each of its transaction points with the rest of its layer. The hidden activation that is protected is called the cell state. The protection of the cell state is undertaken by the three LSTM gates, namely, the forget, input, and output gates. The principle of the LSTM hidden activation structure is shown below. f(t), i(t), and o(t) denote the values of the forget gate, input gate, and output gate at time t, respectively. a(t) denotes the initial feature extraction of h(t-1) and x(t) at time t [1].

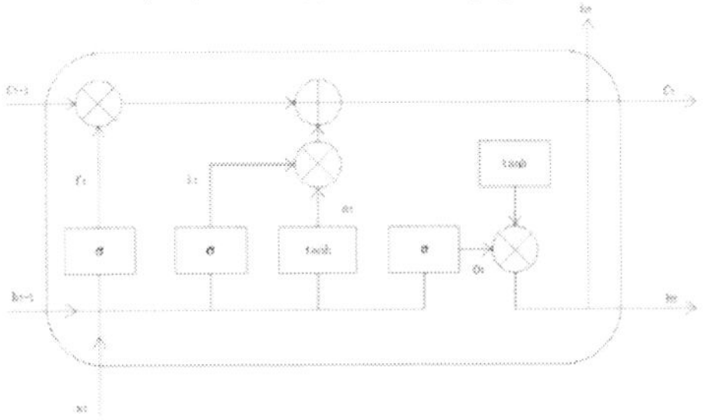

Fig. 2 Structure of LSTM neural network

The LSTM model can be described by following equations

$$f(t) = \sigma(W_f h_{t-1} + U_f x_t + b_f)$$
$$i(t) = \sigma(W_i h_{t-1} + U_i x_t + b_i)$$
$$a(t) = tanh(W_a h_{t-1} + U_a x_t + b_a)$$
$$o(t) = \sigma(W_o h_{t-1} + U_o x_t + b_o)$$
$$tanh(x) = \frac{1 - e^{-2x}}{1 + e^{-2x}}$$
$$\sigma(x) = \frac{1}{1 + e^{x}}$$

Where, x_t is the input at current timestamp
h_{t-1} is the output from the previous LSTM block (at time stamp t-1)
W_x is the weight for the respective gate(x) neurons
U_x is the weight
b_x is the bias for the respective gate(x)
σ is sigmoid function
The results of the forget gate and input gate act on c(t-1), hence the cell state at timestamp t c(t), which can be expressed by following equation

$$c(t) = c(t-1) \odot f(t) + i(t) \odot a(t)$$

Where $\odot$ is Hadamard multiplication
The Bi-LSTM network has two independent LSTMs. The input sequence is sent to the two LSTM neural networks in positive and reverse order for feature extraction, and the feature vector formed by splicing the two output vectors is used as the final feature expression of the instance [2].

$$\overrightarrow{h_n} = LSTM(x_{n,}\overrightarrow{h_{n-1}})$$
$$\overleftarrow{h_n} = LSTM(x_{n,}\overleftarrow{h_{n+1}})$$
$$\sigma_n = \sigma(W_{\overrightarrow{h_0}}\overrightarrow{h_n} + W_{\overleftarrow{h_0}}\overleftarrow{h_n} + by)$$

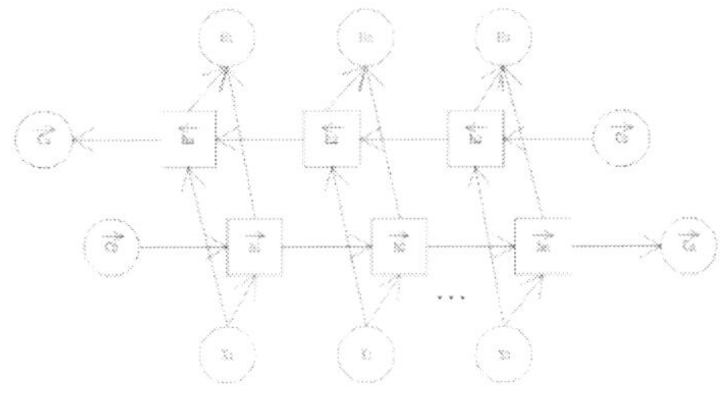

Fig. 3 Structure of Bi-LSTM neural network

3. Bi-LSTM neural network in neck calibration test

In the calibration process of neck frontal flexion and neck extension, the main factors affect peak upper neck My are number of honeycomb aluminum holes, front cable force, rear cable force, neck condition and temperature, considering the instability of the five factors mentioned above during the calibration process, the traditional data-based fitting approach utilizes less information, which makes it cannot learn the features of Thor 50th dummy neck properly, while no significant correlation could be observed between these five factors, and the effect of these factors on the PUN My is non-linear, hence, Bi-LSTM neural network is selected for this study. This neural network is a deep recurrent neural network, which can make good nonlinear approximation using forward and backward data information. In addition, the sensor data collected from the calibration test are considered as time-series data, and the data were simplified into 3 stages for the purpose of the study, namely, $M_y(t_0)$, $M_y(t_+)$, and $M_y(PUN)$, $M_y(t_0)$ and $M_y(t_+)$ were entered simultaneously with the information of the five influencing factors asked above to make predictions on the neck calibration results.
The neck of the Thor dummy is constructed by rubber and metal, in this case rubber has the greatest impact to the overall performance of the neck. However, evaluation against the rubber part of the neck is relatively hard to perform and the neck assembly is not a disposable part, hence the overall performance of the neck in calibration scenario is irregular and cannot be evaluated intuitively. Hence, a subjective evaluation against the neck is conducted based on the neck calibration test results, in order to do so, velocity, peak upper neck My, peak upper neck Fz, Peak head angular rate and peak head angle are chosen as criterions to assess the current state of the specific neck assembly. Based on neck calibration test data, velocity, peak upper neck My, peak upper neck Fz, peak head angular rate and peak head angle are selected to conduct overall evaluation. Each criterion has its upper and lower limit, it can be considered that the neck assembly is in decent condition when the calibration test data is evenly distributed between upper and lower limit in each criterion, Therefore, it is necessary to calculate the proportion of the difference between the actual value of each criterion and the middle value of the upper and lower limit in the limit length interval(expressed in decimal), as shown in the table below the closer the value is to zero, the better the neck condition is, the neck condition value

 https://doi.org/10.1051/matecconf/202236301010

of the test can be described as Neck Goal and shorted as NG in this study.

$$NG = [NG(PV) + NG(My) + NG(Fz) + NG(\omega y) + NG(deg)]/5$$

$$NG(PV) = \frac{\left|4.96 - \frac{5.05 + 4.95}{2}\right|}{5.05 - 4.95} = 0.4$$

Table 1 Neck calibration test data

Description	Lower limit	Value	Upper limit	Unit
Velocity	4.95	4.96	5.05	m/sec
Peak Upper Neck My	-24.9	-22.48	-20.4	N-m
Time To Peak Upper Neck My	N.A.	51.8	N.A.	mSec
Peak Upper Neck Fx	N.A.	566.07	N.A.	N
Time to Peak Upper Neck Fx	N.A.	63.2	N.A.	mSec
Peak Upper Neck Fz	-3103	-2864.05	-2539	N
Anterior Neck Cable Fz first Peak Before 60ms	N.A.	2342.8	N.A.	N
Peak Head Angular Rate Wy	1857	2156.2	2270	deg/s
Peak Head Angle (Relative to Pendulum)	57.1	67.98	69.8	deg
Time To Peak Head Angle	N.A.	58.1	N.A.	mSec
Decay Time to 0deg from Peak Angle	N.A.	69.5	N.A.	mSec
Time to Posterior Neck Cable Fz First Peak	N.A.	53.2	N.A.	mSec
Pendulum Velocity at 10ms after T0	1.66	1.85	2.03	m/sec
Pendulum Velocity at 20ms after T0	3.23	3.56	3.95	m/sec
Pendulum Velocity at 30ms after T0	4.45	4.94	5.44	m/sec
Temperature	20.6	21.1	22.2	degC
Humidity	10	35	70	%

The Neck goal is determined as the latest 2 calibration test results.

3.1 Model inplementation

Number of honeycomb aluminum holes, front cable force, rear cable force, neck condition, temperature, $M_y(t_0)$, $M_y(t_+)$, and $M_y(PUN)$ were entered into the model to analyze the neck calibration test with the following model.

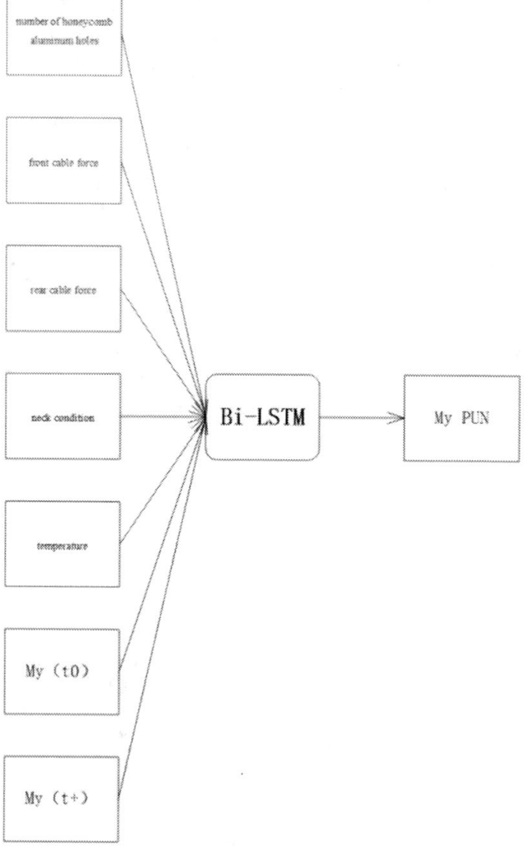

Fig. 4 Model flow chart

3.2 Nomalization process

To improve the stability of the calculation, the data are normalized and the range of value is incorporated between [0, 1] with the formula below.

$$Y = \frac{X - X_{min}}{X_{max} - X_{min}}$$

Where, X is the actual value of a variable, Xmax is the maximum value of a variable, Xmin is the minimum value of a variable, and Y is the normalized value. After the prediction model, the predicted value needs to be denormalized to facilitate error analysis with the true value, and the formula is:

$$X_{BiLSTM} = (X_{max} - X_{min})Y + X_{min}$$

3.3 Bi-LSTM neural network model set-up

The structure of the Bi-LSTM regression network designed in MATLAB is as follows:

layers = [...sequenceInputLayer(numFeatures)
bilstmLayer(numHiddenUnits2,'OutputMode','sequence
')
dropoutLayer(0.2)
fullyConnectedLayer(numClasses)

regressionLayer];

3.3.1 Parameter settings

In general, parameters need to be set including, network structure, number of units, learning rate, training algorithm and training times, etc.

The number of model layers is 2, with a dropout random deactivation layer in the middle to prevent overfitting, and a full linkage layer in the last.

Basically, it is adequate to use same number of neurons for all hidden layers. For some datasets, having a large first layer and following it up with smaller layers will lead to better performance, because the first layer can learn many lower-level features that can feed into a few higher order features in the subsequent layers.

In addition, performance boost is more likely to be acquired from adding more layers rather than adding more neurons in each layer. Hence, adding too many neurons in one single hidden layer is not recommended[3].

$$N_h = \frac{N_s}{(\alpha * (N_i + N_o))}$$

Where,

N_i is the number of neurons in the input layer;

N_o is the number of neurons in the output layer;

N_s is the number of samples in the training set;

α is an arbitrary value, usually in the range of 2 to 10

The number of hidden neurons should be between the size of the input layer and the size of the output layer. The number of hidden neurons should be 2/3 of the size of the input layer, plus 2/3 of the size of the output layer. The number of hidden neurons should be less than twice the size of the input layer[4].

In conclusion, the optimal number of neurons in the hidden layer should be obtained through continuous experiments. It is recommended to start with a small value such as 1 to 5 layers and 1 to 100 neurons, and then slowly add more layers and neurons if underfitting, and reduce the number of layers and neurons if overfitting. In addition, it is also available to introduce Batch Normalization, Dropout, Regularization and other methodologies to reduce overfitting in the actual process.

The two neck calibrations in this study were influenced by the same factors, and the number of hidden neurons was calculated from the data to be between 2-20, and the appropriate number of hidden layer neurons was calculated by selecting 3 points per interval between 2-20, i.e., [1 5 8 11 14 17 20], and the results were as follows, and the number of hidden layer neurons was 5.

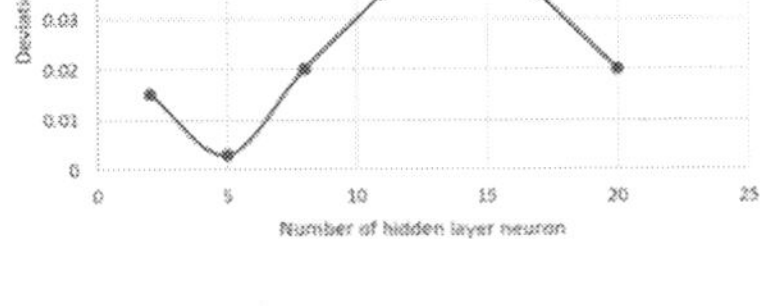

Fig. 5 Deviation of hidden layer neuron number on Neck Extension

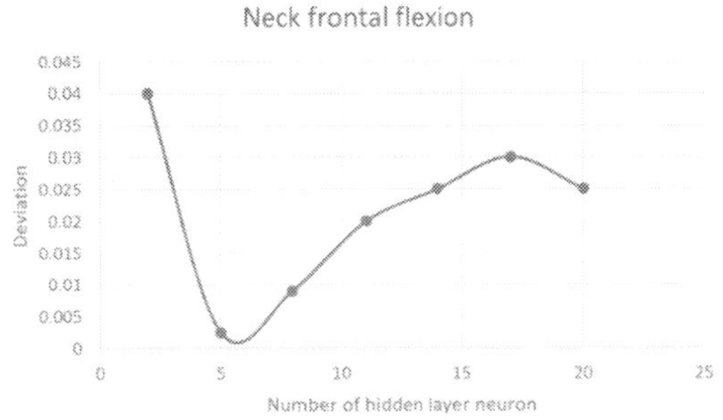

Fig. 6 Deviation of hidden layer neuron number on Neck Frontal Flexion

Adam optimization algorithm was applied as model learning algorithm, which has advantage of fast convergence and high accuracy, and the mean square error function was selected as loss function, which can be expressed by following equation:

$$E_{MSC} = \frac{1}{N}\sum_{i=1}^{N}(y_{pred} - y_i)^2$$

With a learning rate of 0.005 and 500 training times.

300 sets of neck test data were selected in this study, among which, 270 sets of data were chosen randomly as training samples for model training, and the remaining 30 sets of samples were used as prediction samples for outcome prediction. The RBF neural network is determined as the comparison model. For implementation, MATLAB statements are as follows.

RBFnet = newrb(inputn_train, outputn_train, goal, spread, N)

The deviation tolerance of the model (GOAL) is 0.005, the diffusion factor (SPREAD) is 10, and the number of neurons (N) is 5.

3.4 Prediction results

3.4.1 Neck frontal flexion test

The prediction results in neck frontal flexion tests are as follows.

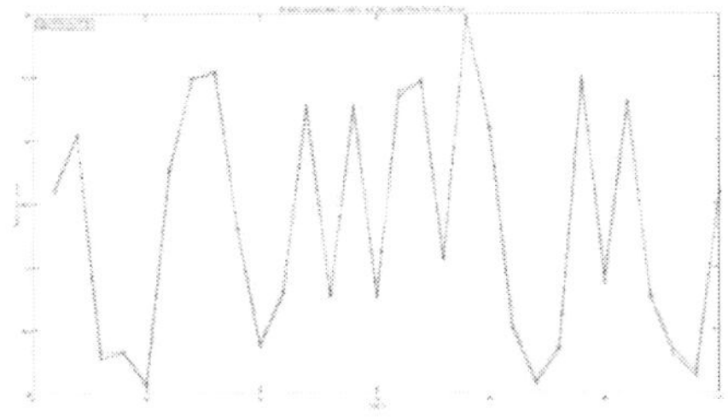

Fig. 7 Neck frontal flexion predicted values vs. actual values using Bi-LSTM neural network

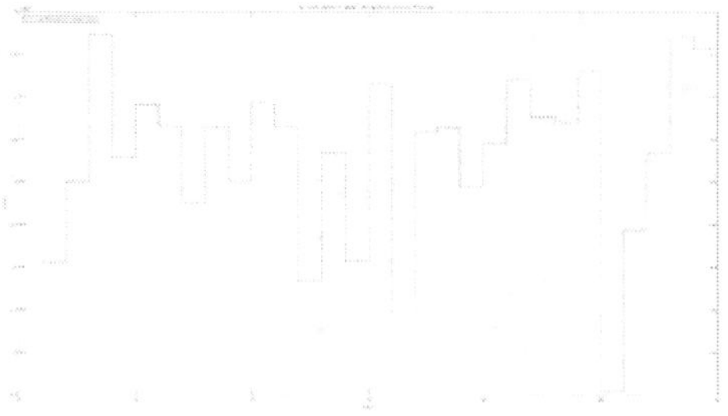

Fig. 8 Deviation of neck frontal flexion predicted values using Bi-LSTM neural network

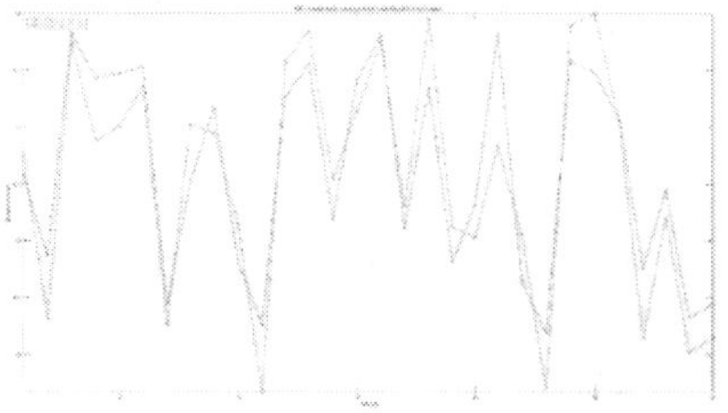

Fig. 9 Neck frontal flexion predicted values vs. actual values using RBF neural network

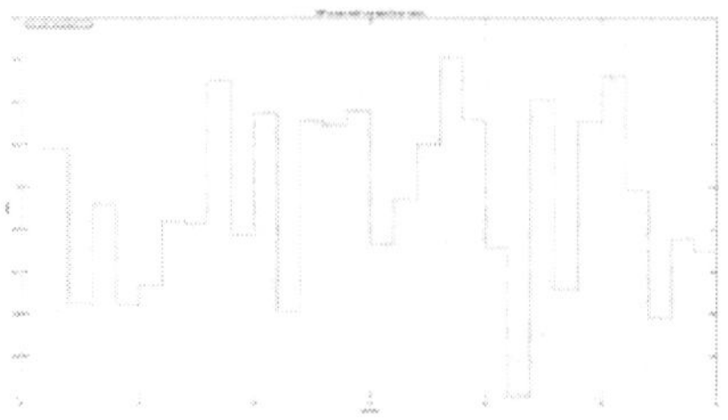

Fig. 10 Deviation of neck frontal flexion predicted values using Bi-LSTM neural network

The prediction error of the Bi-LSTM neural network ranged from -0.25% to 0.2% with good prediction results, proving that the selected factor was indeed the key factor affecting the neck test. While, using RBF neural network for prediction resulted in a deviation of -10%-6%, which indicates significantly lower suitability than Bi-LSTM neural work.

3.4.2 Neck extension test

The prediction results in neck frontal flexion tests are as follows.

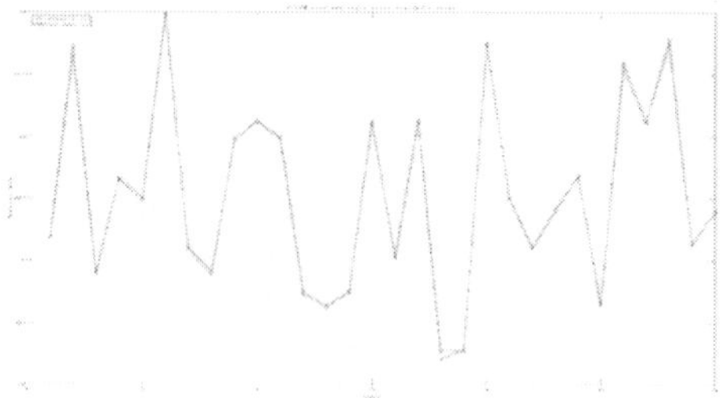

Fig. 11 Neck extension predicted values vs. actual values using Bi-LSTM neural network

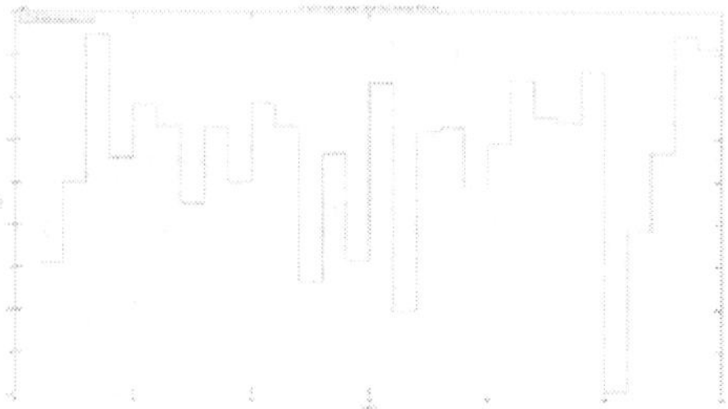

Fig. 12 Deviation of neck extension predicted values using Bi-LSTM neural network

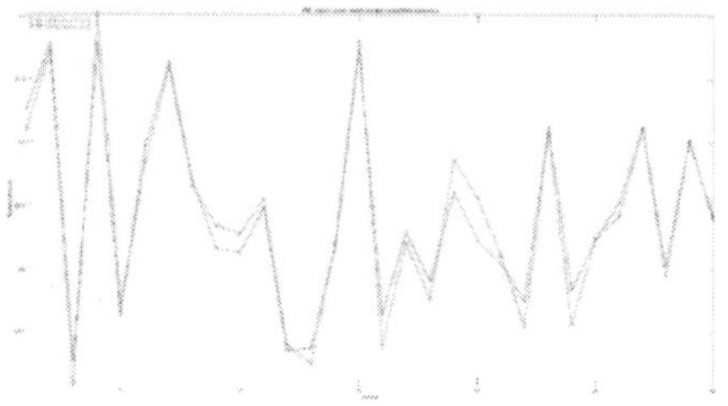

Fig. 13 Neck extension predicted values vs. actual values using RBF neural network

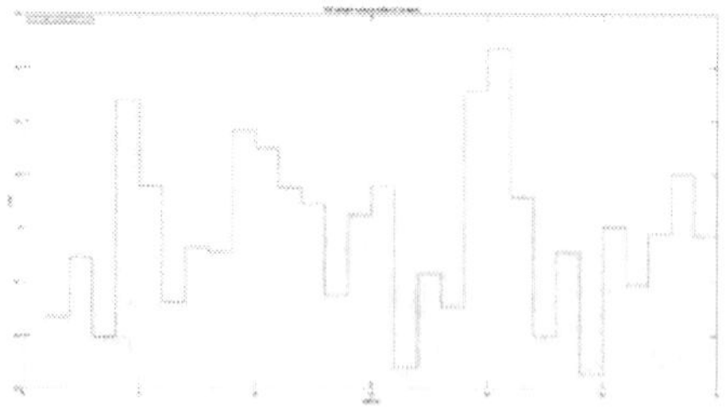

Fig. 14 Deviation of neck extension predicted values using RBF neural network

The prediction error of the Bi-LSTM neural network ranged from -0.3% to 0.4% with good prediction results, proving that the selected factor was indeed the key factor affecting the neck test. While, using RBF neural network for prediction resulted in a deviation of -3%-3%, which indicates significantly lower suitability than Bi-LSTM neural work.

https://doi.org/10.1051/matecconf/202236301010

4. Summary

In this study, a Bi-LSTM neural network based Thor50th neck prediction model was developed, trained the model by inputting the number of honeycomb aluminum holes, front cable force, rear cable force, neck state, temperature, $M_y(t_0)$ and $M_y(t_+)$, which came up with a prediction error lower than 0.5%.

Compared with the RBF neural network, the Bi-LSTM model has higher prediction accuracy due to the special recurrent mechanism and the partial timing factor of the braiding results included in the input of this study

Thor50th neck status evaluation index is proposed, which helps quantitative and intuitive judgment of the current neck status.

References

1. A. Iosifidis, Deep Learning for Robot Perception and Cognition. Elsevier Science, 2022, p. 101-115.
2. S. Abuqran. Arabic Multi-Topic Labelling using Bidirectional Long Short-term Memory. 2021 12th International Conference on Information and Communication Systems (ICICS), 2021. p. 492–494
3. Information on: Http://towardsdatascience.com/designing-your-neural-networks-a5e4617027ed
4. S. Xu, L. Chen, A novel approach for determining the optimal number of hidden layer neurons for FNN's and its application in data mining, 5th International Conference on Information Technology and Applications (ICITA '08), June 2008, p. 683–686.

Crack study of recasting layer on workpiece surface in Micro-EDM

Chunmei Wang [1], Xuyang Chu [2], Yunxiang Lu [1], Hong Li [1]

[1] School of Mechanical and Electrical Engineering, Guizhou normal university, Guiyang, 550025, China
[2] School of Aerospace Engineering, Xiamen University, Xiamen, 361102, China.

Abstract: In this paper, the crack phenomenon of different materials was studied according to the surface morphology of recasting layer of micro EDM workpiece. In this paper, high-speed steel and titanium alloy are used as workpiece materials, and the RC pulse power supply is used for electric discharge machining. The surface and profile of the recast layer of the workpiece are examined by scanning electron microscopy. The difference between the two in the crack phenomenon of the recast layer in micro electric spark machining is analyzed. Finally, some improvement measures are put forward to solve the problem of recasting layer crack on the surface of micro EDM workpiece.

Key words: Micro-EDM; Recast Layer; pulsed power supply; crack.

1. Introduction

The environment and method of the fine EDM machining of the workpiece and the physical properties of the workpiece material will cause cracks on the surface of the workpiece. Moreover, it can be seen from the surface morphology of the EDM workpiece materials that different materials have different degrees of microcracking, and the surface microcracks are basically present in the recast layer and extend to the surface of the substrate[2-4]. Cracks caused by the physical properties of the material are unavoidable, and the surface of the workpiece cannot be completely crack-free, but the method of machining the workpiece can be changed. Therefore, for such a situation, the method of processing the workpiece can be improved, and the occurrence of such a crack phenomenon is relatively improved as much as possible.

2. Crack formation of surface recast layer

Due to the micro-EDM process is carried out in kerosene, the molten metal rapidly cools and solidifies after the end of the pulse discharge, and the rapid cooling shrinkage of the melt generates a large tensile stress, so that the original stress field is redistributed. When the stress is greater than the tensile stress, microcracks will occur on the metamorphic layer. Finally, a melting metamorphic layer with a thickness of 0~10 m is formed on the workpiece surface. The crack of micro EDM is mainly affected by material and discharge energy. The surface crack will reduce the corrosion resistance and fatigue strength of the parts, and seriously affect the service life and performance of the parts.

The surface microcracks of EDM workpiece are caused by the non-uniform stress in the process of EDM. On the one hand, the surface layer of the heated workpiece material produces high temperature and high pressure, which causes the surface of the workpiece material to undergo phase transition (melting and gasification), and the phase transition stress leads to the generation of microcracks. On the other hand, the surface layer of the workpiece will generate thermal stress in the process of thermal corrosion removal, and the thermal stress is caused by the unequal rate of crystallization of the material. Rapid heating and cooling in EDM results in a large temperature gradient, which is an important factor causing various internal stresses and surface cracks. In addition, in the process of electric spark machining, kerosene is used as the processing medium. Due to the distortion around the carbon atoms produced by the kerosene medium cracking, the local structural stresses are generated. These local structural stresses result in microcracks in the thin shell and microcracks can also occur where bubbles exist and where mutations occur.

3. Surface crack pattern

The surface of the micro-EDM workpiece has micro-cracking due to the influence of heat. In different processing materials and different processing parameters, the surface crack phenomenon is different. Radial crack, arc crack and local surface crack are the main forms of

https://doi.org/10.1051/matecconf/202236301011

surface crack in recasting layer. When the energy is large, all kinds of cracks appear at the same time, which is called reticular crack. As shown in Figure 1.

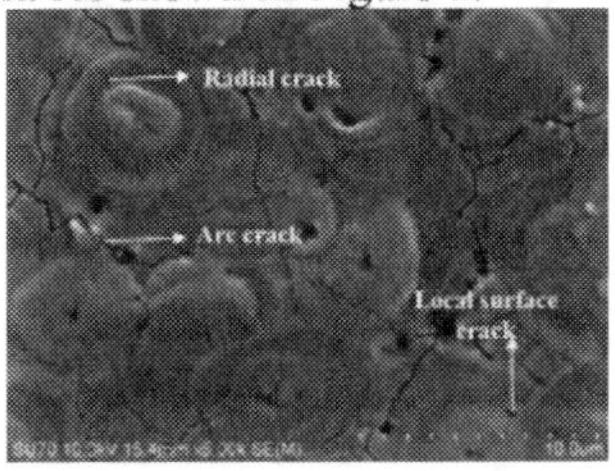

Figure 1 Surface crack representation

Radial cracks appear in the pit marks formed by a single spark, extending from the center of the pit to the edge of the pit. This is due to the fact that during the single-pulse discharge heat process, the heat source model exhibits a time-varying Gaussian distribution in the spark. The central energy density is large; the temperature of the surface of the workpiece material is high; the energy density away from the spark center is small, and the temperature of the surface material of the workpiece is low. Because the temperature of each part of the thermal effect of a single spark is different, a temperature gradient is formed, which in turn generates thermal stress, thus forming radial surface microcracks.

Arc cracks appear at the edge of the single spark discharge traces, approximately the circular arc in shape. This is because at the end of each discharge, part of the higher temperature of heat action produces the phase change processes, such as melting and gasification. Since the workpiece material beyond the boundary of the heat source model is not subjected to thermal action, the characteristics of the matrix are maintained. Thus, different or different simultaneous tissue changes occur in the process of workpiece material processing, and greater tissue stress is generated. As a result, an arc-shaped crack is formed at the junction of tissues, so the arc-shaped crack is thicker and longer than the radial crack and has a greater impact on the mechanical characteristics of the surface.

Local surface cracks appear in the local distortion of the workpiece surface, because the workpiece material, tool material and kerosene medium will appear fine particles during the processing. These fine particles are affected by the thermal effect of the second discharge, whose temperature is not the same, thus forming a temperature gradient, crack phenomenon. The cracks caused by local fine particles are local surface cracks.

4. Surface cracks in different materials

Since surface cracks are not easily quantified in terms of width, length, depth, and number of cracks, H.T. Lee et al [1]. defined "surface crack density", that is, the total length (cm)/unit area (cm2) of cracks, to assess the severity of cracking. P. Govindan et al [3]. used the number of surface cracks and the average length of surface cracks to characterize the cracks. As shown in the formula:

Surface crack density:

$$\lambda_a = \frac{\sum_{i=1}^{n} L_i}{A}$$

In the formula:

λ_a ——Surface crack density (cm/cm2)

n ——Select the number of surface cracks

A ——Assessed surface area (cm2)

L_i ——Each crack length (cm)

Surface crack number density:

$$\rho_a = \frac{n}{A}$$

In the formula:

ρ_a ——Average crack number density

n ——Select the number of surface cracks

A ——Assessed surface area (cm2)

The average length of Surface crack:

$$L_a = \frac{\sum_{i=1}^{n} L_i}{n}$$

In the formula:

L_a ——Mean crack length (cm)

n ——Select the number of surface cracks

L_i ——Each crack length (cm)

4.1 Crack phenomenon on Titanium alloy surface

Due to the difference in the energy of a single pulse in EDM, the surface crack formation process of the workpiece is different, and the number and width of cracks on the surface of the workpiece are also different. Figure 2 shows the crack condition of titanium alloy workpiece material under the surface of EDM milling with different energy. It can be seen from the figure that the surface crack of workpiece mainly includes radial crack, arc crack and local surface crack. The variation of surface crack parameters with discharge parameters is shown in table 1.

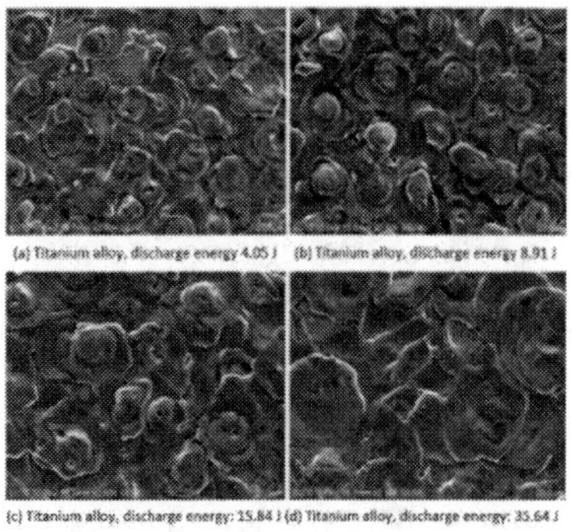

Figure 2 micro-EDM surface cracks in titanium alloy materials

https://doi.org/10.1051/matecconf/202236301011

Table1 RC pulse power supply titanium alloy crack parameters with discharge parameters change data table

Capaitance pF	electric V	Energy μJ	The number of crack	To evaluate the area um²	Total crack length μm	The crack density μm/um²	Average crack length μm	Crack number density
2200	45	2.23	48	483.87	170.16	0.35	3.55	0.10
2200	60	3.96	30	483.87	152.21	0.31	5.07	0.06
2200	90	8.91	33	483.87	120.44	0.25	3.65	0.07
2200	120	15.84	32	483.87	100.24	0.21	3.13	0.07
4000	45.00	4.05	36	483.87	101.36	0.21	2.82	0.07
4000	60.00	7.20	34	483.87	154.47	0.32	4.54	0.07
4000	90.00	16.20	2	483.87	3.70	0.01	1.85	0
4000	120.00	28.80	0	483.87	0	0	0	0
8800	45	8.91	39	483.87	139.2	0.29	3.57	0.08
8800	60	15.84	34	483.87	121.47	0.25	3.57	0.07
8800	90	35.64	26	483.87	106.9	0.22	4.11	0.05
8800	120	63.36	0	483.87	0	0	0	0
16000	45.00	16.20	8	483.87	37.18	0.08	4.65	0.02
16000	60.00	28.80	3	483.87	10.89	0.02	3.63	0.01
16000	90.00	64.80	2	483.87	6.63	0.01	3.32	0
16000	120.00	115.20	0	483.87	0	0	0	0

When using RC pulse power supply for micro EDM, the surface crack of titanium alloy workpiece is very obvious.As shown in figure 3 (a) ,When the same evaluation area is selected, when the discharge capacitance is constant, the number of cracks decreases remarkably with the increase of the discharge voltage, and the change trend of the number density of cracks is the same as the change trend of the number of cracks and this is because most of the cracks occur around a single molten pool. As the discharge voltage increases, the molten pool generated by individual pulse energy increases. In the same assessment area, the number of molten pool decreases correspondingly, so the number of cracks on the workpiece surface decreases correspondingly.

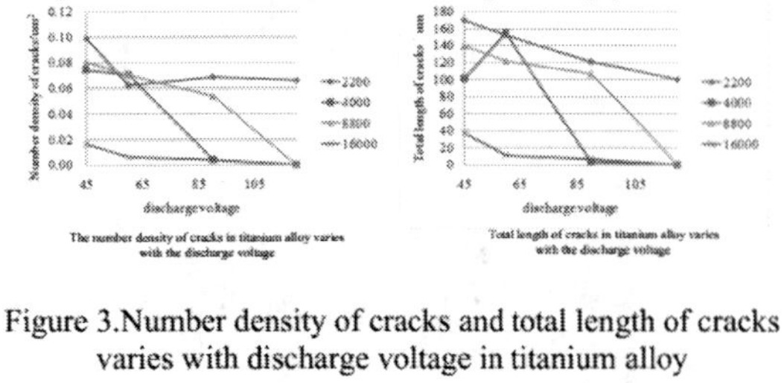

Figure 3.Number density of cracks and total length of cracks varies with discharge voltage in titanium alloy

As shown in figure 3(b), the total length of the surface cracks on the selected evaluation surface gradually decreases with the increase of discharge voltage and this is because the number of cracks decreases with the decrease of the occurrence of cracks. When the capacitance is fixed, the critical gap of discharge breakdown increases with the increase of discharge voltage, which results in the decrease of the temperature gradient acting on the workpiece surface. As a result, the thermal stress decreases. The crack phenomenon decreases accordingly.

As shown in Figure 4, on the surface of micro EDM workpiece, the physical properties of titanium alloy determine the inevitability of surface cracks. When the discharge capacitance remains unchanged, the surface crack density gradually decreases with the increase of discharge voltage. When the discharge voltage remains unchanged, the surface crack appearance gradually decreases with the increase of discharge capacitance. This is different from previous studies that the crack phenomenon increases with the increase of pulse energy and this is because when the discharge voltage is constant, with the increase of discharge capacitance, the monopulse discharge time gradually increases, and the same voltage drops to zero when the discharge time is longer, resulting in a smaller discharge gradient and a smaller thermal stress, resulting in less crack phenomenon. Therefore, as the energy increases, the number of cracks does not increase. On the contrary, the number of cracks, crack density, average crack length, and crack number density all decrease. Therefore, in order to reduce the crack phenomenon, the energy of single pulse cannot be reduced blindly. In these tests, when the capacitance is 4000pF and the voltage is 120V, the surface has no crack phenomenon and the surface quality is relatively good.

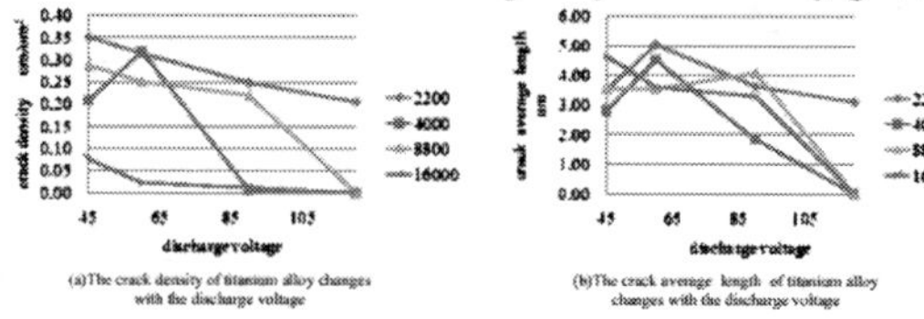

Figure 4 The crack density and crack average length of titanium alloy changes with discharge voltage

4.2 Surface cracks in high speed steel

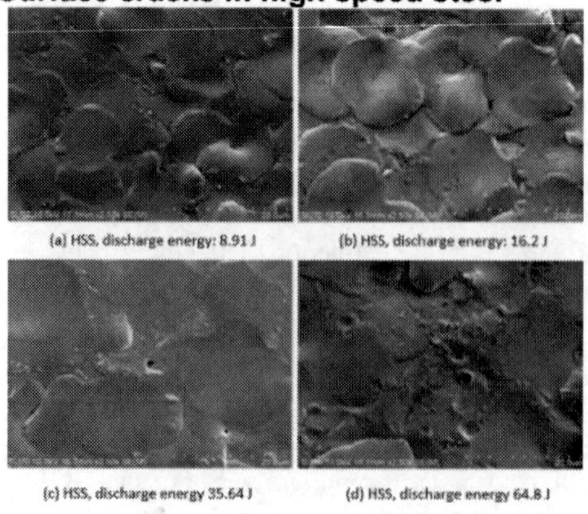

Figure 5. Micro-EDM surface cracks in HSS materials

As can be seen from figure 5, the surface cracks of high-speed steel workpieces are much smaller in size and number than those of titanium alloy materials. There are almost no cracks on the surface of high-speed steel workpieces, and the surface quality is much higher than that of titanium alloy and this is because the specific heat capacity of the titanium alloy is higher than that of the high-speed steel, and the thermal conductivity is lower than that of the high-speed steel. The heat transfer coefficient of the workpiece material is larger, and the heat source transmits more energy to other places, resulting in a smaller temperature gradient and less generated stress. As a result, the surface crack phenomenon is also reduced. That is, the larger the heat transfer coefficient is, the less obvious the crack phenomenon is, so the crack number and crack density of the high speed steel are smaller than those of the titanium alloy under the same energy.

Through a large number of experimental studies on high-speed steel and titanium alloy, it is found that at the same energy level, there is basically no crack in the high-speed steel, and there are a large number of cracks on the surface of the micro-EDM of the titanium alloy. This is because high speed steel has a high thermal conductivity and can quickly transfer heat from the heat source area during the processing. So the heat is far away from the heating area, which will reduce the formation of cracks, indicating that the thermal conductivity has an important effect on the crack. Therefore, the physical properties of the material itself determine that the hot crack of the high speed steel is less than that of the titanium alloy material. Whether it is high-speed steel or titanium alloy, the crack on the surface of the workpiece is deepened with the increase of EDM parameters.

5. Measures to improve surface cracks

Although the physical properties of the material determine that material cracking is unavoidable, surface cracking can be minimized by reasonable selection of processing parameters.

In order to reduce cracks, it is necessary to select a material having a small coefficient of thermal expansion and a small volume change ratio of solid-liquid phase change. The use of electrode materials with better tensile strength will reduce surface cracks and improve the quality of the machined surface.

Adopt workpiece materials with good heat transfer property. Because of good heat transfer performance, the temperature gradient between layers of materials is small in the process of thermal action, resulting in less thermal stress and less crack. For example, high-speed steel with a large thermal conductivity can be selected as the workpiece material in the material titanium alloy and high-speed steel.

The crack is caused by the temperature gradient of each part of the material during the processing, so EDM is used as much as possible with the processing parameters with lower temperature gradient, which can effectively reduce the crack phenomenon. During the processing of RC pulse machining, in order to reduce the temperature gradient, a large capacitor can be used to increase the discharge time and reduce the temperature gradient, thereby reducing the surface crack phenomenon; Similarly, when the RC pulse power supply with the same energy is used for EDM that once the voltage is increased, the critical discharge gap in the discharge circuit will become larger, so that the temperature gradient generated by the workpiece material will be smaller, resulting in less thermal stress and reducing the chance of crack formation. Therefore, when the RC pulse is discharged, the processing parameters of the larger capacitance and larger voltage are selected as much as possible, so that the crack phenomenon can be effectively reduced.

6. Conclusion

In this paper, the crack of recasting layer on the workpiece surface was studied. The results show that different materials have different crack phenomena in EDM. To reduce the crack phenomenon, the processing parameters of large current and short pulse are selected as far as possible for EDM in the processing of equal energy thyristor pulse power supply. In RC pulse discharge, the crack phenomenon can be effectively reduced by selecting machining parameters with larger capacitance and higher voltage.

Acknowledgements

Fund Code: Contract Number: [2022] No. 74
Ideological and political exploration and practice of material processingCAD/CAM course

References

1. Lee Hwa-Teng, Hsu Fu-Chuan, Tai Tzu-Yao, et al.Study of surface integrity using the small area EDM process with a copper-tungsten electrode. Materials Science and Engineering A, 2004, 364(1-2): 346-356.
2. Murray J. W., Clare A. T, et al. Repair of EDM induced surface cracks by pulsed electron beamirradiation. Journal of Materials Processing Technology, 2012, 212(12): 2642-265
3. Govindan P., Joshi Suhas S, et al. Analysis of micro-cracks on machined surfaces in dry electrical discharge machining. Journal of Manufacturing Processes, 2012, 14(3): 277-288.
4. Wujun Feng, Xuyang Chu*, Yongqiang Hong and Li Zhang, et al. Studies on the surface of high-performance alloys machined by micro-EDM. Materials and Manufacturing Processes. 2018, 33(6):616–625.

GPK-YOLOv5s: Content-Aware Reassembly of Features and Self Attention for High Altitude Parabolic Detection and Tracking

Jie Ying [1,*], Han Li [1], Haima Yang [1] and Leqian Zheng [1]

[1]University of Shanghai for Science and Technology, School of Optical Electrical and Computer Engineering, 200093 Shanghai, China

Abstract. High altitude parabolic is difficult to identify because of its small size, fast speed and changeable state, which makes it difficult for subsequent forensics and accountability. This paper proposes a high-altitude parabolic detection and tracking method GPK-YOLOv5s, which integrates Content-Aware Reassembly of Features (CARAFE) and self-attention to realize parabolic detection and tracking. For the detection network, the backbone integrates C3Ghost module to extract effective features and simplify the network. C3Transformer module is embedded in the feature extraction and fusion layer to pay attention to the global context information. The feature fusion layer uses CARAFE module for up sampling to perceive effective features, and integrates shallow features and deep features to form a new large-scale detection layer (Output4) to further obtain smaller receptive fields. Improved multi-scale detection heads are embedded with CBAM to enhance the expression ability of targets. To overcome the frame loss of real-time detection, improved multi-scale detection heads are externally connected with Kalman filter to track targets. This experiment verifies that the detection Precision, Recall and F1 value of GPK-YOLOv5s reached 99.0%, 98.6% and 98.8% respectively, which are 2.8%, 4.1% and 3.5% higher than YOLOv5s respectively. And GPK-YOLOv5s is lighter, and the calculation consumption is reduced by 0.4 GFLOPs.

1. Introduction

High altitude parabolic detection based on computer vision is one of the applications of intelligent video surveillance, which has very important realistic significance. High altitude parabolic is difficult to identify because of its small size, fast speed and variable state.However, there are few literatures on the detection and tracking of high altitude parabolic. Xu W et al. [1] proposed a multi-target tracking algorithm for high altitude parabolic, the improved sorting algorithm and adaptive Gaussian mixture background model are used for detection, and the joint intersection (IOU) combined with correlation filter is used for tracking. Liang X et al. [2] proposed a tracking method based on AprilTa recognition, which can track the color features around the label and improve the tracking accuracy when the target is blocked. Feng W K et al. [3] proposed a moving target detection method based on improved fuzzy C-means clustering (FCM) algorithm. The traditional FCM algorithm is combined with genetic algorithm and Kalman filter algorithm to track and detect moving targets. Murate T et al. [4] proposed a fast target tracking method of learning moving convolutional neural network (CNN) feature extraction. The introduction of self attention mechanism can enhance the expression of target information. Lu X et al. [5] proposed an attention map neural network (AGNN) to complete pixel level target segmentation. The task is described as an iterative information fusion process on the data map, and information is captured from relational visual data through parametric message passing. However, detection speed of above methods is not fast enough, and the recognition accuracy is not high enough, especially a lack of research for the recognition and tracking of high-altitude parabolic.

One-stage target detection algorithms have high detection accuracy, fast speed, and meet requirements of real-time. In particular, YOLOv5 developed by ultralytics is used in wide fields because of its lightweight, high detection accuracy and fast detection speed. Therefore, this paper adopts one stage target detection algorithms to realize high-altitude parabolic detection and tracking. The main contributionsare as follows.

(1) A high altitude parabolic detection and tracking method GPK-YOLOv5s is proposed, which integrates CARAFE and self attention. Improved multi-scale detection CBAMheads are externally connected with Kalman filter to detect and track high-altitude parabolic. (2) C3Ghost is integrated in the backbone to obtain effective features and make the network lightweight. C3Transformer is embedded between feature extraction and fusion layers to capture global context information. (3) For feature fusion layers, CARAFE is used for upsampling to fuse effective features and reduce redundancy. For multi-scale detection layers, further fuse shallow features and deep features to form a new large-scale feature map to obtain smaller receptive fields. Improved multiscale detection CBAMheads could enhance expression ability of targets. (4) The effectiveness and applicability of the proposed GPK-

* Corresponding author: yingjsh@163.com

YOLOv5s are verified by using the self-made high-altitude parabolic dataset OTFH.

2. Methodology

The overall scheme of high-altitude parabolic detection and tracking in this paper is shown in Fig.1.

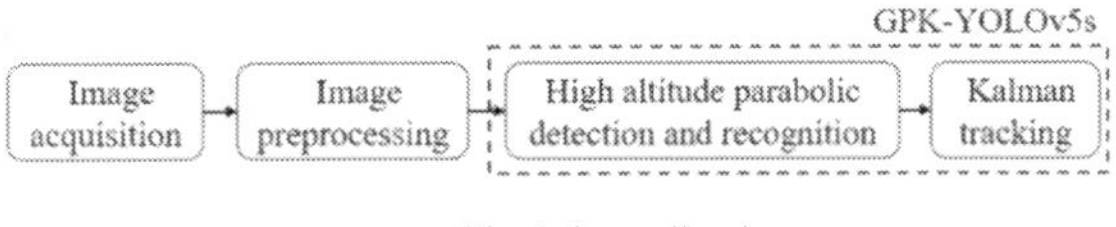

Fig.1 Overall scheme

For data acquisition, we take high altitude parabolic images by CMOS infrared surveillance camera. For image preprocessing, we use Mosaic data enhancement, spatial chrominance transformation, random clipping and left/right flipping to enhance the expression of effective information. The proposed GPK-YOLOv5s realizes the recognition and tracking of high-altitude parabolic.

2.1 Data acquisition

Considering the method of testing and cost, two surveillance cameras with zoom range of 4.7mm-94mm, 200W pixels and 100 m infrared imaging are used in residential or office areas. The installation and layout of high-altitude parabolic surveillance cameras need to consider the installation location, the distance between buildings, household privacy and safety. In this paper, two cameras are installed on the monitoring support 20m away from the building to monitor the high and low floors.

2.2 Data acquisition

Aiming at the problem of difficult identification and few research of high-altitude parabolic, this paper designs the GPK-YOLOv5s network for high-altitude parabolic detection and tracking. As shown in Fig.2, the overall network is composed of detection network and Kalman tracking module.

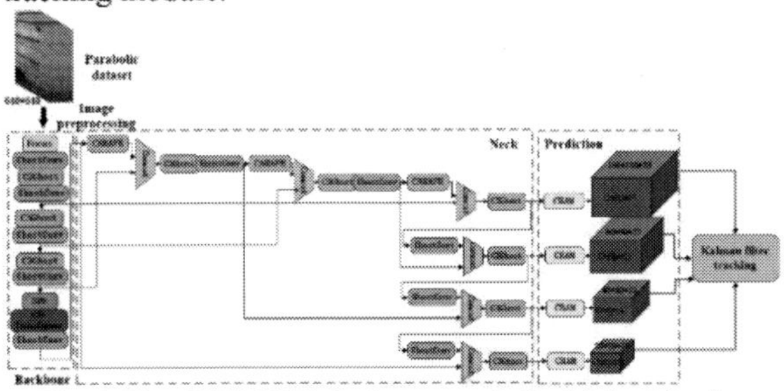

Fig. 2 The GPK-YOLOv5s network for high altitude parabolic detection and tracking

The detection network is composed of backbone, feature fusion layers (Neck) and multi-scale prediction heads. Firstly, the Focus module is used to slice the image, in order to reduce the amount of calculation and improve the speed. Secondly, GhostConv and C3Ghost modules are used to extract features, SPP module adopts convolution kernels with uniform step size but different sizes to realize feature fusion, and embeds C3Transformer module to enhance the context information of backbone. Feature fusion layers adopt CARAFE module [6] for upsampling to fuse deep features and shallow features. Multi-scale detection heads fuse CBAM [7], and feed back the detected target position to the Kalman filter to realize the tracking of high-altitude parabolic.

2.2.1 C3Ghost

Due to the feature extraction network plays an important role in detection results. This paper combines the CSPBottleneck of the original backbone network and three convolution layers to form C3. C3 fuses GhostBottleneck to form C3Ghost module, which is used to extract effective features and simplify the network. As shown in Fig.3 and Fig.4, the GhostBottleneck is composed of Ghost modules [8]. Given input data $X \epsilon R^{c \times h \times w}$, where c is the number of input channels, h and w are the height and width of the input image, respectively. An arbitrary convolutional layer used to generate n feature maps.

$$Y = X * f + b \qquad (1)$$

In formula (1), X and $Y \epsilon R^{h' \times w' \times n}$ denote the input and output of images respectively. $f \epsilon R^{c \times k \times k \times n}$ represents the filter, b denotes the offset. The convolution filter in h' and w' are the height and width of the output data. The backbone of this paper integrates Ghost module, which reduces the redundant calculation of feature map. These intrinsic feature maps are usually small and produced by ordinary convolutional filters. Specifically, m intrinsic feature maps $Y' \epsilon R^{h' \times w' \times m}$ are generated, as formula (2).

$$Y' = X * f' \qquad (2)$$

In formula (2), $f' \epsilon R^{c \times k \times k \times m}$ is the utilized filter, $m \leq n$ and the bias term is omitted for simplicity. In order to get n feature maps, a series of cheap linear operations are used on each intrinsic feature in Y', to generate s ghost features as formula (3).

$$y_{i,j} = \Phi_{i,j}(y_i'), \ \forall i = 1, 2, \dots, m, \quad j = 1, 2, \dots, s, \qquad (3)$$

In formula (3), y_i' is the i-th intrinsic feature map in Y', $\Phi_{i,j}$ is the j-th (except the last one) linear operation for generating the j-th ghost feature map $y_{i,j}$. $n = m \cdot s$ feature maps $Y = [y_{11}, y_{12}, \dots, y_{ms}]$ as the output of a Ghost module as shown in Fig.3.

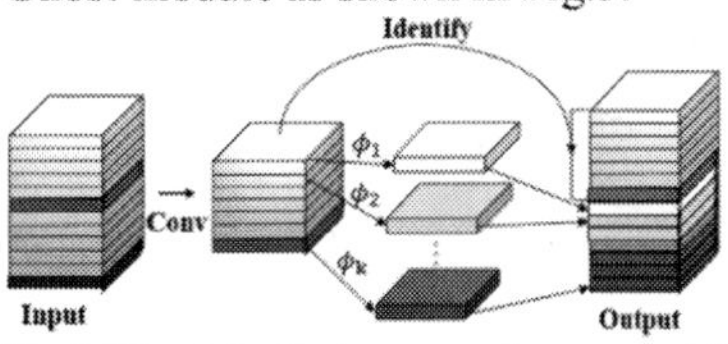

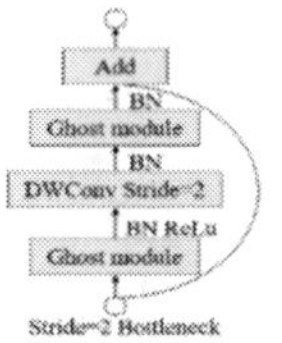

Fig.3 Ghost for feature mapping Fig.4 GhostBottleneck

The step size of the GhostBottleneck module used in this paper is 2, which is formed by stacking two Ghost modules, as shown in Fig.4. The first Ghost module acts as an extension layer, increasing the number of channels. The second Ghost module reduces the number of channels to match the shortcut path. The shortcut is connected between the input and output of the two Ghost modules. BN and ReLU nonlinearity are applied after each layer.

2.2.2 C3Transformer

Considering the diversity of igh altitude parabolic, this paper embeds C3Transformer in the backbone, which can capture global information and context information. The self attention of its module models the dependencies of input and output sequences, which can establish dynamic and longer distance dependencies. Multi heads can learn different tasks respectively, which makes the model more capable of parallel computing. C3Transformer is composed of CSPBottleneck, three convolution layers and Transformer module [9]. The Transformer module is shown in Fig.6.

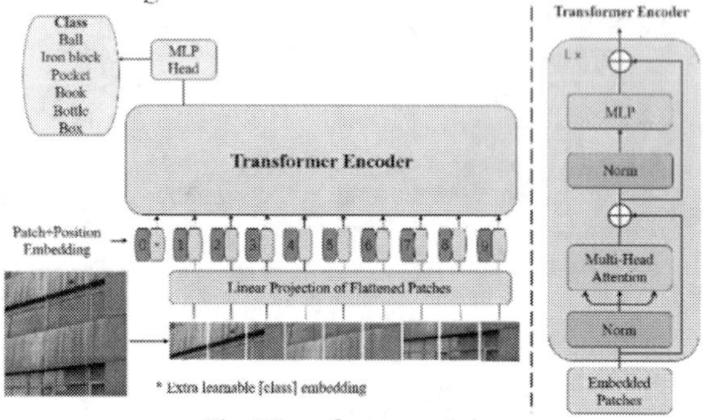

Fig.5 Transformer module

As depicted in Fig.5. The Transformer receives as input a 1D sequence of token embeddings, and reshape the image $X \in R^{H \times W \times C}$ into a sequence of flattened 2D patches $X_p \in R^{N \times (P^2 \cdot C)}$. $H \times W$ is the resolution of the original image, C is the number of channels, P^2 the resolution of each image path, $N = (H \times W)/P^2$ is the resulting number of patches, which also serves as the effective input sequence length for the Transformer. The Transformer uses constant latent vector size D through all of its layers, and flatten the patches and map to D dimensions with a trainable linear projection as formula (4), the output of this projection is patch embeddings. A learnable embedding to the sequence of embedded patches ($z_0^0 = X_{class}$), whose state at the output of the Transformer encoder (z_L^0) serves as the image representation y as formula (7). Both during pre-training and fine-tuning, a classification head is attached to z_L^0. The classification head is implemented by a MLP with one hidden layer at pre-training time and by a single linear layer at fine-tuning time.

$$z_0 = \left[X_{class};\ X_p^1 E;\ X_p^2 E;\ \ldots;X_p^N E\right] + E_{pos,}$$
$$E \in R^{N \times (P^2 \cdot C) \times D}, E_{pos} \in R^{(N+1) \times D} \quad (4)$$
$$z_l' = MSA\left(LN(z_{l-1})\right) + z_{l-1,} \quad l = 1 \ldots L \quad (5)$$
$$z_l = MLP\left(LN(z_l')\right) + z_l', \quad l = 1 \ldots L \quad (6)$$
$$y = LN(z_L^0) \quad (7)$$

Position embeddings are added to the patch embeddings to retain positional information. The resulting sequence of embedding vectors serves as input to the encoder. The Transformer encoder consists of alternating layers of multiheaded self-attention and MLP blocks in formula (5)-(6). Layer norm (LN) is applied before every block, and residual connections after every block. The MLP contains two layers with a GELU non-linearity.

2.2.3 CARAFE

Considering that the traditional upsampling module is easy to lose important feature information and computational redundancy, this paper uses CARAFE module for upsampling to avoid the loss of feature information.

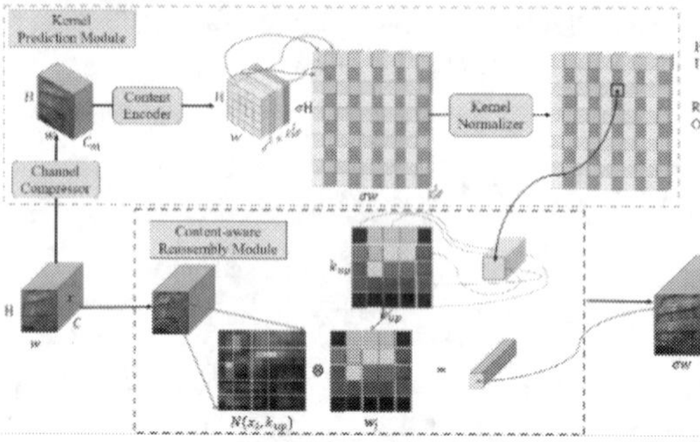

Fig.6 CARAFE module (A feature map with the size $C \times H \times W$ is upsampled by a factor of $\sigma = 2$)

As shown in Fig.6, CARAFE is composed of a Kernel Prediction Module and a Content-aware Reassembly module. The Kernel Prediction Module generates a recombination kernel, which is composed of a Channel Compressor, a Content Encoder, and a Kernel Normalizer. The Channel Compressor reduces the channels of the input feature map, and the content encoder takes the compressed feature map as the input, encodes the content to generate recombination kernels, and the kernel normalizer normalizes each recombination kernel.

CARAFE module predicts recombination kernels according to the underlying content information, uses adaptive and optimized recombination kernels in different positions, and recombines features in predefined nearby areas. The reorganization of the content aware kernel of this module includes two steps. The first step is to predict the reorganization core of each target location according to the content of the target location, and the second step is to use the expected features for reorganization. Given a feature map of size $C \times H \times W$ and an upsample ratio σ (supposing it is an integer), CARAFE will produce a new feature map x' of size $C \times \sigma H \times \sigma W$. For any target location $l' = (i', j')$ of the output x', there is a corresponding source location $l = (i, j)$ at the input x, where $i = \lfloor i'/\sigma \rfloor, j = \lfloor j'/\sigma \rfloor$. Here we denote $N(x_l, k)$ as the $k \times k$ sub-region of x centered at the location l,i.e., the neighbor of x_l.

In the first step, the kernel prediction module ψ predicts a location-wise kernel $w_{l'}$ for each location l', based on the neighbor of x_l, as shown in formula (8). The reassembly step as formula (9), where ϕ is the content-aware reassembly module that reassembles the neighbor of x_l with the kernel $w_{l'}$.

$$w_{l'} = \psi(N(x_l, k_{encoder})) \quad (8)$$
$$x_{l'}' = \phi(N(x_l, k_{up}), w_{l'}) \quad (9)$$

For the kernel prediction module, each source location on x corresponds to σ^2 target locations on x'. Each target locations requires a $k_{up} \times k_{up}$ reassembly kernel, k_{up} is the reassembly kernel size. This module outputs the reassembly kernels of size $C_{up} \times H \times W$, where $C_{up} = \sigma^2 k_{up}^2$. For each reassembly kernel $w_{l'}$, the Content-aware reassembly module will reassemble the features within a local region via the function $\emptyset$. $\emptyset$ is just a weighted sum operator. For a target location l' and the corresponding square region $N(x_l, k_{up})$ centered at $l = (i, j)$, the reassembly is shown in formula (10), where $r = \lfloor k_{up}/2 \rfloor$.

$$x'_{l'} = \sum_{n=-r}^{r} \sum_{m=-r}^{r} w_{l'}(n, m) \cdot x(i+n, j+m) \quad (10)$$

With the reassembly kernel, each pixel in region of $N(x_l, k_{up})$ contributes to the upsampled pixel l' differently, based on the content of features instead of distance of locations.

2.2.4 CBAMheads

Considering that the high-altitude parabolic is generally far from the camera, the multi-scale detection layers need to detect small targets. In this paper, CARAFE is used for upsampling, the shallow features and deep features are further fused to form a new large-scale feature map detection layer (Output4) to obtain smaller receptive fields. Moreover, CBAM is embedded in improved multi-scale detection layers to form multi-scale detection CBAMheads. CBAM module sequentially infers attention maps along two separate dimensions, channel and spatial, then the attention maps are multiplied to the input feature map for adaptive feature refinement.

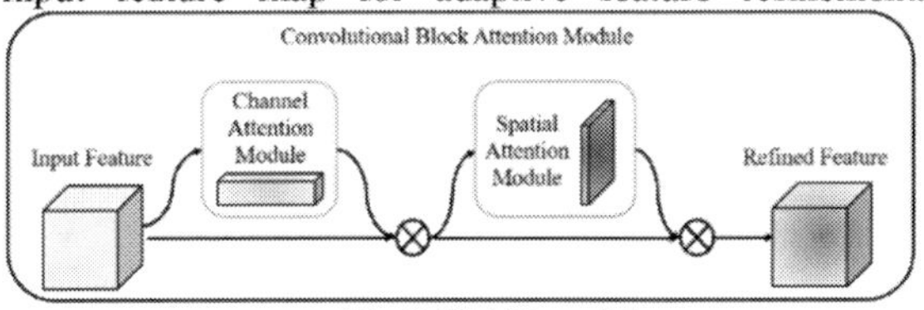

Fig.7 CBAM module

Given an intermediate feature map $F \in R^{C\times H\times W}$ as input, CBAM sequentially infers a 1D channel attention map $M_C \in R^{C\times 1\times 1}$ and 2D spatial attention map $M_s \in R^{1\times H\times W}$ as illustrated in Fig.7. The overall attention process as depicted in formula (11)-(12).

$$F' = M_c(F) \otimes F \quad (11)$$

$$F'' = M_s(F') \otimes F' \quad (12)$$

Where $\otimes$ denotes element-wise multiplication, F'' is the final refined output. The channel attention module uses the average-pooling and maximum-pooling operations to aggregate the spatial information of the feature map and generate two different spatial context descriptors: F_{avg}^{c} and F_{max}^{c}. These descriptors are sent to the shared network to generate the channel attention map $M_c \in R^{C\times 1\times 1}$. The shared network consists of an MLP with a hidden layer. The calculation of channel attention is shown in formula (13).

$$M_c(F) = \sigma\left(MLP\left(AvgPool(F)\right) + MLP\left(MaxPool(F)\right)\right) = \sigma\left(W_1\left(W_0\left(F_{avg}^{c}\right)\right) + W_1\left(W_0\left(F_{max}^{c}\right)\right)\right) \quad (13)$$

Where σ denotes the sigmoid function, the MLP weights, W_0 and W_1. Spatial attention module uses convolution layer to generate spatial attention map $M_s(F) \in R^{H\times W}$, which encodes where to emphasize or suppress, aggregate channel information of a feature map by using two pooling operations, generating two 2D maps: $F_{avg}^{s} \in R^{1\times H\times W}$ and $F_{max}^{s} \in R^{1\times H\times W}$. These images are connected and convoluted through convolution layers to generate 2D spatial attention maps. The calculation of spatial attention is shown in formula (14).

$$M_s(F) = \sigma(f^{7\times 7}([AvgPool(F); MaxPool(F)])) = \sigma(f^{7\times 7}([F_{avg}^{s}; F_{max}^{s}])) \quad (14)$$

σ denotes the sigmoid function, $f^{7\times 7}$ represents a convolution operation with the size of 7×7.

2.2.5 Kalman tracking

To overcome the frame loss of real-time detection, this paper realizes the tracking of high-altitude parabolic by using the improved multi-scale detection headers and external Kalman filter. The target prediction box value and center coordinate value are obtained by the detection network, and the Kalman filter is used to predict the next frame position according to the previous frame position of the target, so as to realize high-altitude parabolic tracking and trace. Kalman filter includes two parts: target state and measurement.

$$x_k = A_{k,k-1}x_{k-1} + Bu_{k-1} + w_{k-1} \quad (15)$$

$$z_k = H_k x_k + v_k \quad (16)$$

As formula (15)-(16), x_k、x_{k-1} are actual state vector of the system at time t_k、t_{k-1}. $A_{k,k-1}$ is the state change matrix t_{k-1} to t_k, B is system control matrix, z_k is state measurement vector at time t_k, H_k represents the measurement matrix, u_{k-1} represents input of the system. w_k denotes process noise of the system, v_k denotes measurement noise of the system.

The Kalman filtering process includes prediction and correction, and its realization of target tracking is divided into three steps. Firstly, the detection network predicted the target position, and initialize the Kalman filter x_{k-1}^{l} and P_{k-1}, record the current frame time t_k. Secondly, state estimation as shown in formula (17) - (18), using the current frame time t_k and the previous frame time t_{k-1} for Kalman prediction, an get predicted target status x_k^- and predicted covariance P_k^-, and search for the best match in the region to obtain the real target state value z_k. Lastly, the status update is as follows (19) - (21), using the real state vector z_k as the input of Kalman filter to update, and get updated status x_k^l and corrected covariance P_k.

$$x_k^- = A_{k,k-1}x_{k-1}^{l} + Bu_{k-1} \quad (17)$$

$$P_k^- = A_{k,k-1}P_{k-1}A_{k,k-1}^{T} + Q_{k-1} \quad (18)$$

$$K_k = P_k^- H_k^T (H_k P_k^- H_k^T + R_k)^{-1} \quad (19)$$

$$x_k^l = x_k^- + K_k(z_k - H_k x_k^-) \quad (20)$$

$$P_k = P_k^- + K_k H_k P_k^- \quad (21)$$

x_k^- represents a priori state estimate, x_k^l and x_{k-1}^l is the posterior state estimate, x_k^l is the optimal estimation after filtering. P_k^- represents the covariance of a priori estimation, and the Kalman filter gain K_k is obtained through the minimum error. P_k represents the covariance of a posteriori estimation, which will be used as the input of the next iteration after updating. R_k is the measurement noise covariance, Q_{k-1} is the process noise covariance. $z_k - H_k\, x_{k,k-1}^{l}$ reflects the error between predicted value $H_k\, x_{k,k-1}^{l}$ and actual measurement value z_k.

3. Experiments and Analysis

3.1 OTFH self-made dataset

Because there is no public dataset for high-altitude parabolic. This paper uses OTFH self-made high-altitude parabolic dataset for experiment. OTFH dataset is made for daily life and office areas, including images of six categories of objects: bottles, books, boxes, balls, iron blocks and pockets. There are 1000 samples in each category, a total of 6000 samples, sample images of this dataset as Fig.8. The experimental dataset is divided into training set: verification set: test set is 6:2:2.

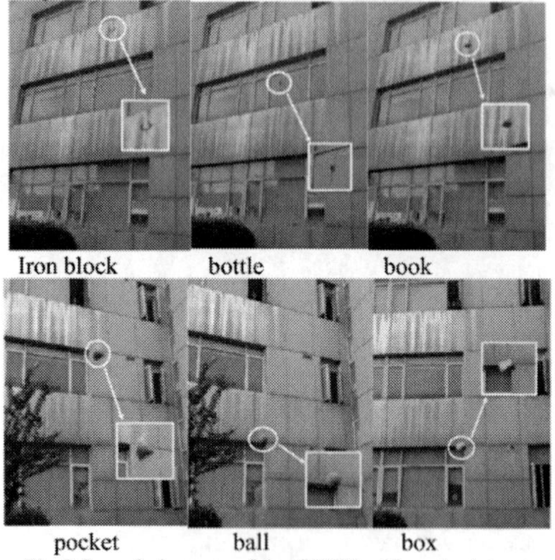

Fig.8 Sample images of the OTFH self-made dataset

3.2 Implementation details

The experimental environment of this paper is based on Windows 10 system. Graphics card: NVIDIA GeForce GTX 1080Ti, learning framework is Pytorch. The programming language is Python 3.7, the programming tool is Pycharm. This experiment uses stochastic gradient descent (SGD) with a mini-batch size of 16 for 200 epochs. Warmup is used to warm up the learning rate during training. Set the learning rate of 0.01 to warm up the training for 10 epochs, so as to improve the stability of deep model training. After Warmup, cosine annealing learning algorithm is used to update the learning rate [10]. This experiment uses CIOULoss function to calculate position loss, BCEWithLogitsLoss function to calculate confidence loss, and FocalLoss function to calculate target class loss.

3.3 Evaluation metrics

In order to quantitatively evaluate the detection performance of the network, precision (P), recall (R), average precision (AP) and F1 value are usually used as indicators of the measurement model as formula (22) - (25). The positive samples of OTFH dataset refer to high-altitude parabolic targets, and the image background and different types of targets are regarded as negative samples.

$$P = \frac{TP}{TP+FP} \quad (22)$$

$$R = \frac{TP}{TP+FN} \quad (23)$$

$$AP = \int_0^1 P(R)dR \quad (24)$$

$$F1 = \frac{2*P*R}{P+R} \quad (25)$$

Where TP is the real value, FP is the false positive value, and FN is the false negative value. When the intersection union ratio (IOU) of the prediction box and the truth box is greater than 0.5, the target is detected.

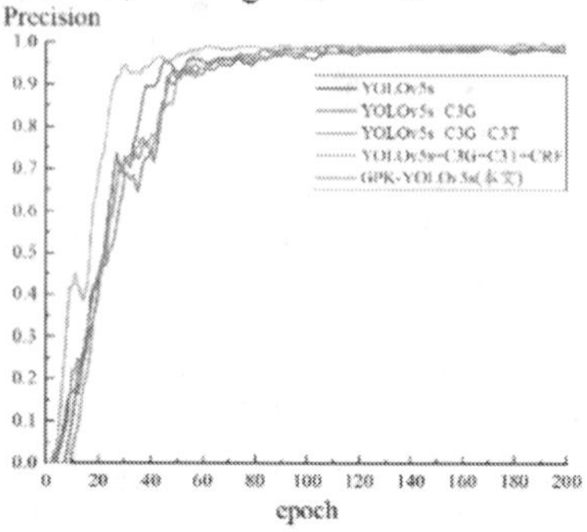

Fig.9 Comparison of Precision of the validation set

For convenient expression of the experiment, C3Ghost module, C3Transformer module, CARAFE module and multi-scale detection headers with CBAM are recorded as C3G, C3T, CRF and CBAMHeaders respectively. Therefore, the detection network of GPK-YOLOv5s includes C3G、C3T、CRF and CBAMHeaders. To verify the effectiveness of each module in GPK-YOLOv5s, ablation experiments of the model were carried out on OTFH dataset.

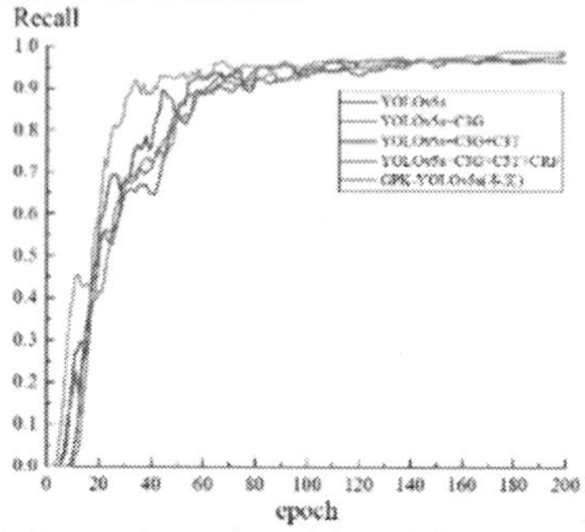

Fig.10 Comparison of Precision of the validation set

Fig.9 and Fig.10 show detection precision and recall curves of OTFH validation set for each model. Through ablation experiments, the detection precision and recall rate of GPK-YOLOv5s in can be rapidly improved, which is significantly higher than other models. The detection precision and recall rate are as high as 99% and 98.6% respectively. Compared with above curves, the embedding of each module can optimize the detection network. The detection error convergence speed of this method is faster, the model stability speed is faster, and the generalization ability of the model is enhanced.

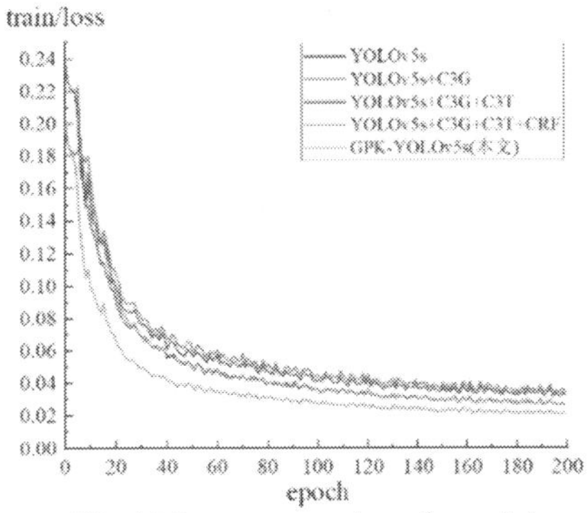

Fig.11 Loss comparison for training

Fig.11 shows the loss values of 200 epochs trained by each model on OTFH dataset. Through a series of ablation experiments, the loss value of GPK-YOLOv5s was significantly reduced to 0.0210, and its loss value was 0.0117 lower than that of YOLOv5s.

4. Experiments results

4.1 Detection and tracking results

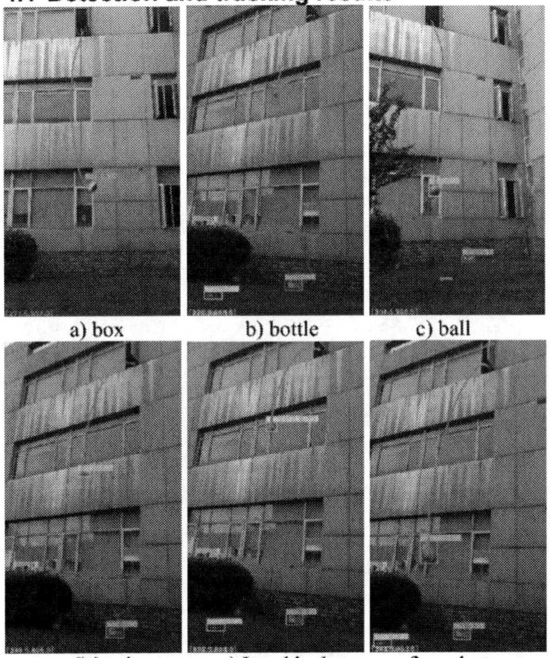

a) box b) bottle c) ball

d) book e) Iron block f) pocket

(1) Office building scene

a) box b) Iron block

c) pocket d) ball

Fig.12 Detection and tracking results of GPK-YOLOv5s ((1) Office building scene; (2) Residential building scene)

Fig.12 (1) (2) show detection and tracking results of GPK-YOLOv5s for high-altitude parabolic in this paper. The high-altitude parabolic is captured from different angles and heights by the surveillance camera, and six types of parabolic such as box, bottle, ball, book, iron block and pocket are detected and tracked in real time, with an average detection confidence of more than 93%, and the tracked target position is displayed in the lower left of the image in real time. This shows that GPK-YOLOv5s is robust to high-altitude parabolic detection and tracking in real scenes.

4.2 Test results

Table 1 shows test results of OTFH dataset by each model, mAP/0.5 represents the average accuracy with a threshold of 0.5. GFLOPs refers to billions of floating-point operations per second. Considering the real-time performance of the detection, one-stage detection such as SSD, YOLOv3, YOLOv3-tiny and YOLOv5 are used for comparison. YOLOv3-tiny is a lightweight version of YOLOv3.

Table 1. Test results of OTFH dataset (%).

Models	P	R	F1	mAP/0.5	GFLOPs
SSD	-	-	-	74.3	-
YOLOv3	88.5	82.1	85.2	81.9	155.1
YOLOv3-tiny	87.7	85.3	86.5	86.2	13.3
v5s(baseline)	96.2	94.5	95.3	95.6	17.1
v5s+C3G	92.3	93.8	93.0	93.6	10.4
v5s+C3G+C3T	96.6	96.3	96.4	96.2	10.7
v5s+C3G+C3T+CRF	98.2	97.4	97.8	97.6	11.2
GPK-YOLOv5s	**99.0**	**98.6**	**98.8**	**98.5**	16.7

In Table1-3, v5s represents YOLOv5s. Considering the complexity of the model, through the ablation experiment, GPK-YOLOv5s realizes the high-precision and rapid detection of high-altitude parabolic. Compared with v5s, v5s+C3G consumes less calculation, has higher detection precision, and reduces 6.7 GFLOPs. The detection precision, recall rate, F1 score, map/0.5 of v5s+C3G+C3T were increased by 0.4%, 0.8%, 1.1%, 0.6% respectively compared with v5s.C3G module lightens the backbone and extracts effective feature information. C3T module can capture global context information. The detection precision, recall rate, F1 score, map/0.5 of v5s+C3G+C3T+CRF were increased by 2.0%, 2.9%, 2.5%, 2.0% respectively compared with v5s, and the consumption of calculation was significantly reduced.

Table 2 shows the detection precision of various high-altitude parabolas of OTFH dataset by each model. SSD and YOLOv3 have low detection accuracy and speed for high-altitude parabolic. In addition, most of bottles are transparent, the iron block moves fast, and the shape changes greatly, so they are difficult to identify. Through

Table 2. Precision of each target in OTFH dataset (%).

Models	bottle	book	box	ball	Iron block	Pocket
SSD	75.1	72.7	79.7	69.8	59.2	89.2
YOLOv3	86.0	81.1	97.0	98.3	73.5	94.8
YOLOv3-tiny	74.5	89.0	97.5	94.7	75.8	94.7
v5s (baseline)	90.6	98.0	98.1	98.9	93.9	97.9
v5s+C3G	86.7	98.9	98.4	97.2	75.4	96.9
v5s+C3G+C3T	92.0	98.0	98.5	97.0	96.1	98.1
v5s+C3G+C3T+CRF	97.0	98.9	98.8	98.8	96.5	99.0
GPK-YOLOv5s	**98.5**	**99.2**	**98.5**	**99.1**	**99.6**	**99.1**

ablation experiments, GPK-YOLOv5s has made a great breakthrough in these difficult to identify targets. The detection precision of bottles, books, boxes, balls, iron blocks and pockets is 7.9%, 1.2%, 0.4%, 0.2%, 5.7% and 1.2% higher than that of YOLOv5s respectively. From ablation experimental results, each optimization module plays an important role in this method.

Table 3. Recall of each target in OTFH dataset (%).

Models	bottle	book	box	ball	Iron block	pocket
YOLOv3	50.0	93.2	97.5	93.7	75.8	82.3
YOLOv3-tiny	63.3	94.3	97.3	92.4	78.8	85.5
v5s (baseline)	94.1	97.2	97.4	92.0	93.7	92.5
v5s+C3G	85.0	96.0	97.0	95.0	96.6	93.0
v5s+C3G+C3T	94.1	98.2	98.0	97.3	97.0	96.1
v5s+C3G+C3T+CRF	94.4	98.2	99.2	98.0	97.3	97.4
GPK-YOLOv5s	**97.6**	**99.1**	**99.4**	**98.5**	**98.4**	**98.3**

Table 3 shows the recognition recall rate of various high-altitude parabolas of OTFH dataset by each model. YOLOv5s has a higher recall rate than YOLOv3 and YOLOv3-tiny. In this paper, GPK-YOLOv5s has increased the recognition recall rate of bottles, books, boxes, balls, iron blocks and pockets by 3.5%, 1.9%, 2.0%, 6.5%, 4.7% and 5.8% respectively compared with YOLOv5s.

Due to the high-altitude parabolic target captured by the camera is generally small. Comparing with the latest small target detection methods, this paper verifies the effectiveness of GPK-YOLOv5s for small target detection in complex background. Comparison results with other small target detection methods are shown in Table 4.

Table 4. Other small target detection methods

Other methods	Pixels	mAP/0.5	FPS
RSRGAN [11]	640*512	98.0%	55.0
YOLOv4-KCF [12]	1920*1080	78.0%	17.8
Improved SSD [13]	-	96.9%	19.0
Improved YOLOv3[14]	-	94.1%	30.0
GPK-YOLOv5s	**640*640**	**98.5%**	**41.7**

Table 4 shows the comparison between GPK-YOLOv5s and other small target detection methods. RSRGAN [11] is faster to detect small targets in lower resolution images, and the mAP reaches 98.0%, which is 0.5% less than that of GPK-YOLOv5s in this paper. The accuracy of improved SSD [13] for small target detection can reach 96.9%, but its detection speed is slow. YOLOv4-KCF [12] has lower value of mAP for small targets. The detection accuracy of proposed method GPK-YOLOv5s for high-altitude parabolic in low resolution images reaches 98.5%, the detection frame rate reaches 41.7, which are better than other small target detection methods mentioned.

5. Conclusion

For intelligent detection network, the backbone integrates C3Ghost to extract effective features and simplify the network. C3Transformer is embedded between feature extraction and fusion layers to strengthen feature extraction and capture global context information. CARAFE is used for upsampling, which has a good effect on pixel level prediction. The new large-scale feature map detection layer can obtain smaller receptive fields, which is convenient for detecting small targets. Multiscale detection heads fused with CBAM can enhance the expression ability of targets. This paper verifies the effectiveness of the proposed method through ablation experiments. The detection results show that the detection precision, recall rate and F1 value of GPK-YOLOv5s for high-altitude parabolic reached 99.0%, 98.6% and 98.8% respectively. The calculation consumption is reduced by 0.4 GFLOPs, and the detection frame rate is 41.7, which is significantly better than other models. To overcome the frame loss during real-time detection, Kalman filter is connected to CBAMheaders of the detection network to realize the tracking of high-altitude parabolic. In future research, the way of training model needs to be optimized, considering the prediction of limited samples, and trying to avoid the dependence of network on big data.

References

1. Xu W, Zhang B. A multi-target tracking algorithm for objects thrown from high[C]//2021 IEEE 3rd International Conference on Civil Aviation Safety and Information Technology (ICCASIT). IEEE, 2021: 1077-1081.
2. Liang X, Chen G, Zhao S, et al. Moving target tracking method for unmanned aerial vehicle/unmanned ground vehicle heterogeneous system based on AprilTags[J]. Measurement and Control -London- Institute of Measurement and Control-, 2020, 53(1):002029401988907.
3. Feng W K, Chun S X. Moving target detection and tracking based on improved FCM algorithm[J]. International Journal of Cognitive Informatics and Natural Intelligence (IJCINI), 2020, 14(1): 63-74.
4. Murate T, Watanabe T, Yamada M. Learning Mobile CNN Feature Extraction Toward Fast Computation of Visual Object Tracking[J]. 2021.
5. Lu X, Wang W, Shen J, et al. Segmenting objects from relational visual data[J]. IEEE Transactions on Pattern Analysis and Machine Intelligence, 2021.
6. Wang J, Chen K, Xu R, et al. CARAFE: Content-Aware ReAssembly of FEatures[J]. IEEE, 2020.
7. Woo, S.; Park, J.; Lee, J.Y.; Kweon, I.S. CBAM: Convolutional block attention module. In Proceedings of the European Conference on Computer Vision (ECCV), Munich, Germany, 6 October 2018; Springer: Munich, Germany, 2018; pp. 3–19.
8. Han K, Wang Y, Tian Q, et al. GhostNet: More Features From Cheap Operations[C]// 2020 IEEE/CVF Conference on Computer Vision and Pattern Recognition (CVPR). IEEE, 2020.
9. Dosovitskiy A, Beyer L, Kolesnikov A, et al (2020) An image is worth 16x16 words: Transformers for

image recognition at scale. arXiv preprintarXiv:201011929

10. LOSHCHILOV I, HUTTER F. SGDR: Stochastic gradient descent with warm restarts[EB/OL]. [2021-05-25]. https://arxiv.org/pdf/1608.03983.pdf.
11. Ren K, Gao Y, Wan M, et al. Infrared small target detection via region super resolution generative adversarial network[J]. Applied Intelligence, 2022: 1-13.
12. Chi Y, Bai C, Guo J. Dynamic Small Target Detection and Tracking Based on Hierarchical Network and Adaptive Input Image Stream[C]//International Conference on Autonomous Unmanned Systems. Springer, Singapore, 2021: 3297-3307.
13. Huang L, Chen C, Yun J, et al. Multi-Scale Feature Fusion Convolutional Neural Network for Indoor Small Target Detection[J]. Frontiers in Neurorobotics, 2022, 16.
14. Cheng Z, Lv J, Wu A, et al. YOLOv3 Object Detection Algorithm with Feature Pyramid Attention for Remote Sensing Images[J]. Sensors and Materials, 2020, 32(12):4537.

Research on straightness calibration method of straight edge working face based on electronic level

Yi Yang*, Tao Wang, Liwei Dai

Hubei Institute of Measurment and Testing Technology, Wuhan, Hubei, China

Abstract. A straightness calibration method of straight edge working face based on electronic level is proposed. According to the proposed calibration method, the evaluation model of measurement uncertainty is established, and the source of uncertainty is analysed. The expanded uncertainty is less than one third of the maximum allowable error required by the calibration specification, which verifies the feasibility and effectiveness of this calibration method.

1. Introduction

The straight edge is a measuring tool used to measure the straightness and flatness of the workpiece when the working surface is flat [1]. The measuring characteristics of the straight edge mainly include surface roughness, straightness of the working face, parallelism of the upper and lower working faces, etc. Among them, the main measuring instrument for the straightness calibration of the working face is the electronic level. Electronic level is a small angle measuring instrument, which is mainly used to measure the inclination of the relative horizontal plane. It can also be used to measure the parallelism of two parts, the straightness of the guide rail and the flatness of the working surface. Other measuring instruments can also be used to calibrate the straightness of the working surface of the straight edge. In reference [3] the author uses the image level to measure the straightness of the straight edge and evaluate its measurement uncertainty. In reference [4] the author studies the straightness measurement system of the straight edge and its experimental instruments, and proposes a fully automatic straightness measurement method of the straight edge. The common calculation methods of straightness of a straight edge are two-point method, least square method and minimum condition method. In reference [5] the author makes a comparative study on the application of these three methods in straightness calibration of a straight edge. This paper uses the electronic level proposed in JJF 1097-2021[1] to calibrate the straightness of the working surface of the straight edge, processes the measurement data with the minimum condition method, and analyzes the measurement uncertainty. The expanded uncertainty of the measurement results meets the requirements of the calibration specification.

* Corresponding author: alex_to@126.com

2. Calibration method

The straightness of the working surface of the straight edge is measured by the pitch method with an electronic level. When measuring, fix the electronic level on the bridge slab, then move the bridge slab from one end of the leveling straight edge to the other end, and record the readings of the electronic level in turn. According to the principle of minimum condition evaluation, data processing is carried out through calculation to evaluate the straightness of the working face of the measured straight edge. During measurement, pay attention to supporting the measured leveling straight edge (except bridge type leveling straight edge) at both ends of the leveling straight edge 2l/9 (l is the length of the leveling straight edge) with equal height blocks, and select appropriate bridge slabs according to the length of the leveling straight edge, generally not less than 8 ~ 10 positions, with a span of (50 ~ 500) mm. The schematic diagram of the measurement process is shown in Figure 1.

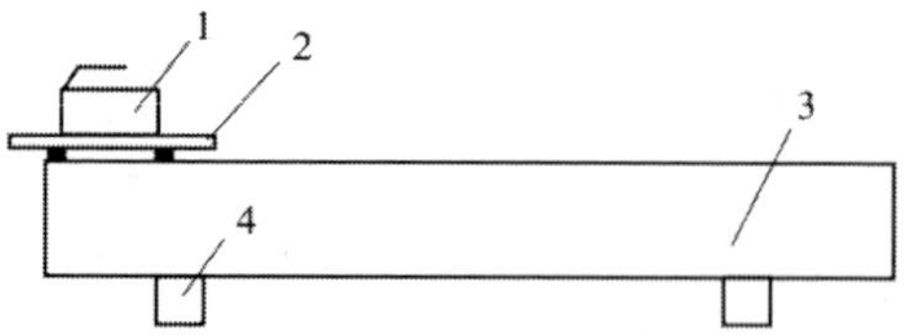

Figure 1. Schematic diagram of straightness calibration of straight edge working face based on electronic level

1 –electronic level; 2 –bridge slab; 3 –straight edge; 4 –block;

After obtaining the measurement data of all electronic level instruments, according to the principle of minimum conditions, the straightness of the working surface of the straight edge can be calculated according to the following

$$F = CL\Delta_i \quad (1)$$

$$F = LC\left[\sum_0^k a_i - \frac{k-g}{p-g}\left(\sum_0^p a_i - \sum_0^g a_i\right) - \sum_0^g a_i\right] = LC\left(\frac{p-k}{p-g}\sum_{i=g+1}^{k} a_i - \frac{k-g}{p-g}\sum_{i=k+1}^{p} a_i\right) \quad (2)$$

Where F is straightness of the working surface of the straight edge, C is graduation value of electronic level, L is span of bridge slab, Δi is straightness reading conforming to the principle of minimum condition, g and p is sequence number of two low points (or two high points), k is sequence number of high point (or low point) contained in two low points (or two high points), n is the reading number of the electronic level at the pitch of section.

3. Evaluation of measurement uncertainty

3.1 Measurement model

Given

$$F_a = \frac{p-k}{p-g}\sum_{i=g+1}^{k} a_i - \frac{k-g}{p-g}\sum_{i=k+1}^{p} a_i \quad (3)$$

Then the straightness measurement model of the working surface of the straight edge based on the electronic level is

$$F = CLF_a \quad (4)$$

Considering that the uncertainty of a single measurement on each dimension segment is the same, then

$$u^2(a_1) = u^2(a_2) = \cdots = u^2(a_i) = \cdots = u^2(a_n) = u^2(a) \quad (5)$$

$$u^2(F_a) = \left[\left(\frac{p-k}{p-g}\right)^2 (k-g) + \left(\frac{k-g}{p-g}\right)^2 (p-k)\right] u^2(a) = \frac{(p-k)(k-g)}{p-g} u^2(a) \quad (6)$$

When g=0, p=n, k=n/2, the u(Fa) is max, then

$$u_{\max}^2(F_a) = \frac{n}{4} u^2(a) \quad (7)$$

$$u_{\max}(F_a) = \frac{\sqrt{n}}{2} u(a) \quad (8)$$

The uncertainty component of the evaluation is composed of the indication error difference of the electronic level, the quantization error of the electronic level, the measurement repeatability, and the influence of the positioning error, then

$$u^2(a) = u_1^2(a) + u_2^2(a) + u_3^2(a) + u_4^2(a) \quad (9)$$

Where u1(a) is the uncertainty introduced by indication error difference of the electronic level, u2(a) is the uncertainty introduced by the quantization error of the electronic level, u3(a) is the uncertainty introduced by the measurement repeatability, u4(a) is the uncertainty introduced by the influence of the positioning error.

3.2 Calculation of standard uncertainty

3.2.1 Calculation of u1(a)

According to the regulation JJG103-2005[2], the indication error shall not exceed (1+A×2%) unit, where A is the change in the number of resolution actually used. When calibrating the level 0 straight edge, the difference between the indicated values of the electronic level from head to tail with the pitch method is not greater than 50 unit. Then the indication error component is

$$A_s = 1 + 50 \times 2\% = 2 \quad (10)$$

It is estimated that the indication error is uniformly distributed in an interval with a half width of 2 unit, then

$$u_1(a) = 2/\sqrt{3} = 1.15 \quad (11)$$

3.2.2 Calculation of u2(a)

Resolution of electronic level is 1 unit (0.001mm/m), which is uniformly distributed, then

$$u_2(a) = \frac{\delta_x}{2\sqrt{3}} = \frac{1}{2\sqrt{3}} = 0.29 \quad (12)$$

3.2.3 Calculation of u3(a)

The standard deviation of measurement repeatability is to install an electronic level with the same span of bridge slab and place it at the same measurement position for 10 times. The reading changes within 1 unit, and the standard deviation is taken as the maximum standard deviation

$$s_x \leq 0.5 \quad (13)$$

Taking the maximum value, then

$$u_3(a) = 0.5 \quad (14)$$

Since the uncertainty component introduced by measurement repeatability is larger than that introduced by quantization error, only the uncertainty component u3(a) is taken and u2(a) is omitted.

3.2.4 Calculation of u4(a)

Due to the influence of the straightness of the leveling straight edge, the bridge plate position is required to be connected head to tail during calibration, but there is a positioning error in the actual operation, which causes the change of the indication value of the electronic level. Practice has proved that for the level straight edge of grade 0, the maximum change of the indication value of the electronic level caused by the above operation deviation is not more than 2 unit. The value is equally probability distributed in an interval with a half width of 2 unit, then

$$u_4(a) = 2/\sqrt{3} = 1.15 \tag{15}$$

3.3 Combined standard uncertainty

3.3.1 The calculation results of standard uncertainty

The source of standard uncertainty components of measurement results is shown in Table 1.

Table 1. The source of standard uncertainty components of measurement results

$u(x_i)$	The source of standard uncertainty	coefficient of sensitivity c_i	results of $u(x_i)$
$u_1(a)$	The uncertainty introduced by indication error difference of the electronic level	1	1.15
$u_2(a)$	The uncertainty introduced by the quantization error of the electronic level	1	0.29
$u_3(a)$	The uncertainty introduced by the measurement repeatability	1	0.5
$u_4(a)$	The uncertainty introduced by the influence of the positioning error	1	1.15

3.3.2. The calculation results of combined standard uncertainty

The above standard uncertainty components are not related to each other, so the combined standard uncertainty is

$$u(a) = \sqrt{u_1^2(a) + u_3^2(a) + u_4^2(a)} = 1.7 \tag{16}$$

3.4 Expanded uncertainty

Taking the inclusion factor k = 2, the expanded uncertainty of the straightness measurement of straight edge working face based on electronic level is

$$U = 2u(F) = 2LCu_{\max}(F_o) = 2LC\frac{\sqrt{n}}{2}u(a) = 1.7\sqrt{n}LC \tag{17}$$

According to the provisions of JJF1097-2021[1], different spans L of the bridge slab are selected for measurement. Select the length of 300mm, 1000mm, 2000mm and 3000mm respectively straight edge, the evaluation results are shown in Table 2.

Table 2. The uncertainty of straightness measurement of different length straight edge

length of straight edge (mm)	spans *L* of the bridge slab (mm)	number of measurement sections	expanded uncertainty *U*(k=2) (μm)	straightness requirements of grade 00 straight edge[1] (μm)
300	50	6	0.2	1.3
1000	100	10	0.6	3.0
2000	200	10	1.1	5.4
3000	300	10	1.7	7.8

According to the evaluation results in Table 2, the expanded uncertainty of this method is less than one third of the maximum allowable error required by the specification.

4. Conclusion

A straightness calibration method of straight edge based on electronic level is proposed. The evaluation model of measurement uncertainty is established, and the source of uncertainty is analyzed, and the extended uncertainty of straightness calibration of straight edge based on electronic level is obtained. The expanded uncertainty is less than one third of the maximum allowable error required by the calibration specification, which shows that the method in this paper is effective and feasible.

References

1. JJF1097-2021 Calibration Specification for Straight Edges. China Standards Press, 2021.
2. JJG103-2005 Verification Regulation of Electronic Levels and Coincidence Levels. China Metrology Press, 2005.
3. Wu Huling. Uncertainty Evaluation to Ruler Straightness Based on Combined Diagram Level. Mechanical Research & Application, 2017, 30(2):129-132.
4. Chen Gang, Zhai Qiongjie, Hu Zhen. Straightness measurement system and experimental instrument of straight edge, 2014, 40(z1):105-108.
5. Dong Zhijiao, Wang Xue.Three data processing methods of straightness of ruler. Measurement Technique, 2012, 10:38-42.

Research of controller of Permanent magnet linear synchronous motors via model predictive direct thrust control

Chunmei Wang [1], Hong Li [1], Yunxiang Lu [1], zhang Song [2]

[1] School of Mechanical and Electrical Engineering, Guizhou Normal University, Guiyang, 550025, China
[2] Suiyang County Secondary Vocational School, Zunyi, Guizhou, 563300, China

Abstract: How to achieve high-precision control is the core issues of the linear motor research. To solve this problem, this paper proposes a new control algorithm. on the one hand, through the sliding mode control for the unknown torque observations, in the process of observation, select a specific based function to approximate the sign function in order to eliminate the jitter of the vicinity of the switching surface; On the other hand, the use of model predictive control method (Model Predictive Control) to achieve the torque compensator track to achieve the velocity and displacement tracking. The constant sampling time of the model predictive control in the new method can effectively avoid the problems that the number of switching per unit time of the inverter can't be expected.

Keywords: Permanent magnet synchronous linear motors; sliding mode control; model predictive direct thrust control.

1. Introduction

Linear motor technology is widely used in CNC machining, industrial production and so on due to the advantages of simple structure, fast response. Displacement tracking control of linear motor is a core issues of the linear motor applied research. The linear motor mostly used displacement tracking control method: PID control, adaptive control, direct thrust control, sliding mode control, preview control, fuzzy control, neural network control. The MPC current tracking technology developed in recent years, has the advantages of fast dynamic response and high tracking accuracy, the algorithm is simple, model predictive control sampling time constant, can effectively circumvent the direct thrust control sampling time is not a constant problem. Kennel R. et al [1, 2] used generalized predictive control method, established linear controller based on Controlled Autoregressive Integrated Moving Average Model (CARIMA) by data analysis, and used in motor control.
Based on the above analysis, the authors proposed a new method of sliding mode model predict direct thrust control. In the new approach, on the one hand, to observe the unknown torque in the observation process by sliding mode control, select a specific function approximation sign function to eliminate jitter in the vicinity of the switching surface. On the other hand, using model predictive control to achieve the torque compensator to achieve precise control of speed and displacement. The method is an improvement for the conventional direct torque control method, the sampling time is constant, reducing the technical requirements of the inverter.

2. Mathematical Model of the PMLSM

Mathematical Model of the PMLSM as follows:

$$\begin{cases} u_d = R_a i_d + L_d \dfrac{di_d}{dt} - \dfrac{\pi}{\tau} v_m L_q i_q \\ u_q = R_a i_q + L_q \dfrac{di_q}{dt} + \dfrac{\pi}{\tau} v_m L_d i_d + \dfrac{\pi}{\tau} v_m \psi_f \\ F_e = \dfrac{3}{2}\dfrac{\pi}{\tau}[\psi_f i_q + (L_d - L_q) i_d i_q] \end{cases} \quad (1)$$

Where u_d and u_q are the d−q voltages; i_d and i_q are the d−q currents; L_d and L_q are the d−q axis inductance. R_a is phase winding resistance of the stator; v_m is mechanical speed of the rotor; ψ_f is permanent magnet flux linkage.
The mover dynamic equation of the PMLSM is

$$F_e = F_l + B \cdot v_m + M\frac{dv_m}{dt} \quad (2)$$

where F_e is the output electromagnet thrust, M is the mass of moving inertia, B is the friction factor and F_l is the external disturbance term.
It can be seen that the basic characteristics of the permanent magnet linear synchronous motor like with permanent magnet synchronous motor, it is a high-order, nonlinear, strong coupling of multi-variable drive control

system, and therefore it is the actual dynamic mathematical model is a complex high-order differential equations, it is difficult to accurate solution for their model. Therefore, in the actual application is often simplified into a simple second-order differential equations.

3. Linear motor controller study

In this paper, the sliding mode model to predict direct thrust control method with constant sampling time, and the advantages of easier to realize in the engineering applications. Its control block diagram is shown in Figure 1 : the reference displacement and the actual displacement as input of sliding mode control, whose output is the reference thrust and reference flux, reference thrust、reference flux、estimate the flux and estimates thrust as inputs of the direct thrust model predictive control. the output of Model predictive control is the switching state of the voltage source inverter. Finally voltage source inverter output voltage to the linear motor.

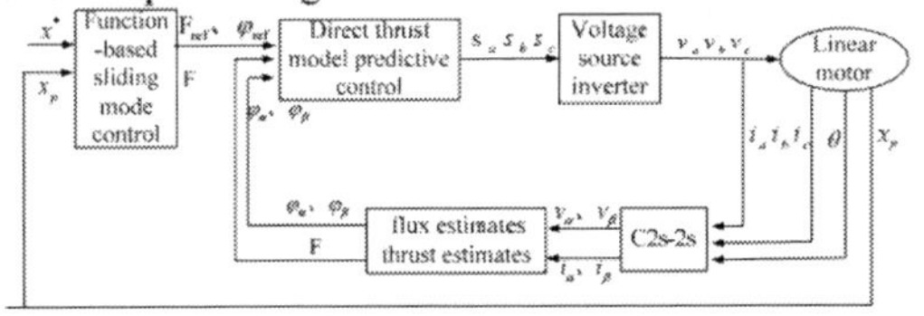

Figure 1. Linear motor SMMP-DTC control block diagram

3.1 Research on sliding mode controller

The linear motor has strong nonlinearity and uncertainty, There is no any the buffer of the middle part, the transformation of the any parameters in the control system will seriously affect the static performance and dynamic performance of the control system of linear motor. sliding mode variable structure control apply to nonlinear and uncertainties of the linear motor. This control method through switch of the control force to make the system state sliding along the sliding surface, the system invariant by parameter perturbation and external interference.

Consider the system: $\dot{x} = Ax + Bu + f(t,x)$ (3)

where $x \in R^n$, $A \in R^{n\times n}$, $B \in R^{n\times n}$, $u \in R^{n\times 1}$,

If there exists $g(t,x) \in R^{n\times 1}$, such that $f(t,x) = Bg(t,x)$, n>m and B have the full rank and (3)meets the demands of the matching conditions.

By(2),The dynamics of the PMLSM can be represented by the following second-order differential equation:

$$F_e = M\ddot{x} + B\dot{x} + F_L \tag{4}$$

where x is the position output and $\|F_L\| \le E$ is the bounded external uncertainty and disturbance term that may comprise dry and viscous friction, as well as any other unknown forces. Here, $\|F_L\|$ is satisfy the matching conditions of formula (3). It can be seen that the robustness of the SMC method makes this intensive approach suitable for use in a motor control system[5].

formula (4) can be written as:

$$\dot{x} = \begin{bmatrix}\dot{x}_1\\ \dot{x}_2\end{bmatrix} = \begin{bmatrix}0 & 1\\ 0 & -\frac{B}{M}\end{bmatrix}\begin{bmatrix}x_1\\ x_2\end{bmatrix} + \begin{bmatrix}0\\ \frac{1}{M}\end{bmatrix}(u - F_L) \tag{5}$$

where: $x_1 = x_p, \dot{x}_1 = x_2$, and, $u = F_e$ is control input.

In this paper ,we selected shift tracking error as the system state variables:

$$e = x^* - x_p \tag{6}$$

Where, x* is reference displacement; x_p is actual displacement. PMSLM system error equation:

$$\dot{e} = \dot{r} - \dot{x}_1 = \dot{r} - x_2 \tag{7}$$

$$\ddot{e} = \ddot{r} - \dot{x}_2 = \ddot{r} + \frac{1}{M}\left[B\dot{r} - \dot{B} + F_L\right] - \frac{1}{M}u \tag{8}$$

This paper select sliding surface: $s = ce + \dot{e}$ (9)

Where, c is a positive real number.

Using sliding mode equivalent control method to achieve Sliding mode control of linear motor:

$$u = u_{eq} + u_s \tag{10}$$

u_{eq} is sliding equivalent control system parts, namely, required the control amount when the sliding surface and all of the uncertain part is the zero, i.e. the average value of switch control;

$$u_{eq} = M\left[\ddot{r} + \frac{B}{M}x_2 + c\dot{e}\right] \tag{11}$$

u_s is sliding mode switching control part, the system state towards the sliding surface by the high-frequency switching control, and slide into steady-state status along the lines of the sliding mode, in order to limit and compensation uncertainty of the servo control system.

$$u_s = k\,\mathrm{sgn}(s) \tag{12}$$

Where, k is the scale factor of switching, $\mathrm{sgn}(s)$ is the sign function, It can be obtained $s\dot{s} < 0$ that existence conditions and capacity conditions of the sliding mode by Lyapunov stability theorem. In order to ensure at any point in the phase plane tends reached Sliding line with in a limited time, the switching control must be above conditions. By the formula (8), (9), (10), (11) and (12) can be solved the selection scope of the scale factor:

$k > \frac{F_{L\max}}{M} + \varepsilon$, Where, $F_{L\max}$ is maximum generalized disturbance. ε is a positive real number.

Based on the above principle, to build linear motor matlab simulation model of the sliding mode control [6].

3.2 The sliding mode control of the linear motor based on the function

In this paper, based on function instead of sign function sgn (s), can greatly weaken the speed control system buffeting phenomenon. Its function is expressed as[9]:

$$\alpha = \frac{2}{1 + e^{-f\cdot s}} - 1 \tag{13}$$

Where, f is a positive real number, so that:

$$u = u_{eq} + \alpha K \quad (14)$$

3.3 The linear motor model predict direct thrust control

The optimizer based cost function as nonlinear model predictive control method [7]. This method will begin the initial value through the system model to predict the predicted outputs, and then compared the predicted value with a reference value to obtain the future errors. The prediction error based on the cost function and other constraints optimizer to calculate future input. The next sampling time system model through past inputs and outputs and the predicted future input state to re-calculate the output state, so the cycle.

$$s_a = \begin{cases} 1, if & s_1 \quad on \quad and \quad s_4 \quad off \\ 0, if & s_1 \quad off \quad and \quad s_4 \quad on \end{cases}$$

$$s_b = \begin{cases} 1, if & s_2 \quad on \quad and \quad s_5 \quad off \\ 0, if & s_2 \quad off \quad and \quad s_5 \quad on \end{cases}$$

$$s_c = \begin{cases} 1, if & s_3 \quad on \quad and \quad s_6 \quad off \\ 0, if & s_3 \quad off \quad and \quad s_6 \quad on \end{cases}$$

The vector expression: $s = \frac{2}{3}(s_a + as_b + a^2 s_c) \quad (15)$

Where, $a = e^{j\frac{2\pi}{3}}$ 。

Load voltage: $v = V_{dc} s \quad (16)$

Future reference thrust and reference flux can be respectively given from the current thrust value and the flux value and the last thrust value and flux values by second-order extrapolation [5].

$$F_{ref} = 3f_{ref1} - 3f_{ref2} + f_{ref3} \quad (17)$$

$$\varphi_{ref} = 3\varphi_{ref1} - 3\varphi_{ref2} + \varphi_{ref3} \quad (18)$$

Where, F_{ref1}, φ_{ref1} are the current values, F_{ref2}, φ_{ref2} are the values of the last time, F_{ref}, φ_{ref} are the values of the last two sample time.

The expression of the linear motor in the two phase static coordinates:

$$\begin{bmatrix} u_\alpha \\ u_\beta \end{bmatrix} = R \begin{bmatrix} i_\alpha \\ i_\beta \end{bmatrix} + p \begin{bmatrix} L_\alpha & 0 \\ 0 & L_\beta \end{bmatrix} \begin{bmatrix} i_\alpha \\ i_\beta \end{bmatrix} + \begin{bmatrix} -\phi_f w_e \sin\theta \\ \phi_f w_e \cos\theta \end{bmatrix} \quad (19)$$

According to the principles of direct force and model predictive control can know that: In the stationary coordinate $\alpha - \beta$, the discrete magnetic chains within the k+1 sampling interval can be expressed as:

$$\begin{cases} \varphi_\alpha(k+1) = \varphi_\alpha(k) + [u_\alpha - Ri_\alpha(k)]T_s \\ \varphi_\beta(k+1) = \varphi_\beta(k) + [u_\beta - Ri_\beta(k)]T_s \end{cases} \quad (20)$$

So winding flux vector:

$$\varphi_s(k+1) = \sqrt{\varphi_\alpha^2(k+1) + \varphi_\beta^2(k+1)} \angle \tan^{-1}\left(\frac{\varphi_\beta(k+1)}{\varphi_\alpha(k+1)}\right) \quad (21)$$

in the stationary coordinate $\alpha - \beta$, the force of the linear motor within the k+1 sampling interval can be expressed as:

$$F_e(k+1) = \frac{3\pi}{2\tau}[\varphi_\alpha(k+1)i_\beta(k+1) - \varphi_\beta(k+1)i_\alpha(k+1)]$$

The general expression of the thrust can be expressed as:

$$F_e(k+1) = k_s |\varphi_s(k+1)| \times \sin\delta(k+1) \quad (22)$$

Where, $k_s = \frac{3}{2}\frac{\pi}{\tau}\frac{1}{L_d}\varphi_f$

This paper set g function in order to flux and thrust can be achieved good tracking:

$$g_1 = \left|\varphi_{ref} - \sqrt{\varphi_\alpha^2(k+1) + \varphi_\beta^2(k+1)}\right| \quad (23)$$

$$g_2 = \left|F_{ref} - [F(k) + k_s(u_\beta - Ri_\beta)T_s]\right| \quad (24)$$

$$g = a * g_1 + b * g_2 \quad (25)$$

Where, a、b is respectively flux estimation and thrust estimated weight coefficient. It can be calculated the most appropriate a、b value according to the proposed method [8].

The g function is set, so that the minimum value of the function g conducted inverter switch is selected so that a current can track known current, thereby a good tracking of the displacement of a given signal.

4. Simulation realization based on direct thrust control and sliding model predictions

Using simulation software, and established a permanent magnet synchronous linear motor displacement control system simulation model based on the fuzzy sliding mode and model predict direct thrust control. The selected permanent magnet synchronous linear motor parameters: resistance R = 4.6 Ω, L_d = 0.0035 H, L_q = 0.0035 H; permanent magnet flux ψ_f = .618794 Wb; thrust coefficient k_f = 48.6, the pole pair number p = 2, τ = 0.06 m, load mass M = 1.5Kg, viscosity coefficient B = 120 N • m • s. Sliding parameter c = 2000, k = 600, f = 90; A = 100; B = 1; sampling time is 10us, load torque F_d = 10 N, cogging is estimated in accordance with the sine function $F_{ripple} = 10\sin(\frac{\pi}{\tau}x)$: where x is a straight line the displacement of the motor.

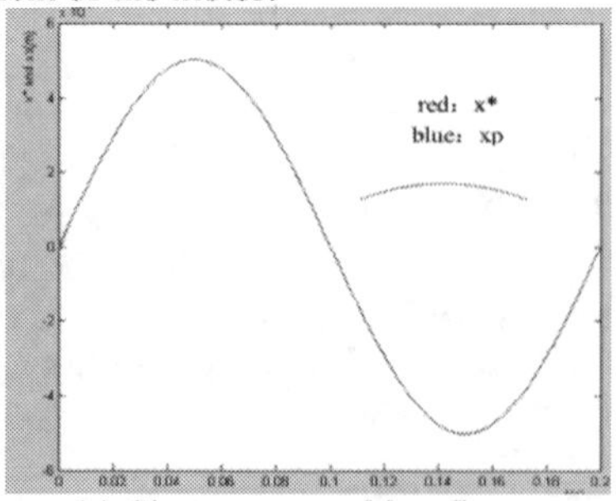

(a) Sine wave tracking figure

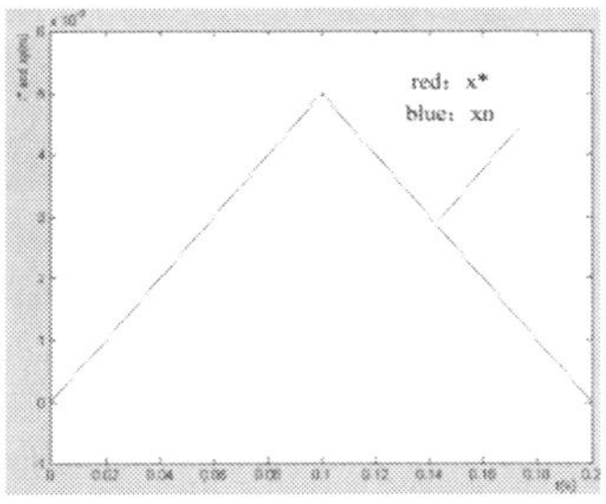

(b)Triangular-wave tracking figure

Figure 2 Displacement tracking diagram of the linear motor the direct thrust predict model

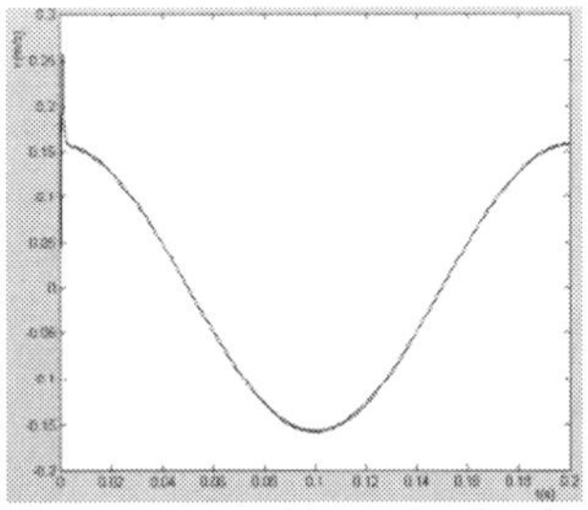

(a) Sine wave tracking velocity figure

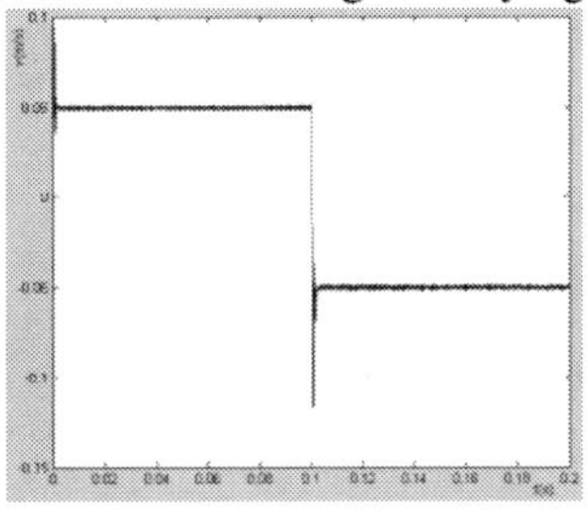

(b) Triangular-wave tracking velocity figure

Figure 3 Velocity tracking diagram of the linear motor the direct thrust predict model

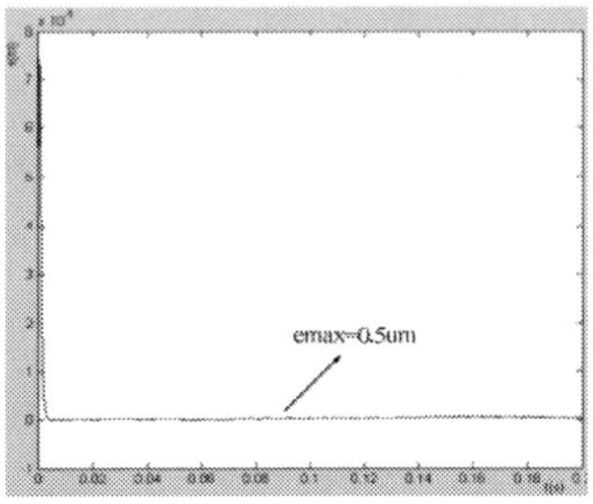

(a) Sine wave tracking error figure

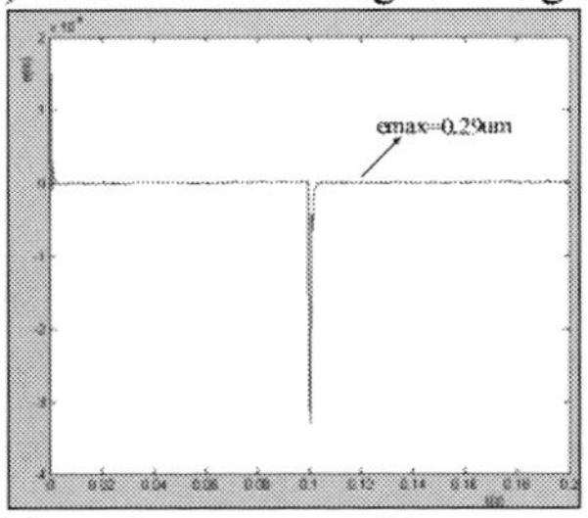

(b) Triangular-wave tracking error figure

Figure 4 Displacement tracking error diagram of the linear motor the direct thrust predict model

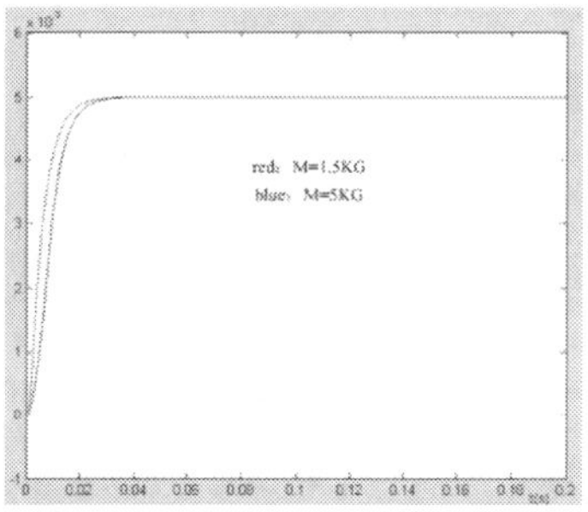

a. Linear motor rotor quality changes

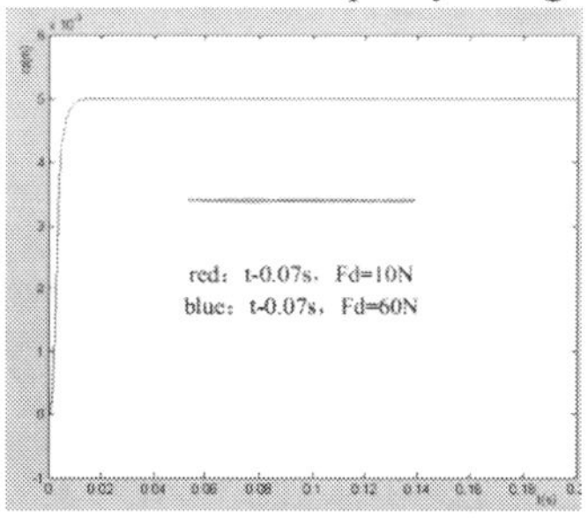

b. Linear motor load changes

Figure5.Displacement tracking figure

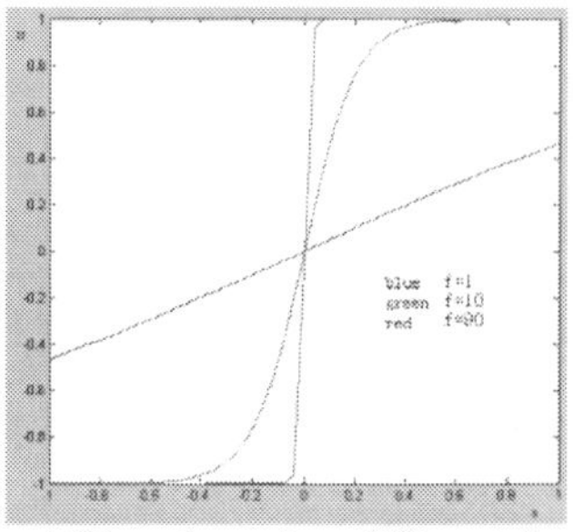

(a)Function when different f

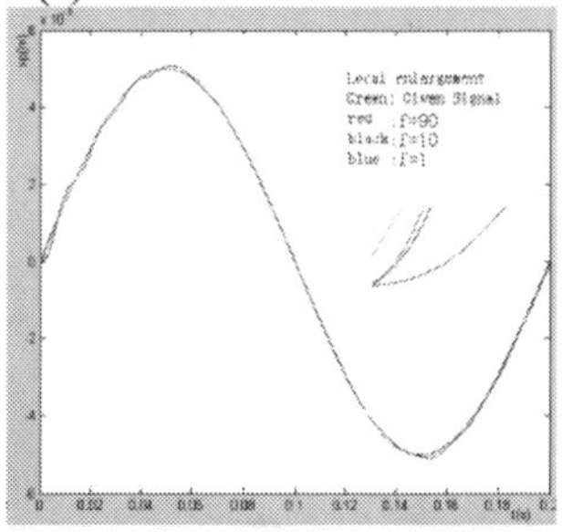

(b) Displacement tracking figure when different f

Figure 6. Linear motor displacement tracking map when different f values

The first case: when a given signal types, respectively, a sinusoidal signal $x_p = 0.005\sin(10\pi t)$ m, triangular signal (when $t = [0 \quad 0.1 \quad 0.2]$, $x_p = [0 \quad 0.005m \quad 0]$), The simulation result shown in Figure 2, Figure 3 and Figure 4, The simulation results indicate that the new method of based predictive control track all kinds of displacements of linear motor, and its tracking accuracy is high, the tracking error can reach 0.5um level tracking effect in the sampling period of 10us, and comparison with Yi-Sheng Huang[9] proposed based on the function of the sliding mode direct thrust control, its accuracy is improved by an order of magnitude.

The second situation: When the mass M of the linear motor transform, it can be seen that the response characteristics of the tracking control process will not have a significant impact by the simulation results of Figure5(a) when its quality changed greatly.

Third case: when the load of the linear motor system due to the influence of the external environment, from Figure 5(b) shows, When uncertain load varied from 10N to 60N, the response will not be affected as long as the load does not exceed the maximum of the controller design, this paper, the new control method can better track its displacement.

Fourth case: when f of the function $\alpha = \frac{2}{1+e^{-f \cdot s}} - 1$ transform ,it can be seen by the simulation results of Figure 6, when f from the change from 0 to 100 ,the sinusoidal displacement tracking simulation results can be seen, the larger the value of f, its tracking better higher accuracy.

5. Summary

This article put forward sliding mode model predicts direct thrust control system, for shortcomings of the linear motor nonlinearities and uncertainties as well as the control cycle non-constant of the traditional direct torque control, the number of switching inverter in unit time is not expected, complex frequency components .which control performance has been improved and perfected. The model predicts direct thrust control using discrete control design, to solve the problem SVPWM control assumes that the system as a continuous system. Moreover, the sampling time of the model predictive control constant, reducing the requirements of the computing unit and the implement unit. On the one hand, sliding mode control of function-based to meet the linear motor nonlinear and uncertainties requirements. The other hand, the model predicts direct thrust control algorithms, software and hardware are simple, fast dynamic response, constant control cycle, the control system makes the project easier to realize, lay the foundation for widely using of the direct thrust control. In summary, model predictive direct thrust control system opened up another new direction for linear motor, provides the foundation and help for the latter part of the experimental study.

Acknowledgements

Fund: Study on the design of penetration meter for soil hardness measurement, Department of Education of Guizhou Province (qianjiehe KYword [2021]060)

References

1. Kennel R., Linder A., Linke M, et al. Generalized predictive control (GPC)-ready for use in drive applications. IEEE, 2001: 1839-1844 vol. 1834.
2. Linder A., Kanchan R., Kennel R., Stolze P. Model-Based Predictive Control of Electric Drives. Cuvillier. 2010.
3. Zhu Xiaodong, Cheng Zhiping, Jiao Liucheng, et al.permanent magnet linear synchronous motor simulation model study. Mining Machinery. 2006, 34 (7): 84-87.
4. Wang Fuzhong, Wang Xudong,Jiao Liucheng, et al.synchronous motor electromagnetic force angle and the maximum thrust control strategies. Coal Society. 2001, 26 (3), p. 307-312.
5. Y. S. Huang and C. C. Sung, et al. Implementation of sliding mode controller for a linear synchronous motor based on direct thrust control theory. IET Control Theory Appl. 2010, p. 1–13.
6. Liu Jinkun sliding model variable structure control MATLAB simulation. Beijing: Tsinghua University Press, 2005.
7. José Rodríguez, Jorge Pontt, César A. Silva, et al. Predictive Current Control of a Voltage Source Inverter. IEEE transaction on industrial electronics. vol. 54, NO. 1, february, 2007.
8. S. Alireza Davari, Davood Arab Khaburi, and Ralph Kennel, et al. An Improved FCS–MPC Algorithm for an Induction Motor With an Imposed Optimized Weighting Factor. IEEE transaction on industrial electronics, vol. 27, NO. 3, march 2012.
9. Yi-Sheng Huang, Cheng-Chung Sung, et al. Function-Based Controller for Linear motor Control Systems . IEEE transaction on industrial electronics, vol.57, NO. 3, march, 2010.

https://doi.org/10.1051/matecconf/202236301015

Effect of anodic oxidation of 2024-T3 aluminum alloy on interface properties of metal fiber laminates

Zhiqiang Zhou [1], Changjiang Qian [2], Zonghao Hu [1], Feiyuan Yang [3], Meiduo Lou [4], Guangan Xie [2]

[1] Shenyang Aircraft Design Institute, Shenyang 110000, China
[2] College of Aerospace Engineering, Shenyang Aerospace University, Shenyang 110136, China
[3] Shenyang Tancho Industrial Co. LTD, Shenyang, 110041, China
[5] School of Electric Power, Shenyang Institute of Engineering, Shenyang 110136, China

Abstract: Interlaminar fracture failure is the main factor restricting the development and application of fiber metal laminates. In this paper, the interlayer bonding properties of 2024-T3 aluminum alloy plates with different surface treatments were compared and studied through lap-shear experiments. Compared with the aluminum plate treated with acid and alkali corrosion, the interlaminar shear strength of the aluminum plate after sulfuric acid anodization increased by 34.3%. After different surface treatments, experimental stress-strain curves of 2024-T3 aluminum alloys reveal the enhancement mechanism and changing trend of interlaminar shear strength.

Keywords: Metal fiber laminate; Surface treatment; Aluminum anodizing; lap-shear experiment.

1. Introduction

Metal fiber laminate is a new type of composite material formed by alternately bonding metal plates and fiber materials. Composite materials have the advantages of strong corrosion resistance, high specific strength, high specific modulus, light weight, non-magnetic, and good designability compared with traditional steel [1]. Although the many advantages of metal fiber laminate composites will make the design of auxiliary equipment in aerospace and other fields more abundant and diverse, but the unique directionality and layering of composite materials will also make their structural characteristics different from traditional metal structures [2]. Metal fiber laminates have a completely different damage form from general isotropic metal materials due to the high anisotropy of each ply, the bonding method between component materials, and the bonding method between plies and plies. Damage can occur within a ply or between two plies called intra-laminar and inter-laminar. Interlaminar damage generally refers to delamination damage. The bonding performance between the layers of composite laminates is not enhanced so delamination damage can occur under lower loads, and the stiffness, strength and stability of the structure will be reduced. Delamination damage is one of the most easily accumulated damage forms in composite materials, and severe delamination damage can directly lead to the overall failure of the structure. Therefore, the study of delamination damage is one of the most popular topics in composite damage mechanics. The occurrence and expansion of delamination mean the degradation of the interlayer performance of adjacent materials and the interlayer delamination significantly reduces the structural integrity, which will lead to a significant reduction in the structural strength and stiffness of the composite laminates, which seriously restricts the application in spacecraft structure. There are many reasons for the delamination failure of metal fiber laminates, including delamination caused by defects in the manufacturing process, cracks in the matrix during impact, etc. [3]. Mohammad Ahsin Muflikun investigated the stress-strain failure mode behavior, and mechanical properties of carbon fiber reinforced plastic-steel cold commercial (CFRP-SPCC) hybrid thin laminates through axial tensile and bending tests. The tensile test results show that premature failure of CFRP laminates can be delayed by up to 58.5% by adding SPCC to hybrid laminates[4]. Miguel Angel Caminello conducted tensile and three-point bending tests to study the different behavior of composites in tension and compression to evaluate their effect on failure modes and to determine the strength of different CFRP laminates and mechanical responses of stiffness[5]. Pagano and Piper combined the interlaminar stress analysis and experiment to propose the maximum tensile interlaminar normal stress criterion for resin matrix composites [6,7]. Dong Lei studied the effect of anodizing on the bonding properties of 6061 aluminum alloy and peeked through lap shear experiments [8]. V. FioreF. Di Franco studied the effect of anodizing surface treatment on the mechanical strength of 5083 aluminum alloy on fiber reinforced composite bonded joints[9]. In this paper, the surface of 2024-T3 aluminum alloy was corroded with sodium hydroxide solution and hydrochloric acid solution for 2 minutes, respectively, and anodized in the sulfuric acid solution for 15 minutes. The maximum interlaminar shear strength of the acid-base corrosion aluminum plate is 5.86MPa, and the maximum interlaminar shear strength of the hydrochloric acid anodized aluminum plate is 7.87MPa, which is 34.3% higher than that of the acid-base corrosion specimen.

2. Materials and Experimental Procedures

2.1 Materials

In this experiment, the main materials are aluminum alloy plate, glass fiber and epoxy resin. The metal material is 2024-T3 magnesium-aluminium alloy, and the fiber layer is the EW100 glass fiber woven material which is 0.11 mm thick, produced by Shanghai Yaohong Glass Fiber Co. Ltd. China. The matrix material is epoxy resin produced by Ciba Geigy in Australia, and the curing agent is Jeffamine D230 produced by Huntsman in the United States.

2.2 Material Pretreatment and Forming

First, put the 120*25mm 2024-T3 aluminum alloy sheet into acetone for ultrasonic degreasing for 2 minutes, then rinse with deionized water and dry it for later use. Take a part of the aluminum sheets and put them into the 5wt% NaOH solution for corrosion for 2 minutes, rinse them with deionized water after taking them out, dry them and put them in the 11 wt% HCL solution and rinse them with deionized water after taking them out and wipe them dry. Take the remaining aluminum sheets as the cathode and the graphite plate as the anode, put them into a 180g/L H2SO4 solution and connect to the DC power supply, and the power parameters are adjusted to 15V and 2A/dm2. Take them out after 20 minutes of power on and rinse them with deionized water and wipe them dry. During the anodization process, the circulating flow rate of the oxidation bath directly affects the temperature uniformity in the bath. Whether the anodized thick film on all aluminum workpieces in the whole anodization bath is uniform depends to a large extent on the inside of the bath. temperature is uniform[10]. Usually, the bath temperature is controlled within ±2°C for ordinary anodizing requirements; and the bath temperature is controlled within ±1°C for hard anodizing requirements.

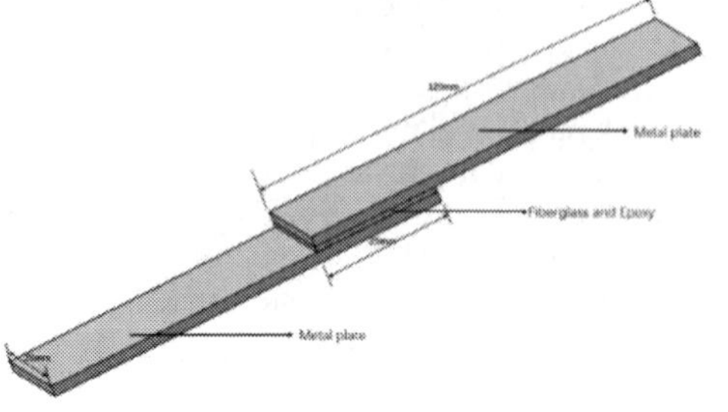

Figure 1 Schematic diagram of bonding

Then, cutting the glass fiber in 25*25mm as the bonding intermediate layer,and mixing the epoxy resin and D230 curing agent in a ratio of 3.3/1. Stiring that evenly and vacuuming to remove air bubbles as the adhesive. According to ASTM D5868-01, the surface-treated aluminum alloy sheets are bonded with epoxy resin according to the bonding area of 25*25mm, and the bonding area of the two aluminum alloy sheets is sandwiched with a layer of glass fiber to simulate the interlayer of the metal fiber laminate. After applying adhesive, fix it with dovetail clips and put it in an oven at 120°C for 2 hours to cure. The sample after bonding is shown in Figure 1.

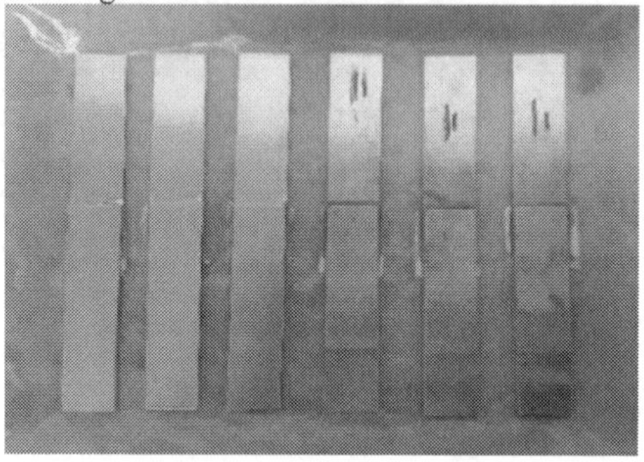

Figure 2 Samples after bonding

2.3 Experiment and performance test

According to ASTM D5868-01, the lap-shear experiment can measure the maximum interlaminar shear strength of the laminate. The samples shown in Figure 2. Each surface treatment sample was tested at a tensile speed of 2 mm/min. The lap-shear experiment is shown in Figure 3. The lap-shear experiment is repeated at least three times, and the maximum tensile stress obtained in the test divided by the bonding area is the maximum interlaminar shear strength of the sample.

Figure 3 The lap-shear expriment

3. Results and Discussion

The cross-section of the sample after stretching is shown in Figure 4. Figure a is the acid-base corrosion sample, and Figure b is the anodized sample. It can be seen from the figure that the cross-section of the acid-base treated sample fails with the degumming of glass fiber and aluminum plate. The main failure mode of the anodized samples is tearing the glass fiber in the middle layer. The maximum interlaminar shear strength of the acid-base corrosion sample measured in the experiment is 5.86MPa, while the maximum interlaminar shear strength of the sulfuric acid anodized sample is 7.87MPa. The maximum shear strength of the anodized sample increased by 34.3% compared with the acid-base corrosion sample, and the stress-strain curve of the lap-shear experiment is shown in Figure 5.

It can be concluded that the interlayer mechanical properties of the 2024-T3 aluminum alloy after sulfuric acid anodizing treatment are significantly improved compared with that of acid-base corrosion, according to the experimental data, because sulfuric acid anodizing can generate a diameter of about 12 nm and a pore wall thickness of 12 nm on the surface of the aluminum alloy. The 0.8nm/V dense oxide film layer has a smaller and denser pore size than the oxide film produced by the acid-base etching process, and it combines firmly with the metal matrix and provides more adhesion area for the adhesive. Therefore, the experiment can withstand more tremendous interlaminar shear stress.

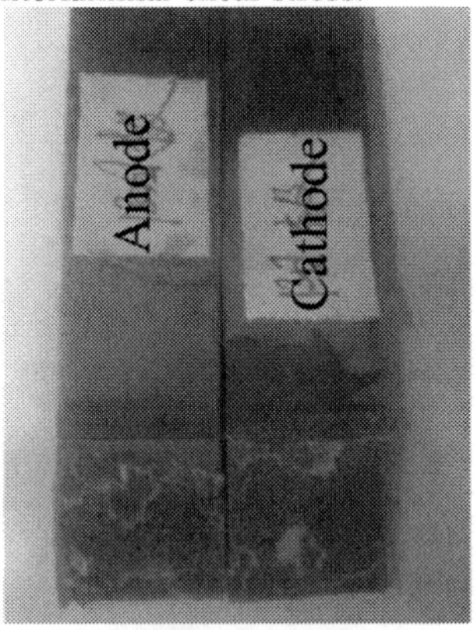

a. Acid-base corrosion

b. Anodizing

Figure 4 Section of the specimen after stretching

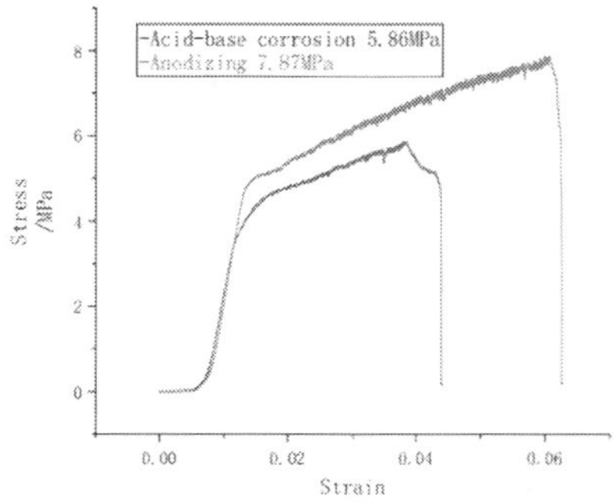

Figure 5 Stress-strain curve of lap-shear experiment

4. Conclusion

The bonding fracture mode can be divided into cohesive failure inside the adhesive layer and interface failure between the adhesive layer and the adherend. Competition relationship. When the cohesive strength is higher than the interface strength, interfacial failure occurs; when the cohesive strength is lower than the interface strength, cohesive failure occurs[11]. The fracture force for cohesive failure is usually higher than for interfacial failure. Therefore, the surface treatment state is essential to obtain a good bonding effect.

This paper mainly studies the effect of 2024-T3 aluminum alloy on the interlaminar mechanical properties of metal fiber laminates after acid-base corrosion and sulfuric acid anodization. Dense, the interlaminar mechanical properties of the metal fiber laminates are greatly improved. The maximum interlaminar shear strength of this group of samples reaches 7.87MPa, which is 34.3% higher than that of the acid-base corrosion samples. Compared with the traditional acid-base corrosion pretreatment method, sulfuric acid anodizing can significantly improve the common delamination phenomenon of metal fiber laminates. To sum up, the anodizing treatment of the inner side of the metal layer of the metal fiber laminate can significantly improve the interlaminar mechanical properties of the laminate, and the anodizing treatment can also be used for the anti-corrosion process on the outside of the metal layer. The overall mechanical properties and reliability of the parts are helpful.

Acknowledgments

This work was financially supported by the Key Research and Development program of the Ministry of Science and Technology of China (2017YFB1104001), Aeronautical Science Foundation of China (2020Z055054002)

References

1. Xiao-yan Huang, Liu Yuan, Liu Bo. Application of composite materials on ships [J]. Jiangsu Ships, 2008(02):13-17. (in Chinese)

2. WU Kai. Structural Analysis and Optimal Design of Composite Laminates [D]. Dalian University of Technology, 2019. (in Chinese)
3. ZHANG Xia. Research on fatigue delamination expansion of composite laminates with fastening [D]. Nanjing University of Aeronautics and Astronautics, 2019. (in Chinese)
4. Muhammad Akhsin Muflikhun,Ryo Higuchi, Tomohiro Yokozeki,Takahira Aoki. The evaluation of failure mode behavior of CFRP/Adhesive/SPCC hybrid thin laminates under axial and flexural loading for structural applications[J]. Composites Part B, 2020,185(C).
5. Miguel Angel Caminero, Gloria Patricia RodrAguez, Jesus Miguel ChacAn, Irene GarcAaaMoreno. Tensile and flexural damage response of symmetric angleaply carbon fiberareinforced epoxy laminates: Nonalinear response and effects of thickness and plyastacking sequence[J]. Polymer Composites, 2019,40(9).
6. Pagano N.J, Pipes R. B. The Influence of stacking sequence on laminate strength [J]. Journalof composite materials, 1971, 5:50-57.
7. Pagano N.J, Pipes R. B. Some Observations on the Interlaminar Strength of CompositeLaminates [J].International journal of mechanical Sciences, 1973, 15(8):679-686.
8. Lei Dong. Effect of anodizing surface morphology on the adhesion performance of 6061 aluminum alloy [J]. International Journal of Adhesion and Adhesives, Volume 113, 2022.
9. V. Fiore. Effects of anodizing surface treatment on the mechanical strength of aluminum alloy 5083 to fibre reinforced composites adhesive joints [J]. International Journal of Adhesion and Adhesives,Volume 108, 2021,
10. Yun-fang Huang. Aluminum anodic oxidation thick film process and control of main influencing factors[J]. Guangzhou Chemical Industry, 2021, 49(12):154-156.
11. Ying-dong Li. Effect of Anodizing Treatment on Bonding Properties of 5754 Aluminum Alloy Sheet[J]. Processing Technology of Light Alloy, 2021,49(08):39-43.

Research on calibration method of indication error of grade III standard metallic scale based on length measuring machine

Tao Wang, Yi Yang, Zou Rong and Xiao Xiang

Hubei Institute of Measurement and Testing Technology, Wuhan, Hubei, China.

Abstract. A calibration method of indication error of grade III standard metallic scale based on length measuring machine is proposed. The existing length measuring machine is reformed to realize the calibration of the indication error of grade III standard metallic scale. According to the proposed calibration method, the evaluation model of measurement uncertainty is established, and the source of uncertainty is analysed. The expanded uncertainty is less than one third of the maximum permissible error specified in the verification regulation, which verifies the feasibility and effectiveness of the calibration method.

Keywords: Metrology; grade III standard metallic scale; length measuring machine; uncertainty in measurement.

1. Introduction

The grade III standard metallic scale is mostly used to verify the level ruler, steel ruler and other line scale[1]. Common grade III standard metallic scale materials include zinc, white copper, stainless steel, brass, etc. The grade III standard metallic scale is a rigid ruler, and its scribed line is on two inclined planes at an angle of 45° to the bottom. 1020mm scribed lines are engraved on both sides, of which the division value of one side is 0.2 mm and the other side is 1 mm. The verification regulation of JJG71-2005 standard metallic scale (Grade III) requires that the maximum permissible error of indication is $\pm (0.03 + 0.02L)$ mm. The verification method of indication error can be verified by laser interferometer or grade II standard metallic scale as standard ruler on length measuring machine or other verification devices. According to the requirements of verification regulations, reference[2] installed a grade II standard metallic scale on a one meter length measuring machine to carry out verification, and analyzed its measurement uncertainty. In reference [3], the measurement accuracy factors of two indication error verification methods in the verification regulation was discussed. The research shows that the laboratory ambient temperature and the flatness of the bottom of the third-order linear ruler are important factors for the measurement accuracy. In reference [4], the author proposed a method to verify grade III standard metallic scale with MLD one meter measuring machine, and analyzed its measurement uncertainty. In this paper, a method of directly using the length measuring machine to verify the indication error of grade III standard metallic scale is proposed, and the influence of the temperature in the measurement process, the Abbe error of the device and the inclination error of the reading device are corrected to improve the measurement accuracy, and the measurement uncertainty is analyzed. The expanded uncertainty result of the indication error meets the requirements of the maximum permissible error of the verification regulation.

2. Calibration method

The grade III standard metallic scale indication error verification device based on the length measuring machine is transformed from the existing length measuring machine, as shown in Figure 1. A fixed support of the scale with two-way micro adjustment is designed near the track axis of the length measuring machine to realize the fixation and position fine adjustment of the scale. The distance between the two fixed supports is equal to the distance of the Bessel support point of the grade III standard metallic scale. A reading device which can read the linear scale of the scale is installed on the tailstock of the length measuring machine, and an attached thermometer is installed on the track of the length measuring machine and the scale to realize the real-time temperature measurement of the length measuring machine and the tested scale.

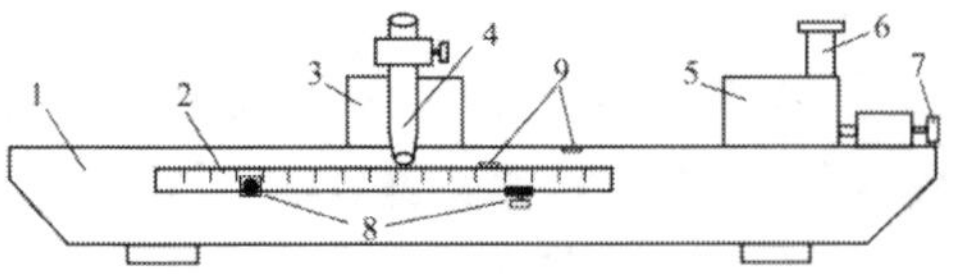

Fig. 1 Schematic diagram of grade III standard metallic scale indication error verification device

1 – length measuring machine; 2 – grade III standard metallic scale; 3 – tailstock; 4 –micro reading device; 5 – measuring seat; 6 – microscope; 7 – adjustment handwheel; 8 – adjustment of the scale; 9 – thermometer

The verification process of the indication error of the grade III standard metallic scale based on the length measuring machine is similar to that in part 6.3.8.2 of JJG71-2005 standard metallic scale (Grade III). The main difference is that this method takes the actual length of the length measuring machine as the standard length. At the same time, because the axis of the length measuring machine is different from the axis of the scale, it does not comply with the Abbe principle, and Abbe error correction is required during measurement data processing. The Abbe error calculation method is as follows: firstly, use the autocollimator to measure the straightness runout angle of the length measuring machine track in the horizontal and vertical directions at the tested interval, and multiply this angle by the distance between the length measuring machine track and the tested scale to obtain the Abbe error in the corresponding direction. In addition, due to the influence of the flatness of the engraving surface of grade III standard metallic scale, the error will be introduced when there is an inclination error of the optical axis of the micro reading device, so it also needs to be corrected.

3. Evaluation of measurement uncertainty

3.1 Measurement model

According to the calibration method, the measurement model of indication error of grade III standard metallic scale based on length measuring machine is

$$L_x = L_s - \Delta L_x + (\alpha_s \delta t_s - \alpha_x \delta t_x)L + \delta l_a + \delta l_b \quad (1)$$

Where L_x is actual length of grade III standard metallic scale, L_s is actual length of the length measuring machine at the inspected interval, ΔL_x is measured deviation of the tested interval of the grade III standard metallic scale, α_s is linear expansion coefficient of length measuring machine, δt_x is deviation between the average temperature of the length measuring machine and the reference temperature of 20°C, α_x is linear expansion coefficient of grade III standard metallic scale, δt_s is deviation between the average temperature of the grade III standard metallic scale and the reference temperature of 20°C, L is nominal length of the tested interval of the scale, δl_a is Abbe error between grade III standard metallic scale and length measuring machine, δl_b is optical axis inclination error of micro reading device.

Because each component is independent of each other, the combined standard uncertainty is calculated by the following formula[5]

$$u_c^2(L_x) = c_1^2 u^2(L_s) + c_2^2 u^2(\Delta L_x) + c_3^2 u^2(\alpha_s) + c_4^2 u^2(\delta t_s) + c_5^2 u^2(\alpha_x) + c_6^2 u(\delta t_x) + c_7^2 u^2(\delta l_a) + c_8^2 u(\delta l_b) \quad (2)$$

Where coefficient of sensitivity c_1 is equal to 1, c_2 is equal to -1, c_3 is equal to $\delta t_s L$, c_4 is equal to $\alpha_s L$, c_5 is equal to -$\delta t_x L$, c_6 is equal to -$\alpha_x L$, c_7 is equal to 1, c_8 is equal to 1. During the actual measurement, the scale and the length measuring machine are fully isothermal, and the actual measurement temperature is about 21.0°C, so δt_s and δt_x are both equal 1.0°C. According to the materials of the grade III standard metallic scale and length measuring machine, α_s is 11.5×10-6°C, α_x is 10.8×10-6°C.

3.2 Calculation of standard uncertainty

3.2.1 *Calculation of u(Ls)*

Expanded uncertainty of indication error of micrometer scale of length measuring machine U is 0.3μm+2.1×$10^{-6}L$ (k=2), then the uncertainty component $u(L_s)$ is

$$u(L_s) = \frac{(0.3\mu m + 2.1\times 10^{-6} L)}{2} = 0.15\mu m + 1.05\times 10^{-6} L \quad (3)$$

3.2.2 *Calculation of u(ΔLx)*

Uncertainty component $u(\Delta L_x)$ introduced by measuring the actual length deviation of the tested ruler is mainly introduced for measurement repeatability. Carry out 10 times repeated alignment measurements on 100mm scribed lines, and the calculated experimental standard deviation s of a single reading is 1.91μm. according to the verification regulation JJG71-2005[1], the indication error of the tested interval is the reading difference between the tested scribed line and the zero scribed line. It is necessary to carry out two times measurements, that is, the average value of the four measured values. Then the uncertainty component $u(\Delta L_x)$ introduced by the actual length deviation measurement of the tested ruler is

$$u(\Delta L_x) = \frac{\sqrt{2}\times 1.91\mu m}{\sqrt{4}} = 1.35\mu m \quad (4)$$

3.2.3 *Calculation of u(αs)*

The linear expansion coefficient of the length measuring machine shall be (11.5±1)×10-^{6}°C^{-1}, the half width range is 1×10^{-6}°C^{-1}, estimated by uniform distribution, uncertainty component $u(\alpha_s)$ introduced by linear expansion coefficient of the length measuring machines is

$$u(\alpha_s) = \frac{1\times 10^{-6}℃^{-1}}{\sqrt{3}} = 0.577\times 10^{-6}℃^{-1} \quad (5)$$

3.2.4 *Calculation of u(Δts)*

The average temperature of the length measuring machine is the average value of the starting temperature and the ending temperature of each measured return. The instrument used is an attached thermometer, and its MPE is ±0.2°C. If it is estimated by uniform distribution, then the uncertainty component $u(\Delta t_s)$ introduced by the average temperature deviation of the length measuring machine is

$$u(\Delta t_s)=\frac{\sqrt{4}\times 0.2^\circ\mathrm{C}}{\sqrt{3}}=0.23^\circ\mathrm{C} \quad (6)$$

3.2.5 Calculation of u(αx)

The tested ruler coefficient of the length measuring machine shall be $(10.8\pm1)\times10^{-6}$°C^{-1}, the half width range is 1×10^{-6}°C^{-1}, estimated by uniform distribution, uncertainty component $u(\alpha_x)$ introduced by linear expansion coefficient of the tested ruler is

$$u(\alpha_x)=\frac{1\times10^{-6}{}^\circ\mathrm{C}^{-1}}{\sqrt{3}}=0.577\times10^{-6}{}^\circ\mathrm{C}^{-1} \quad (7)$$

3.2.6 Calculation of u(Δtx)

The average temperature of the tested ruler is the average value of the starting temperature and the ending temperature of each measured return. The instrument used is an attached thermometer, and its MPE is ±0.2°C. If it is estimated by uniform distribution, then the uncertainty component $u(\Delta t_s)$ introduced by the average temperature deviation of the tested ruler is

$$u(\Delta t_x)=\frac{\sqrt{4}\times 0.2^\circ\mathrm{C}}{\sqrt{3}}=0.23^\circ\mathrm{C} \quad (8)$$

3.2.7 Calculation of u(δla)

Because the tested ruler and the length measuring machine are parallel longitudinal measurement, this method does not conform to Abbe principle, and Abbe error compensation is required during measurement. The Abbe error δl_a is calculated by the following

$$\delta l_a = d_H\varphi_H + d_V\varphi_V \quad (9)$$

Where d_H is the horizontal distance between the axis of the tested ruler and the axis of the length measuring machine, φ_H is the horizontal angle between the guide rail of the length measuring machine at the inspected point and its ideal axis, d_V is the vertical distance between the axis of the tested ruler and the axis of the length measuring machine, φ_V is the vertical clamping angle between the guide rail of the length measuring machine at the inspected point and its ideal axis.

Then the uncertainty component $u(\delta l_a)$ caused by Abbe error between the tested ruler and the length measuring machine is

$$u^2(\delta l_a)=d_H^2\times u^2(\varphi_H)+\varphi_H^2\times u^2(d_H)+d_V^2\times u^2(\varphi_V)+\varphi_V^2\times u^2(d_V) \quad (10)$$

During actual measurement, the distance between the axis of the tested ruler and the axis of the length measuring machine measured with a steel ruler in the horizontal direction is 190mm, and the distance in the vertical direction is 175mm. The expanded uncertainty of distance measurement is U=5mm(k=2). Then there are

$$u(d_H)=u(d_V)=\frac{5\mathrm{mm}}{2}=2.5\mathrm{mm} \quad (11)$$

The maximum horizontal angle between the length measuring machine guide rail and its ideal axis measured by the autocollimator is 4.8×10^{-5}rad, the maximum included angle in the vertical direction is 1.5×10^{-4}rad, expanded uncertainty of included angle measurement is $U=1.0\times10^{-5}$rad(k=2). Then there are

$$u(\varphi_H)=u(\varphi_V)=\frac{1.0\times10^{-5}\mathrm{rad}}{2}=5.0\times10^{-6}\mathrm{rad} \quad (12)$$

Then the uncertainty component $u(\delta l_a)$ caused by Abbe error between the tested ruler and the length measuring machine is

$$u(\delta l_a)=1.30\mu\mathrm{m} \quad (13)$$

3.2.8 Calculation of u(δlb)

When there is deflection deformation on the surface of the tested ruler, and the optical axis of the reading device is not perpendicular to the surface of the tested line, the inclination error will affect the measurement results. Optical axis inclination error of reading device δl_b is calculated as

$$\delta l_b = r\times\varphi_G \quad (14)$$

In the actual measurement, the deflection r of grade III standard metallic scale is taken as 0.2mm, and the expanded uncertainty of the measurement of the included angle between the optical axis and the normal direction of the inspected line surface is $U=2.91\times10^{-3}$rad(k=2). Then the $u(\delta l_b)$ is

$$u(\delta l_b)=0.2\mathrm{mm}\times10^3\times\frac{2.91\times10^{-3}\mathrm{rad}}{2}=0.29\mu\mathrm{m} \quad (15)$$

3.3 Combined standard uncertainty

3.3.1 The calculation results of standard uncertainty

The source of standard uncertainty components of measurement results is shown in Table 1.

Tab.le 1. The source of standard uncertainty components of measurement results

$u(x_i)$	The source of standard uncertainty	coefficient of sensitivity c_i	results of $u(x_i)$
$u(L_s)$	The uncertainty introduced by the length measuring machine in the actual length measurement of the inspected interval	1	0.15μm+ $1.05\times10^{-6}L$
$u(\Delta L_x)$	The uncertainty introduced by the tested ruler in the actual length measurement of the inspected interval	-1	1.35μm
$u(\alpha_s)$	Uncertainty introduced by linear expansion coefficient of length measuring machine	1.0°C×L	0.577×10^{-6}°C^{-1}

$u(\delta t_s)$	Uncertainty introduced in the measurement of average temperature deviation of length measuring machine	11.5×10^{-6}°C $\times L$	0.23°C
$u(\alpha_x)$	Uncertainty introduced by linear expansion coefficient of the tested ruler	-1.0°C$\times L$	0.577×10^{-6}°C^{-1}
$u(\delta t_x)$	Uncertainty introduced by the measurement of the average temperature deviation of the tested ruler	10.8×10^{-6}°C $\times L$	0.23°C
$u(\delta l_a)$	Uncertainty caused by Abbe error between the measured ruler and the length measuring machine	1	1.30μm
$u(\delta l_b)$	Uncertainty introduced by optical axis inclination error of reading device	1	0.29μm

3.3.2 The calculation results of combined standard uncertainty

According to formula (2), the combined standard uncertainty of different tested intervals can be calculated, and the results are shown in Table 2.

Table 2. The combined standard uncertainty of different tested intervals

tested intervals (mm)	0	100	200	300	400	500	600	700	800	900	1000
$c_1\times u(L_s)$ (μm)	0.15	0.26	0.36	0.47	0.57	0.68	0.78	0.89	0.99	1.10	1.20
$c_2\times u(\Delta L_s)$ (μm)	-1.35	-1.35	-1.35	-1.35	-1.35	-1.35	-1.35	-1.35	-1.35	-1.35	-1.35
$c_3\times u(\alpha_s)$ (μm)	0.00	0.06	0.12	0.17	0.23	0.29	0.35	0.40	0.46	0.52	0.58
$c_4\times u(\delta t_s)$ (μm)	0.00	0.26	0.53	0.79	1.06	1.32	1.59	1.85	2.12	2.38	2.65
$c_5\times u(\alpha_x)$ (μm)	0.00	0.06	0.12	0.17	0.23	0.29	0.35	0.40	0.46	0.52	0.58
$c_6\times u(\delta t_x)$ (μm)	0.00	0.25	0.50	0.75	0.99	1.24	1.49	1.74	1.99	2.24	2.48
$c_7\times u(\delta l_a)$ (μm)	1.30	1.30	1.30	1.30	1.30	1.30	1.30	1.30	1.30	1.30	1.30
$c_8\times u(\delta l_b)$ (μm)	0.29	0.29	0.29	0.29	0.29	0.29	0.29	0.29	0.29	0.29	0.29
$u_c(L_x)$ (μm)	1.90	1.95	2.07	2.25	2.48	2.74	3.03	3.34	3.66	4.00	4.34

Linearize uc(Lx) in Table 2 to obtain

$$u_c(L_x)=1.9\mu m+2.5\times10^{-6}L \quad (16)$$

3.4 Expanded uncertainty

Taking the inclusion factor k=2, the expanded uncertainty of the indication error measurement of grade III standard metallic scale is

$$U=k\times u_c(L_x)=3.8\mu m+5.0\times10^{-6}L \quad (17)$$

4. Summary

By reforming the existing length measuring machine and considering the influence of temperature compensation and Abbe error correction during calibration, the calibration of the indication error of grade III standard metallic scale can be realized. The expanded uncertainty of indication error is less than one third of the maximum allowable error of indication required by the verification regulation, which indicates that the method in this paper is effective and feasible.

References

1. JJG171-2005 Standard Metallic Scale (Grade III). China Metrology press, 2005.
2. Yang Kun, Yan Fuchao.Evaluation of uncertainty in Standard Metallic Scale (Grade III). Geomatics Technology and Equipment, 2006, 8(2):47-48.
3. Chi Zhanduo. Discussion on the factors affecting Standard Metallic Scale (Grade III). China Metrology, 2012, 5:82-83.
4. Huang Xiuli. Verification of the Standard Metallic Scale (Grade III) by a MLD Meter Measuring Machine and Analysis Uncertainty in Measurement. Metrology & Measurement Technique, 2008, 35(10):54-56.
5. JJF1059.1-2012 Evaluation and Expression of Uncertainty in Measurement. China Metrology Press, 2012.

Research on differentiated scheduling method of electric vehicle considering random probability

Hanghang Liu, Yuanyua Wang, Sentao Liu, Qi Liu, Yun Xie, Xiaoyi Ma, Yajin Li

Dongying Power Supply Company of State Grid Corporation of China, China

Abstract. On the basis of the charging load of a single electric vehicle, in order to reduce the variance of the equivalent load curve and alleviate the peak plus peak caused by electric vehicles, a local electric vehicle charging cooperative scheduling model with the distribution network node as the optimization unit is set up. The model is composed of node agent and electric vehicle agent. The electric vehicle agent is based on the node load obtained. Rate information, combining with its own state, using random probability charging algorithm to make the decision to start charging or stop charging, and use IEEE-33 node system as an example to use MATLAB software for the designed electricity. The algorithm is verified by dynamic vehicle charging cooperative scheduling strategy. The results show that the proposed algorithm can effectively load the load curve and reduce the variance of load curve.

Keywords: Electric vehicle; collective motion; ordered charging; synegistic dispatch; random probability charging algorithm.

1. Introduction

A large scale electric vehicle access to the power grid, its disordered charging will lead to the peak of the load curve peak in the distribution network, threaten the safe operation of the power system, and reduce the economy of the system operation [1]. At the same time, distributed generation and electric vehicles are widely connected to the power grid, which makes the center of gravity of the power system structure shift to the "load" side, and the amount of information to be collected increases exponentially. Under the traditional "many to one" information transmission mechanism, there are bottlenecks in the information transmission channel and computer processing capacity, making the traditional centralized dispatching method no longer feasible [2-4]. It is necessary to study the cooperative scheduling method suitable for the wide access of distributed resources to the power grid.

Based on the cluster motion theory and aiming at reducing the variance of the equivalent load curve, this paper studies the charging collaborative scheduling strategy of electric vehicles. Each electric vehicle is regarded as an agent with independent decision-making [5]. According to the information of other power generation and consumption objects in a certain neighborhood, the electric vehicle adopts the random probability charging algorithm to decide whether to charge or not. The feasibility and effectiveness of the designed cooperative scheduling algorithm are verified by MATLAB simulation software.

2. EV charging collaborative scheduling process

Assuming that the electric vehicle is connected to the power grid at time t, the required power and the time required to complete charging are calculated according to the driving demand of the electric vehicle, and finally the latest charging time T_{last} of the electric vehicle is obtained. If $t = T_{last}$, no matter the node load rate is high or low, the electric vehicle starts charging immediately, and then it is no longer controlled until the charging is completed. If $t < T_{last}$, it indicates that the electric vehicle has sufficient charging time and can be scheduled. Enter the following scheduling process.

1) Charging process

The electric vehicle receives the load rate information of the access node, and when the node load rate $r < \lambda_1$, the electric vehicle immediately charges; When $\lambda_1 < r < \lambda_2$, the electric vehicle charges immediately with probability p_1 and requests charging again with probability $1 - p_1$ delay time t_d ;When $r > \lambda_2$, the electric vehicle delay time requests charging again.

2) Stop the charging process

The threshold λ_2 indicates that the node is close to full load. When $r > \lambda_2$, if the load continues to increase, the node is at risk of overload. Therefore, in addition to stopping the charging of electric vehicles, it is also necessary to stop the charging behavior of some charging electric vehicles. The process of stopping charging is, the electric vehicle being charged stops charging with the probability of p_2, and tries to charge again after waiting time t_d, and continues charging with the probability of $1-p_1$.

3. Design of random probability charging algorithm

The principle of charging probability threshold generation algorithm is: firstly, ensure that the electric vehicle power meets the owner's requirements before the electric vehicle owner leaves the charging pile; Secondly, we should ensure the economy of power grid operation, improve power grid efficiency, and ensure that the real-time power of all nodes cannot exceed the maximum power supply capacity of the node. Therefore, the goal is to minimize the variance of load curve; Finally, the algorithm should not be based on the accurate prediction of future load information, but only need to obtain real-time load information.

In order to meet the algorithm objectives, first, when the running time t reaches the latest charging time of the electric vehicle, the electric vehicle does not accept any control and starts charging until the charging is completed. Secondly, when the node load rate is greater than the threshold, the schedulable electric vehicle charging behavior connected to the power grid is limited; when the node load rate is greater than the threshold, the stop charging mechanism will start and stop the charging behavior of some charging electric vehicles, so as to ensure that the node load is at a safe and stable level. Therefore, when the electric vehicle is connected to the power grid, the intelligent experience of the electric vehicle will be based on the nodes received at the previous time Load rate information to make a decision to charge or stop charging.

3.1 Start charging function design

According to the requirements of random charging algorithm, when the node load rate r slightly exceeds the threshold λ_1, the starting charging probability $p_1(r)$ should be close to 1, and when it approaches the threshold λ_2, it should quickly drop to the minimum value close to 0.At the same time, the waiting time of electric vehicles should also be considered to ensure the fairness of charging. Therefore, it is designed as an exponential function about r, as shown in formula (1). As shown in the figure 1, with the increase of, the concave of the function curve becomes stronger and stronger.

$$p_1(r) = \begin{cases} e^{-\alpha(r-\lambda_1)+\beta t_w/t_m}, & t < T_{last} \\ 1, & t = T_{last} \end{cases} \quad (1)$$

$$t_w = t_m - t_c$$
$$t_m = t - T_a$$

Where, t_w represents the accumulated time of the electric vehicle without charging, t_m represents the accumulated time of the electric vehicle access time, t_c represents the accumulated charging time after the electric vehicle connected to the power grid, t represents the current operation time, and t_w / t_m represents the waiting factor after the electric vehicle connected to the power grid β; α indicates the rapidity parameter that drops to 0 when the node load rate is greater than λ_1.

When r is greater than λ_1, p_1 decreases exponentially with the increase of r, so there are only a few electric vehicles to charge, so as to ensure that the node load rate will not increase further. When $t < T_{last}$, in order to ensure the formula $p_1(\lambda_1) = 1$, $p_1(\lambda_2) = \varepsilon$ (here represents the positive decimal), α is determined by formula (2).

$$\alpha = \frac{\ln \varepsilon}{\lambda_1 - \lambda_2} \quad (2)$$

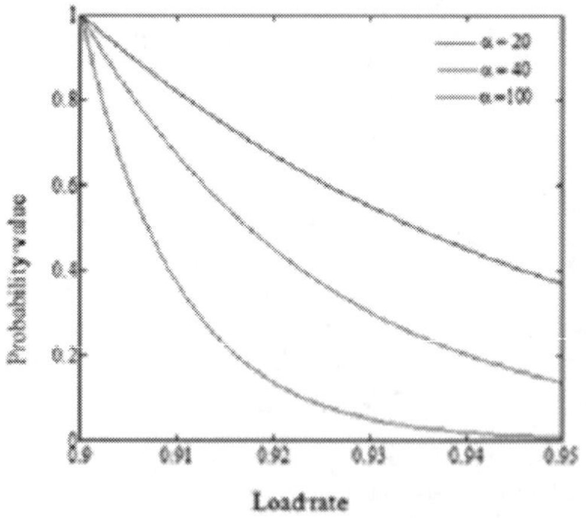

Fig. 1 Different α corresponding curves

According to the random probability charging rule, the probability of electric vehicle charging request being allowed is p_1 and the probability of being rejected is $1-p_1$. When the electric vehicle charging request is rejected, the electric vehicle will send the charging request again after a delay time t_d. The determination rule of delay time is: when $T_{last} - t > t_n$, the delay time is

https://doi.org/10.1051/matecconf/202236301017

randomly determined from the set $\left[0,t_n\right]$, and when $T_{last}-t\le t_n$, the delay time is determined as $T_{last}-t$.

3.2 Stop charging function design

When the node load rate r exceeds the threshold λ_2, the stop charging mechanism starts. p_2 indicates the probability that the electric vehicle stops charging. Obviously, when r is slightly exceeded λ_2, the value p_2 should be small to avoid too many charging electric vehicles stopping charging and affecting the owner's charging experience; When r is close to 1, it shall be increased rapidly to prevent node overload. Therefore, we design the stop charging probability p_2 as an exponential function of r, as shown in formula (3).

$$p_2(r)=\begin{cases} e^{\gamma(r-1)}, & t<T_{last} \\ 0, & t=T_{last} \end{cases} \tag{3}$$

$$\gamma=\frac{\ln\varepsilon}{\lambda_2-1}$$

Figure 2 shows the function curves corresponding to different γ values in the case of $\lambda_2=0.9$. It can be seen that the concave characteristic of the function curve gradually increases with the increase of γ. When $\gamma\ge 30$, it is slightly exceeded λ_2, the probability of stopping charging is small, which closed to 0, and when r is close to 1, the probability of stopping charging increases rapidly, which close to 1. Therefore, the design function can meet the requirements of demand side response.

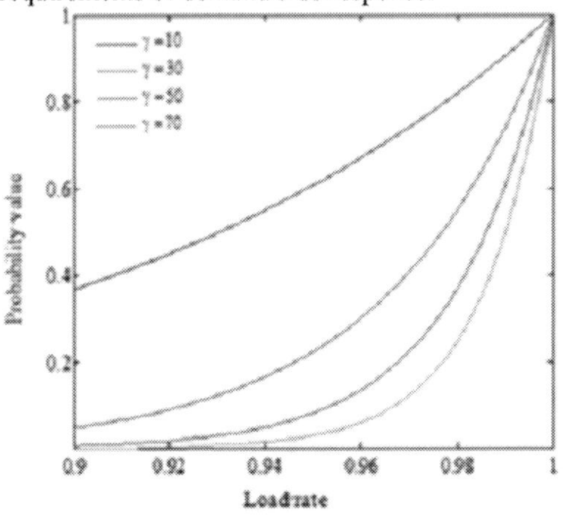

Fig. 2 Function curves under different node load rates

3.3 Example analysis

In this paper, IEEE-33 node example system] is used for simulation analysis. Its wiring is shown in Figure3. Node 1 is the system balance node, the system power reference value is s = 100MVA, the total basic active load of the system is 3.715MW, and the active capacity of distribution transformer is 6500kW. The 24-hour basic load curve of the system is shown in Figure 4.

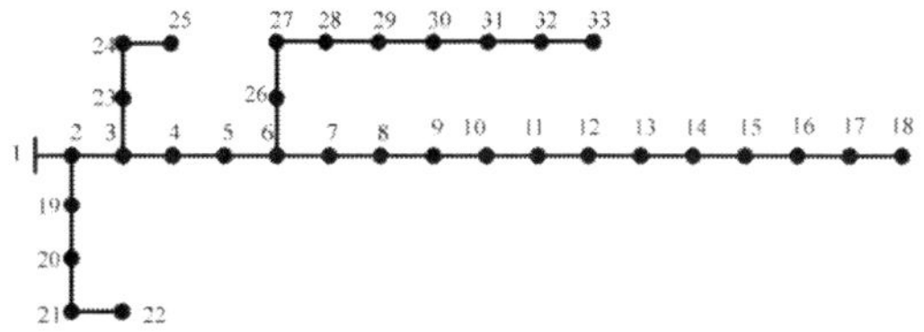

Fig. 3 IEEE-33 node system wiring diagram

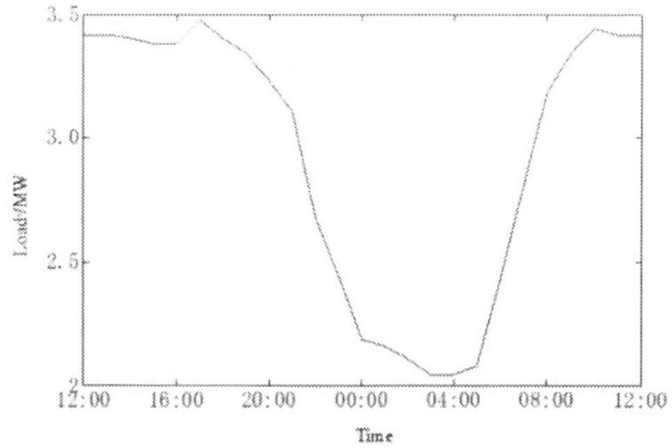

Fig. 4 24-hour basic load curve

The day is divided into 2400 time periods. 12:00 noon is the zero operation time, and 12:00 the next day is the 2400 operation time. The maximum allowable power of each node takes the node active power reference value, for example, the maximum allowable power of node 25. Battery capacity of electric vehicle. The charging mode of electric vehicle is slow charging mode. Refer to the literature for parameters such as charging power, driving habits and daily driving mileage of electric vehicles. Electric vehicles are evenly distributed and connected to each node.

Monte Carlo simulation is conducted for disorderly charging and collaborative dispatching charging of 1000 electric vehicles and 1500 electric vehicles connected to the system. Each case is simulated 100 times, and the average value is taken as the simulation result.

Tab. 1 Parameter values of cooperative scheduling algorithm under different vehicle access conditions

Parameter	λ_1	λ_2	ρ	ε	α	β	γ
N=1000	0.854	0.855	0.5	e^{-1}	1000	0.307	6.90
N=1500	0.938	0.940	0.5	e^{-1}	500	0.307	16.67

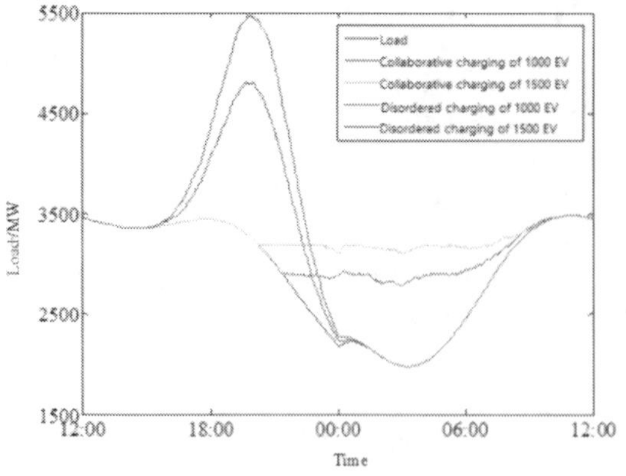

Fig. 5 Total load curve of different number of EV under disordered charging and collaborative dispatching charging modes

4. Conclusion

Aiming at minimizing the variance of load curve, a local cooperative dispatching model for electric vehicle charging is established in this paper. As an independent individual, according to the received node load rate information, combined with its own state information, aiming at the minimum variance of the load curve, the electric vehicle spontaneously makes the decision of charging or stopping charging by using the stochastic probabilistic control algorithm. Through this distributed local decision algorithm, the global expected peak cutting and valley filling function can be realized as much as possible.

Acknowledgements

Fund Project: Science and technology project of State Grid Shandong electric power company (52061618004N).

References

1. Liu Zhenya. Global energy Internet [M]. Beijing: china power press, 2015.
2. Li Wensheng. Distributed resources and power grid interacton mechanism and synergistic dispatch technology research [D]. Shandong University, 2012.
3. Ming Lingling. Study on synergistic dispatch technology of EV charge-swapping station and PV power..Shandong University, 2015.
4. Zu Xiangdong, Zhang Yuhui, Niu Guojun, et al..Supply-demand Coordination Optimal Dispatch of Active Distribution Network With Electric Vehicle Group Flexible Charging /discharging Response [J]. Journal Of Northeast Electric Power University, 2019, 39(03): 29-37.
5. Li Beibei. The Collective Motion Model and Synegistic Dispatch of Electric Vehicle Charging [D]. Shandong University, 2018.

Discussion on safety protection of on-line monitoring in self provided power plant

Lingkai Zhu [1,*], Jiyan Liu[2], Weishuai Wang[2], Wei Zheng[1], Ziwei Zhong[1], Zhiqiang Gong[1], Junshan Guo[1], Panfeng Shang[1]

[1]State Grid Shandong Electric Power Research Institute, Jinan, China
[2]State Grid Shandong Electric Power Company, Jinan, China

Abstract. This paper expounds the necessity of formulating the security protection strategy for the flexible interactive platform between the captive power plant and the power grid, analyzes the overall security protection strategy, and divides the security zone of the platform according to the overall security protection principle.

1. Introduction

Smart grid integrates power, computer, communication, network and other advanced technologies on the basis of traditional power grid, and is the main direction of current power grid development [1]. "Strong intelligent power grid" is strong because it has ultra-high voltage backbone network and multi-level power grid coordination. Intelligent power grid refers to the transformation of traditional power grid into a new modern power grid by using network communication, informatization and automatic control technology [2]. The intelligence and robustness of the smart grid are mainly guaranteed by three systems: the intelligent infrastructure system, the intelligent management system and the intelligent protection system. With the expansion of the power grid scale and the increasingly complex structure, it is increasingly difficult to ensure the safe and stable operation of the power grid. The network security problem is a major challenge [3].

Participating in the power market competition, captive power plants can promote the power supply and demand balance of the power grid from the load side, which can reduce the disadvantages of traditional regulation and generation side to ensure the power balance of the power grid due to the shortage of power coal supply, the increase of temporary repair and output reduction of thermal power units, and the increase of new energy penetration[4]. On the basis of the research on the operation characteristics, evaluation and market mechanism of the captive power plant, combined with the achievements of the demonstration project to improve the theory and technology research, the flexible interactive platform between the captive power plant and the power grid has the functions of archive management, resource adjustable potential analysis, operation status detection, energy consumption analysis, transaction management, etc. of the captive power plant, and has formed an operable and popularized evaluation system and market mechanism[5,6].

The "flexible interactive platform between self owned power plant and power grid" is connected with the State Grid System and becomes a part of the smart grid.In order to strengthen the security protection of the flexible interactive platform between the captive power plant and the power grid, ensure the security of business data in the process of generation, storage, transmission and use, resist malicious damage and attacks on the platform by hackers and malicious codes, and other illegal operations, and prevent the paralysis and out of control of the flexible interactive platform between the captive power plant and the power grid, as well as the resulting transaction interruption or failure, a system accident of the power plant and other accidents, It is necessary to adopt a number of security protection technical means to build a comprehensive security defense system in depth. Through multi-level information security protection measures, the security problems caused by the breakthrough of some protection layers are prevented, so as to effectively ensure the security of the platform. In order to achieve cost control, it is not necessary to deploy the whole system in an all-round way when building the security defense system. Instead, multi-level security protection configuration is adopted for key nodes, and security protection measures with relatively low protection intensity are adopted for common areas[7].

2. Overall protection principle of platform

"Flexible interaction platform between captive power plant and power grid" is a system connection with State Grid Corporation of China, which has certain particularity. The research on flexible interaction security protection strategy between captive power plant and power grid will be completed on the basis of complying with relevant national regulations and information management regulations of State Grid Corporation of China. The

* Corresponding author: zhulingkai@woyoxin.com

general principle of security protection is "security zoning, network dedicated, vertical authentication"[8] .

2.1 Security partition

Safety zoning is the structural foundation of flexible interactive safety protection system between captive power plant and power grid[9]. The internal business system of the captive power plant based on computer and network technology is divided into production control area and management information area in principle. According to the importance of the business system, the production control area can be divided into control area and non control area, and the management information area can be divided into production management area and management information area.

The business system or its functional module in the control area is an important part of power production. It is the focus and core of security protection to directly realize the real-time monitoring of the primary power system and vertically use the power dispatching data network or special channel. The main users of all systems in the control area are dispatchers and operators. The real-time data transmission is at the millisecond level or second level. Its data communication is transmitted through the real-time subnet or special channel of the power dispatching data network.

The business system or its functional module in the non control area is a necessary part of power production. It operates online but does not have control function. It uses the power dispatching data network and is closely connected with the business system or its functional module in the control area. The main users of various systems in the non control area are power dispatchers, hydropower dispatchers, relay protection personnel and power market traders. The data acquisition frequency of non control area is minute level or hour level, and its data communication uses the non real time subnet of power dispatching data network.

The production management area mainly includes the production management system. The typical systems are lightning monitoring system, meteorological information access, etc. the external communication boundary of the area is the power data communication network.

The information management area is the management information area. The area includes office management information system, customer service, etc. the external communication boundary of the area is power data communication network and Internet.

Different safety protection requirements are determined for different safety zones, and the control zone has the highest safety level. On the premise of meeting the general principles of security protection, the setting of security zones can be simplified according to the actual situation of the business system, but the formation of vertical cross connection of different security zones should be avoided.

2.2 Security partition

The power trading module undertakes the online trading business and needs to be linked with the power trading platform of the State Grid. Therefore, it can be combined with the special channel of the power dispatching data network at the power plant end to use independent network equipment for networking. The power dispatching data network is a special data network for the production control area, which carries the power real-time control, online production transaction and other businesses. The security protection isolation strength between the external boundary networks of the security zone should match the security protection isolation strength between the connected security zones. The power dispatching data network shall use independent network equipment for networking on the dedicated channel, and adopt different channels, different optical wavelengths, different fiber cores and other methods based on sdh/pdh to realize the safe isolation from other data networks of power enterprises and external public information networks at the physical level. The power dispatching data network shall adopt the following security protection measures: network route protection, network boundary protection, security configuration of network equipment, layered and partitioned setting of data network security. In addition, the data networks at all levels should be safely isolated through route restriction measures. When a public communication network is used within an area, it is prohibited to interconnect with the dispatching data network to ensure that network failures and safety events are limited within the local area.

2.3 Vertical certification

Vertical encryption authentication is the vertical defense line of the security protection system of the platform. Adopt authentication, encryption, access control and other technical measures to realize the remote secure transmission of data and the security protection of vertical boundaries. The vertical connection between the platform terminal and the platform shall be equipped with a dedicated vertical encryption authentication device or encryption authentication gateway and corresponding facilities that have been tested and certified by the designated department of the state to realize two-way identity authentication, data encryption and access control. Vertical communication in the secure access area shall adopt security measures such as one-way authentication based on asymmetric key technology, and two-way authentication can be adopted for important services.

3. Division of safety zone for interaction platform between captive power plant and power grid

The flexible interactive platform between captive power plant and power grid has functional modules such as file management, resource adjustable potential analysis, operation status monitoring, energy consumption analysis, transaction management, etc. the databases involved are: captive power plant equipment, captive power plant terminal, power grid information, flexible interactive strategy and other databases. According to the above concept of safety zone and the principle of dividing safety zone, the division of safety zone for each business system in the interactive platform between captive power plant and power grid is shown in Table 1.

Table 1. Security zone division of platform business system.

Numble	Module name	Large area	Region
1	File management	Management information region	Information management area
2	Analysis of resource adjustable potential	Management information region	Production management area
3	Operation status monitoring	Production control area	Uncontrolled area
4	Energy consumption analysis	Management information region	Production management area
5	Transaction management	Management information region	Production management area

According to the overall security protection principle and the system security zone division, the security protection methods developed by this platform mainly include border security protection, terminal security protection, network security protection and comprehensive security protection, as shown in Table 2.

Table 2. Platform safety protection method.

Numble	Safety protection classification	Safety protection measures
1	Border security protection	Lateral boundary protection Longitudinal boundary protection Third party border security protection
2	Terminal safety protection	Terminal virus protection Malicious code protection Patch management Terminal security management
3	Network security protection	Network equipment security protection Network information security protection Network channel security protection
4	Comprehensive safety protection	Intrusion detection Host and network equipment reinforcement Safety inspection and analysis Network security management Backup and disaster recovery User interface security protection Data interface security protection

The overall security protection scheme needs to divide the security zones of the involved systems in combination with the management similarity and business similarity of each application system, so as to design the security protection measures of each system more pertinently. The security zone division of the information system can decompose the complex security protection problems, help to realize the hierarchical protection of the information system, implement targeted application boundary protection, and prevent the proliferation of security problems.

4. Summary

During the operation of the smart grid software system, due to business expansion and other reasons, it is necessary to add new information systems or access a large number of intelligent terminal equipment, which makes the original system scale increase, the business dependence become stronger and stronger, and the system application environment becomes more and more complex. The security of each region has an important impact on the safe operation of the entire power grid system. The security control strategy for the power grid system is mainly considered from three aspects: control security, access security and application security. Control security effectively prevents illegal intrusion initiated by external users, access security prevents illegal user intrusion, and application security mainly improves the availability and reliability of the system.

Based on the self-developed system "flexible interactive platform between self-contained power plant and power grid", this paper expounds the overall strategy of system security protection, divides the security zone according to the specific business content of the system, and formulates specific protection rules under the guidance of the overall strategy, effectively ensuring the safe operation of the platform.

This work was supported by The Science and Technology Project of State Grid Corporation of China - Research on low-carbon regulation oriented coal-fired captive power plants' participation in power grid supply and demand regulation and source grid load coordinated operation technology.（5400-202236173A-1-1-ZN）.

References

1. Xiaofeng Zhu, Kejie Cui, Jianwei Huang. Research on integrated security protection of power monitoring systems[J]. Electrical Technology, 2021,(07):114-115+119.
2. Jie Cheng, Zhijie Shang, Wei Hu, etc. Security hidden dangers and countermeasures of smart grid information system[J]. Electrical applications, 2020,39(04):99-102.
3. Han Qian, Jie Gu. Analysis of network security protection strategy of smart grid system[J]. Application of integrated circuit, 2021,38(09):42-43.
4. Shijun Liu, Qing Cai, Yuhui Bi. Research and judgment on restrictive policies and development

direction of enterprise owned thermal power units[J]. Baosteel Technology, 2016,(5):6-12.

5. Yifan Huang, Haijing Zhang, Lei Wang, etc. Evaluation method for flexibility regulation of captive power plants based on production characteristics of enterprises [J]. Power demand side management, 2021,23(1):61-66.
6. Lingkai Zhu, Weishuai Wang, Haijing Zhang, etc. Application prospect of captive power plant in future comprehensive energy field [J]. Shandong Electric Power, 2022,49(2):12-16.
7. Yanwei Shang. Research on security protection of power grid information platform[J]. Automation technology and Application, 2018,37(7):126-129.
8. Xiasheng Gao, Shaoxiong Huang, Xiao Liang. Safety protection elements of power monitoring system [J]. Computer knowledge and technology, 2017,13(8):212-214+222.
9. Weihua Zhang. Discussion on network security protection of power monitoring system [J]. Information recording materials,2017,18(6):46-47.

Analysis of the O-rings that influence the performance of RCP hydrostatic seal based on TEH coupling method

Jianghong Zhang [1], Kan Chen [2], Yijin Ma [1], Mingzhong Feng [2], Wei Liu [2], Junkai Zhang [2], Hebin Ren [2]

[1] Suzhou Nuclear Power Research Institute, China General Nuclear Power Corporation, Shenzhen 518026, China
[2] Department of Nuclear Equipment, Sichuan Sunny seal company Co., Ltd, Chengdu 610031, China

Abstract: A Multi-body contacting Thermal-Elastic-Hydro Model, MTEHM has been established in this study. And which has been applied to analysis the leakage behavior of No.1 seal of the RCP hydrostatic seal. The hydro-model is based on 2D Reynolds equation and programmed by M-code. An interface program written in C has been carried out. And which is used to transit the end face pressure value from hydro-model to elastic-solid model. The FEA multi-body contacting method has been taken into the elastic-solid model. And which is used to obtain the deformation angle of seal rings. An opening force ratio has been carried out as a convergent judgment of the multi-physics iterative coupling process. The influence of position of O-rings (on the back of seal rings and relative to the hydrostatic clamp rings) has been discussed. For test and verify the MTEHM, a testing rig has been established. It can be used to simulate a high temperature high pressure seal media with a temperature range from 22℃ to 100℃ and a pressure range from 0.1MPa to 22MPa. At last, a comparison of pressure-leakage data from Westinghouse, Saint Alban power plant and this study has been presented. And a variance analysis has been attached.

Key words: hydrostatic seal; RCP; RCP seal; multi-physics coupling; TFSI.

1. Introduction

Reactor coolant pump (RCP) is used to transport media for cooling down the reactor core and exchanging heat. The hazardous high pressure media is separated by RCP seal from ambient. The performance of this seal influence safety and stability of whole nuclear power plant directly. A hydrostatic RCP seal which is applied on Chinese generation-III PWR is shown as Fig. 1. It contains three mechanical seals which are arranged tandem. The No.1 seal has two silicon nitride seal rings. One of which is rotating relative to another. The stationary ring has a taper with several minutes. The pressure difference between inner and outer of the end face make the seal rings separated from each other with a 10um gap. Usually, the differential pressure is 15.5MPa and with a 700L/h leakage. The No.1 seal subject to the whole working pressure so it is the most significant sub-part in the whole RCP seal[1-2]. In normal work condition, the No.2 seal subjected to 0.02~0.2MPa differential pressure and has an 18L/h leakage. In emergence work condition, the type of No.2 Seal will be changed from rub seal to non-contacting seal and keep the seal function for at least another 30 minutes[2].

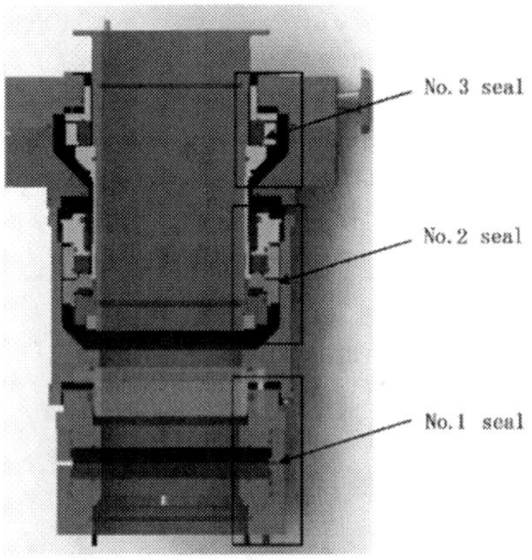

Fig. 1. Hydrostatic RCP seal assemble

The No.3 seal has a dual-dam. The block media is supplied by a high tank and injected into the end face by several orifices. No.3 seal has a leakage of hundreds milliliter per hour which is too small to be measured by general flow meter but it can be measured by weighing[2]. Considered the importance of No.1 seal, an enhanced research is necessary.

A large number of institutes and researchers have published their theoretical and experimental study results. T. Boardman from Rockwell has established a hydrostatic calculation model for dual-taper RCP seal based on Bernoulli's equation[3]. An inlet pressure loss factor has been introduced. But the structure deformation caused by pressure and temperature has not been taken into his model. The pressure-leakage curve carried out by this model is linear. Which is not suit to the real situation from power plants (in fact it is similar to convex curve)[4]. Noël Brunetiere etc. have taken the inertia term into their model [5], but it is pure fluid model also. Wang Heshun has established a brief explicit model but solid deformation is still not being considered[6]. Yasuhiro Sasaki etc. have analyzed No.1 seal rings' deformation and optimized the design by increase axial dimension to increase operational stability[7]. Liao Chuanjun etc. from Tsinghua University have considered the end face deformation caused by pressure[8]. They have deduced this deformation angle by stress analysis of ring's section. And then the gap function with end face deformation has been carried out. A solid-fluid coupled model has been established at last. But in their research, only a single ring's section has been analyzed. The interaction between rings, clamp rings and ring seats has been ignored. Especially, the clamping force induced by differential pressure on the clamp rings has been ignored. WeiFeng Huang etc. from Tsinghua University also has presented their TFS model to analysis RCP No.1 seal. The interaction of rotating parts has been considered, but the clamping force produced by clamp ring's two O-rings with different diameters has been ignored in their research[9].

Therefore, to establish a multi-body contacting thermal-Elastic-hydro-coupling model and use it to analysis No.1 seal leakage behavior is significant.

2. Structures and Function of No.1 Seal

As shown in Fig. 2, the sealing boundary consists with rotating component, stationary component and insert. Rotating ring, hydrostatic clamp ring and rotating ring seat are assembled as the rotating

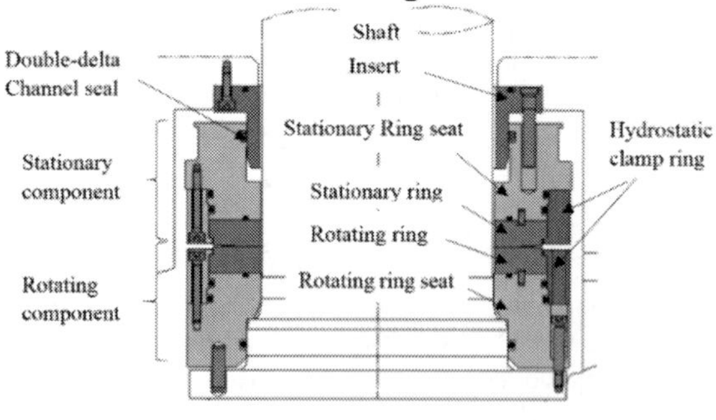

Fig. 2. No.1 seal assemble

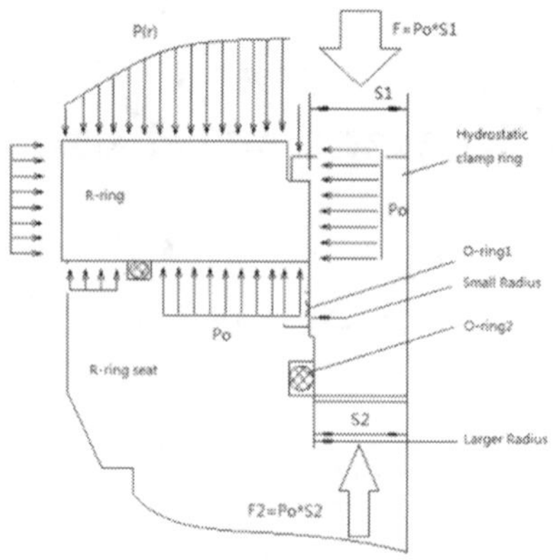

Fig. 3. Ring section and subject to force

component. Stationary ring, hydrostatic clamp ring and stationary ring seat are assembled as the stationary component. Seal rings are made of Si3N4. The stationary ring has a taper approximate equal to 2.8 minutes.

What needs to be emphasized is that the seal rings are not bond with seats by screws and clamp rings mainly. The initial load of these screws is very small about 1.5Nm. In fact, an axial hydrostatic force ($F - F_2$, as shown in Fig. 3) produced by two O-ring with different radius push the rings onto seats tightly. So that the clamp rings are named "hydrostatic clamp rings" strictly. The hydro-load has a range from 0 to 14809.2N when media pressure is between 0~15.5MPa.

3. Hydrostatic Model of the End Face

As shown in Fig. 4. It is assumed that the ring section rotate φ_r, φ_s relative to initial position when subjected to pressure and thermal gradient. Then the end face film thickness will be changed. End face pressure distribution changed as well as. The end face opening force and leakage will be changed at last.

Nomenclature

R_o — Outer radius of rotating ring, mm

R_i — Inner radius of rotating ring, mm

R_g —End face taper transition radius, mm

R —Radial variable, mm

φ_r —Deflection angle of rotating ring, minutes

φ_s —Deflection angle of stationary ring, minutes

$\Delta\varphi$ —total deflection angle of end face, minutes, $\Delta\varphi = \varphi_s + \varphi_r$

φ_t —initial taper of end face, minutes

φ_t' —net angle of end face after deformation, minutes, $\varphi_t' = \varphi_t - \Delta\varphi$

h_a —film thickness of end face parallel region, μm

h_b —film thickness of end face taper region, μm
G—gravity of stationary component, N
F_{open} —end face open force, N
F_{close} —end face close force, N

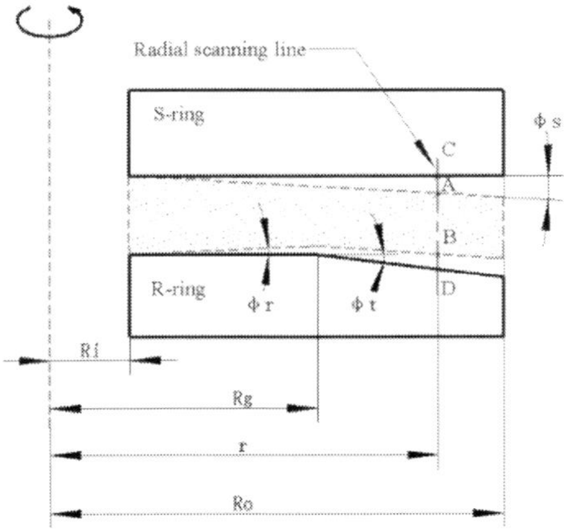

Fig. 4. No.1 seal end face dimensions

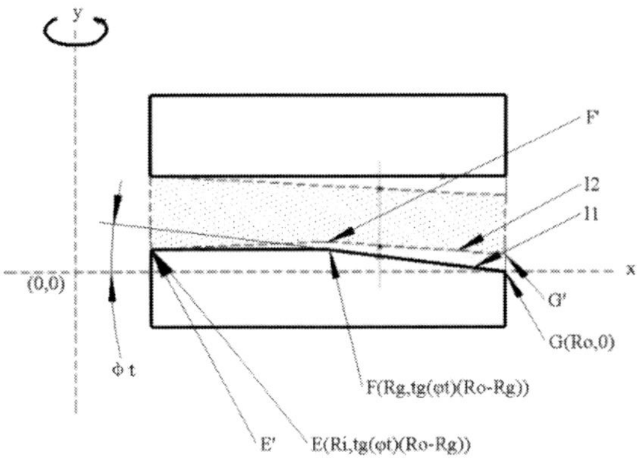

Fig. 5. Hydrostatic model of end face

As shown in Fig. 4 and Fig. 5, Establish a Cartesian coordinate, seal swirl around y axis, and the taper face outer point on the x axis. After the seal deflected (dash line), the points E, F, G will be translated to point E', F', G'. The function of straight line L1 and L2 can be established in two points form:

L1:

$$y = -tan(\varphi_t) \cdot x + R_O \cdot tan(\varphi_t) \quad (1)$$

L2:

$$\frac{y-y_{G'}}{y_{F'}-y_{G'}} = \frac{x-x_{G'}}{x_{F'}-x_{G'}} \quad (2)$$

For L2, $y_{G'}$, $x_{G'}$, $y_{F'}$, $x_{F'}$ in (2) is the horizon position and vertical position of point G and F after seal deflection which can be carried out by rotating coordinates formula simply. Once this four positions has been got. Take them into (2), and then the linear function of L2 is:

$$y = -\frac{1}{sin(\varphi_r)\,tan(\varphi_t) + cos(\varphi_r)}\Big(\big(((R_i - R_O)\,cos(\varphi_r) + tan(\varphi_t)\,(R_g - R_o)\,sin(\varphi_r) + x - R_i)(cos(\varphi_r)\,tan(\varphi_t) - sin(\varphi_t))\big) + (cos(\varphi_t) - 1)(R_g - R_o)\,tan(\varphi_t) + (R_o - R_i)\,sin(\varphi_r) \quad (3)$$

Assume the variable $R \subseteq (R_g, R_o)$ and position of points B and D are known as $(X_{B'}, Y_B)$, $(X_{D'}, Y_D)$. Then L_{BD} the distance between B and D is:

$$L_{BD} = \frac{1}{sin(\varphi_r)\,tan(\varphi_t)+cos(\varphi_r)}(-sin(\varphi_r)(R - R_g)\,tan(\varphi_t)^2 - (cos(\varphi_r)^2 + sin(\varphi_r)^2 - cos(\varphi_r))(R_g - R_i)\,tan(\varphi_t) - sin(\varphi_r)(R - R_i)) \quad (4)$$

Film thickness control function h(R):

The film thickness control function on whole end face can be divided into two parts. One is parallel region $R \subseteq (R_i, R_g)$ and another is taper region $R \subseteq (R_g, R_o)$. "h_1" is the film thickness before seal deflecting and "h'" is that after seal deflecting. Then:

The parallel region film thickness h_a is:

$$ha = h_1 - (R - R_i) \cdot \tan(\varphi_s) - (R - R_i) \cdot \tan(\varphi_r) \quad (5)$$

The taper region film thickness h_b is:

$$ha = h_1 + \tan(\varphi_t) \cdot (R - R_g) - (R - R_i) \cdot \tan(\varphi_s) - L_{BD} \quad (6)$$

Take (4) into (6), the taper region film thickness h_b is:

$$h_b = h_1 + tan(\varphi_t)(R - R_g) - (R - R_i)\,tan(\varphi_s) + tan(\varphi_t)R - tan(\varphi_t)R_o - \frac{1}{sin(\varphi_r)\,tan(\varphi_t)+cos(\varphi_r)}(((R_i - R_o)\,cos(\varphi_r) + tan(\varphi_r)(R_g - R_o)\,sin(\varphi_r) + R - R_i)(cos(\varphi_r)\,tan(\varphi_t) - sin(\varphi_r))) + (cos(\varphi_r) - 1)(R_g - R_o)\,tan(\varphi_t) + (R_o - R_i)\,sin(\varphi_r) \quad (7)$$

The film thickness profile after seal deflected can be plotted as Fig. 6 once φ_r and φ_s are obtained.

Assume the media of RCP seal is uncompressible. The scale of film thickness is much smaller than that in faceplate direction. Therefore, the 2D Reynolds equation in pole coordinate is suitable[10-11]:

$$\frac{1}{R}\frac{\partial}{\partial \theta}\left(\frac{h^3}{12\mu}\frac{\partial p}{\partial \theta}\right) + \frac{\partial}{\partial R}\left(\frac{rh^3}{12\mu}\frac{\partial p}{\partial R}\right) = \frac{R\omega}{2}\frac{\partial h}{\partial \theta} \quad (8)$$

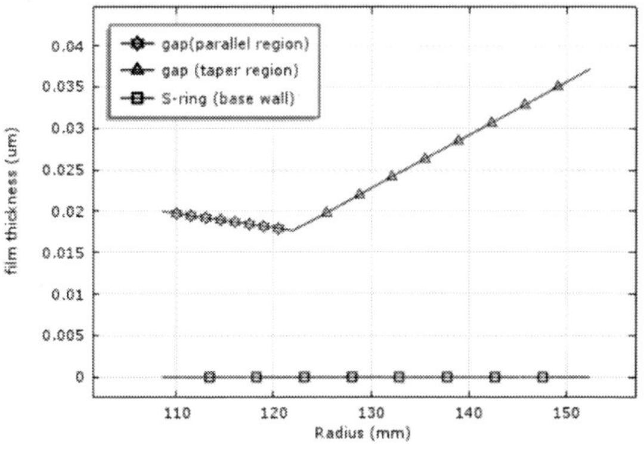

Fig. 6. The film thickness profile of the end face (assume $\varphi_r = 0.367', \varphi_s = 0.046'$)

For RCP hydrostatic seal, the pressure and thickness gradient along pole angle θ is equal to 0. It means $\frac{\partial p}{\partial \theta} = \frac{\partial h}{\partial \theta} = 0$, so that (8) can be simplified into:

$$\frac{\partial}{\partial R}\left(\frac{Rh^3}{12\mu}\frac{\partial p}{\partial R}\right) = 0 \tag{9}$$

Take (5) or (7) into (9), the end face pressure distribution function P(R) with two subsections can be obtained. It is difficult to carry out an explicit solution about the P(R). A program written in Matlab based on FE method is used to solve numerical solutions. The detail procedure of FE method can be found in reference[10]. A comparison between MTEHM and common hydrostatic model is taken out as instance:

The pressure distribution curves of these two models can be plotted as Fig. 7. An obvious difference can be found especially in parallel region ($R \subset (108,125)$ mm). The result of common hydrostatic model is approximate to linear but the MTEHM is a concave line. That means the area below the MTEHM curve is smaller. The end face open force is equal to this area. The film thickness and leakage under balance status is smaller than that of common hydrostatic model also. A further plot about how the taper changing influence end face pressure distribution is shown as Fig. 8

Table 1. Boundary conditions of MTEHM and common hydrostatic model

parameter	MTEHM	Common hydrostatic model	physics
h1	5[um]	5[um]	Film thickness(assumed)
RPM	1500	1500	Rotating speed (RPM)
φ_r	0.367	0	Rotating component deflection angle(minutes)
φ_s	0.0456	0	stationary component deflection angle(minutes)
φ_t	3	3	Initial taper (minutes)
T	55+273	55+273	injection temperature (K)
Rg	122[mm]	122[mm]	End face taper transition radius (mm)
Ri	108.5[mm]	108.5[mm]	Inner radius of end face(mm)
Ro	152.5[mm]	152.5[mm]	Outer radius of end face (mm)
Rb	121.4[mm]	121.4[mm]	Balance Radius (mm)
Roring	119.5[mm]	119.5[mm]	O-ring radius back of rings (mm)
P2	15.8[MPa]	15.8[MPa]	Media pressure (MPa)
P1	0.21[MPa]	0.21[MPa]	Leakage pressure (MPa)

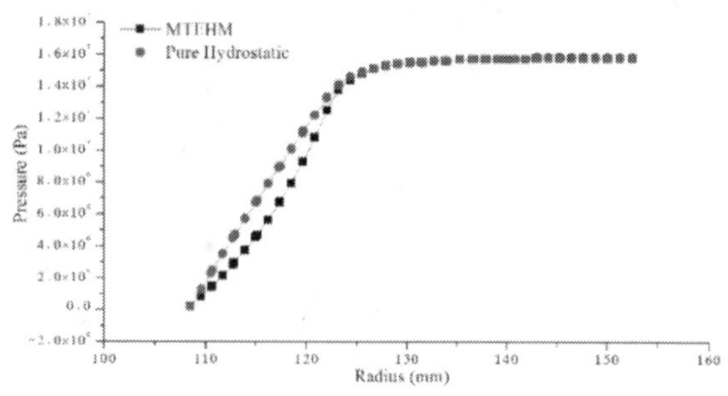

Fig. 7. The P(R) (MTEHM vs. common model)

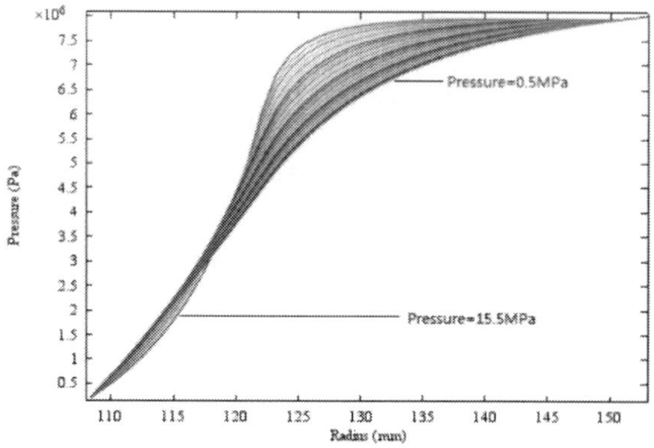

Fig. 8. P(R) with taper changing

In the instance above, the end face deflecting angle φ_r, φ_s are assumed to be known at the beginning. For closing the coupling circle, an Elastic-solid model with thermal condition and multi-body contacting is required to obtain the deflecting angle.

4. Thermal Elastic multi body contacting model

In this section, the interaction of the mechanical parts and temperature gradient are taken into solid model. According to experimental results, the differential temperature along radial direction of the end face is $4^\circ C \sim 8^\circ C$. It has a small variation when injection temperature or leakage changed. A FEA method with multi-body contacting and thermal condition is applied.

A 2D axial symmetry model which includes hydrostatic clamp rings, seal rings, ring seats and virtual shaft is assembled. All contact pairs have a different size of mesh and are set to be frictional as the A~G shown in Fig. 9. The friction coefficient is 0.37. Contact source has a fine size and contact target has a chaos size to prevent nodes from penetrating. The contact force is defined by:

$$F_{normal} = k_{normal} x_{penestration} \tag{10}$$

Where the k_{normal} is contact stiffness and $x_{penetration}$ is penetrating amount.

 https://doi.org/10.1051/matecconf/202236301019

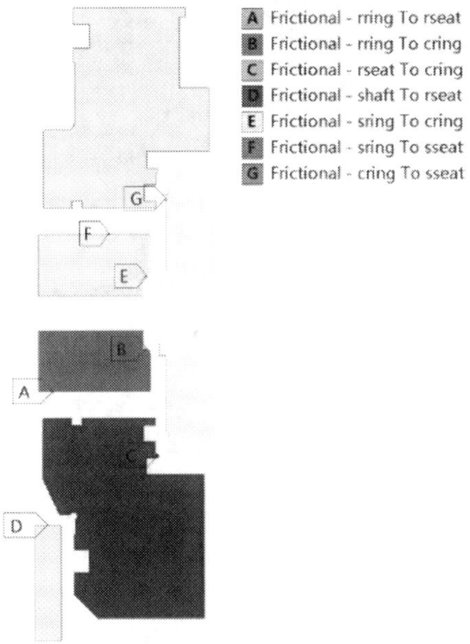

Fig. 9. multi-body contacting model of No.1 seal

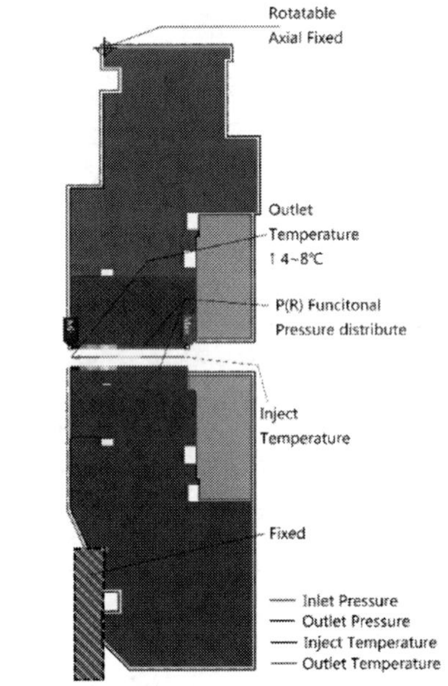

Fig. 10. Boundary conditions and constraint

4.1 Boundary Conditions, Constraints, Materials Property

Table 2. Loads, materials, work conditions

		Young's modular (GPa)	Poisson's ratio	Thermal conductivity (W/m*°C)
Rings	Si3N4	320	0.272	28.4
Clamp rings, ring seats	stainless steel	217	0.27	30
media	water	viscosity 5.1E-4Pa*s	Specific heat 4200J/kg*°C	0.645
Inlet pressure	15.8MPa		Inlet temperature	55°C
Outlet pressure	0.21MPa		Outlet temperature	60°C

The boundary condition applied in solid model is shown as Fig. 10. An interface program written in C language is used to transit the pressure value from fluid zone to solid zone automatically. Therefore, the coupling between fluid and solid zones is archived.

The calculation of temperature gradient of end face can be found in reference[12]. It is the same with reference [12], the solid thermal conductivity, and solid-fluid conjugate heat transfer is considered. Injection temperature, leakage, thermal conductivity, heat transfer coefficient will influence the end face temperature gradient.

Four probes located on end face of seal rings inner corner and outer corner are set to get the axial displacement. They are:

Rin —inner corner displacement of rotating ring, mm

Rout —outer corner displacement of rotating ring, mm

Sin —inner corner displacement of stationary ring, mm

Sout —outer corner displacement of stationary ring, mm

At last, the seal deflection angle φ_r, φ_s can be obtained by solving a triangle function simply.

4.2 Coupling processing

The solving program is beginning from fluid zone.

Step1: assume that the $\varphi_s = \varphi_r = 0$, and a sets of film thickness $|h| \supset h_{eq}$ are taken as input. h_{eq} is the film thickness when seal works in balance status。 Use $F_{open} = F_{close}$ to evaluate whether the seal archives balance status or not. Then the end face pressure distribution P(R) can be obtained as output.

$$F_{open} = \int_0^{2\pi} P(R) d\theta \tag{11}$$

$$F_{close} = G + Po \cdot \pi \cdot (R_o^2 - R_b^2) - Pi \cdot \pi \cdot (R_b^2 - R_i^2) \tag{12}$$

Step2: the value of P(R) can be translated by interface program and be transferred to end face nodes in solid model. Other input containsT_o,T_i,$T(R)$,T_o, T_o is injection temperature, T_i is leakage temperature, and $T(R)$ is temperature gradient on the end face. The linear level of $T(R)$ influence seal deformation less. So the $T(R)$ is treated as linear in this study. The φ_s, φ_r can be obtained in this step as output.

Step3: Take the φ_s, φ_r back into Step1 to calculate the end face pressure distribution $P(R)_n$ again, n=1…n which means the iterative steps。

Step4: repeat Step1~Step3, and output $P(R)_n$ each time. It can be found that the open force by integrating the $P(R)_n$ will not decrease obviously after several repeats. Therefore, the coupling convergence criterion shall be:

$$\frac{\int_0^{2\pi} P(R)_n d\theta - \int_0^{2\pi} P(R)_{n-1} d\theta}{\int_0^{2\pi} P(R)_{n-1} d\theta} \leq 0.01 \qquad (13)$$

Step5: the final solutions can be obtained by iterating n times and which include film thickness h_{eq_n}, leakage Q_n and end face open force F_{open_n} .

$$F_{open_n} = \int_0^{2\pi} P(R)_n \, d\theta \qquad (14)$$

5. Discussion of Engineering Application

In order to confirm the engineering applicability of MTEHM, a test rig as shown in Fig. 11 has been established in Sichuan Sunny Seal Company. The Fig. 12 is the No.1 seal and its stationary ring after 4000 hours life circle experiment.

Fig. 11. Test rig for RCP hydrostatic seal

Fig. 12. No.1 seal disassembled

5.1 The Influence of Temperature

The temperature influences the viscosity and density of seal media mainly. The leakage increase as the temperature increase[12-15]. Except for that, the temperature influence end faces taper also, but much smaller than pressure. A comparison between pure pressure effect (Fig. 13) and pure thermal effect (Fig. 14) is shown as below Table 3 and Table 4:

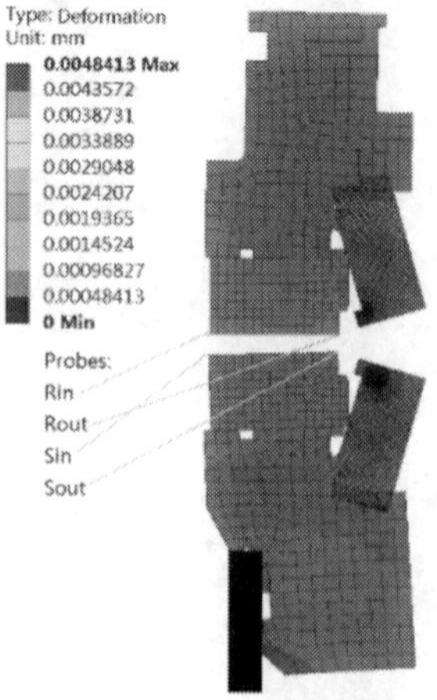

Fig. 13 Pressure effect (15.5MPa)

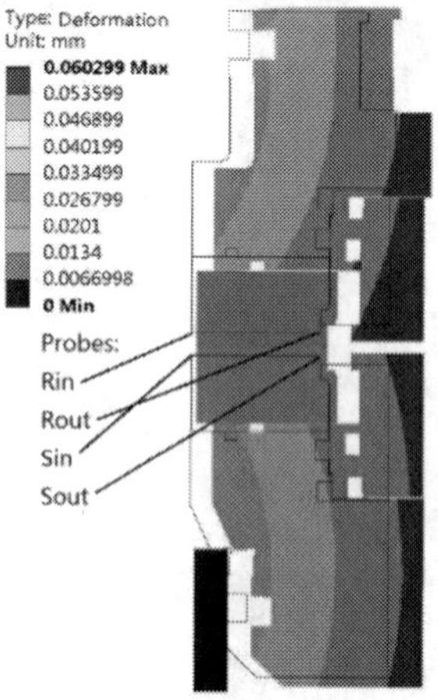

Fig. 14 Thermal effect (with 5°C differential temperature)

Table 3. Seal deflection of single pressure effect (15.5MPa)

Probe		Axial displacement (mm)	End face width (mm)	Deflecting angle (minutes)	Total angle (minutes)
R-ring	Rin	1.34E-03	4.63E-03	0.3678	0.4135
	Rout	5.97E-03			
S-ring	Sin	-1.32E-03	5.75E-04	0.0456	
	Sout	-7.46E-04			

Table 4. Seal deflection of single thermal effect (with 5°C differential temperature)

Probe		Axial displacement (mm)	End face width (mm)	Deflecting angle (minutes)	Total angle (minutes)
R-ring	Rin	1.5971e-2	2.80E-05	0.0022	0.0002
	Rout	1.5943e-2			
S-ring	Sin	-3.0729e-2	2.50E-05	0.0019	
	Sout	-3.0704e-2			

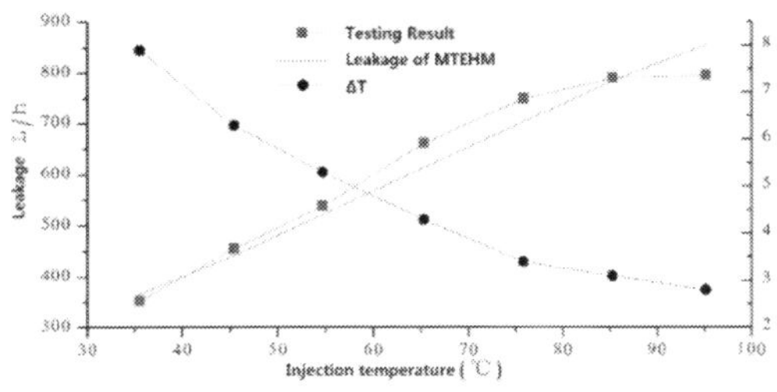

Fig. 15. Injection temperature VS. Leakage and ΔT (15.5MPa)

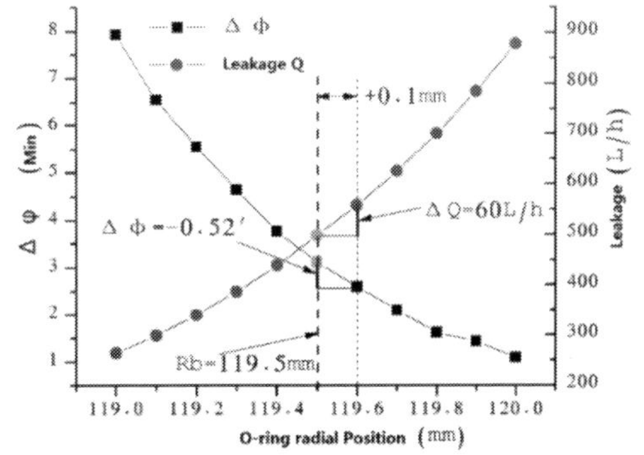

Fig. 16. The influence of radius O-ring on back of seal rings (15.5MPa, 55°C, up)

It can be observed from column "axial displacement" in Table 3 and Table 4 that the taper is divergent form R_o to R_i when seal subject to pressure only. On contrary the taper is convergent when seal subject to temperature only. And the "Total angle" column shows that the influence of pressure is more important than that of temperature.

Injection temperature influence the end face differential temperature and leakage. The higher injection temperature, the larger leakage is. The simulation and experiment results are shown in Fig. 15.

5.1 O-Ring Positions

There are two sets of O-rings influence leakage mainly. One is on back of the seal rings and another is a pair of O-rings on the hydrostatic clamp rings.

It is need to clarify that the O-rings' material is FKM[14]. When it is compressed by seal media with different pressure, the pressure boundary points will changing. In this study, the error caused by the boundary points offset has been ignored because of it is too tiny to influence the final results.

For the O-ring back of seal rings. Once the distributed pressure be treated as concentrated force, the origin and value of this force will be influenced by the radius of O-ring. And then the end face taper changed. Larger the radius of O-ring is, the pressure effective area is smaller, the end face taper variation $\Delta\varphi$ is smaller, the taper after deflecting φ' is larger, and the leakage is larger. With smaller radius, the changes are on contrary. As shown in Fig. 16:

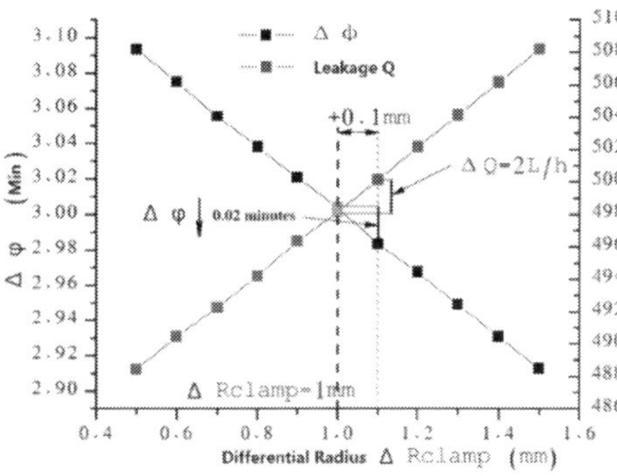

Fig. 17. Influence of differential radius of O-ring on clamp rings (15.5MPa, 55°C,below)

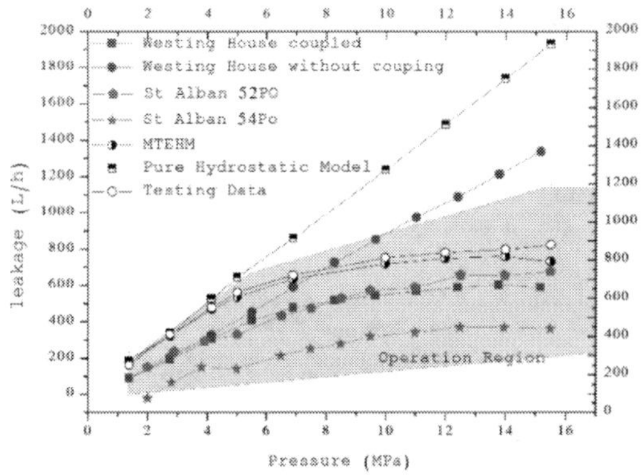

Fig. 18. Comparison of MTEHM, Westinghouse, St Alban and testing data

There are two O-rings with different radius of the hydrostatic clamp rings. The difference of radius of this two O-rings ΔR_{clamp} influence the axial load of clamp rings. The ΔR_{clamp} increased, the axial load increased. The end face taper deflecting angle $\Delta\varphi$ decreased as well. The end face net taper after deflecting φ' increased.

$$\varphi_t' = \varphi_t - \Delta\varphi \tag{14}$$

The leakage increased at last, shown as Fig. 17:

5.2 End Face Differential Pressure

It can be found from Fig. 18 that the leakage (with 55°C and 15.5MPa injection) obtained by MTEHM is close to which obtained by common hydrostatic model when inject pressure is blow 5MPa. But the leakage curve becomes more flat as the pressure increasing. This is because of the end face net taper decreasing. As shown in Fig. 18, this trend is same to Westinghouse's[3] and the real operation data from Saint Alban 52PO, 54PO[16]. There is a 200L/h leakage difference between MTEHM and Westinghouse's result because of different research targets' geometric dimensions. It can be adjusted by changing initial end face taper. The initial taper increased the leakage curve rise up entirely and it is the key parameter that influences the RCP No.1 seal's performance. And the optimize design points can be obtained by this MTEHM now.

6. Conclusions

A MTEHM has been presented in this study to simulate the RCP seal end face pressure-leakage relationship. The results display an obvious difference with pure hydrostatic model. Especially in the inner radius region of the end face. The pressure-leakage trend curve of MTEHM is concave. Different with that, the trend curve is convex of common pure hydrostatic model as shown in Fig. 7. The open force of MTEHM is 4.7% smaller than pure hydrostatic model(15.5MPa, 55°C).
According to the MTEHM results: the changing of end face net taper caused by pressure and changing of viscosity of media caused by injection temperature are key factors that influent RCP seal leakage directly. The changing cause by pressure is more important than by injection temperature as shown in Table 3 and Table 4. The viscosity of media decreased as the temperature increased and this lead to leakage increase directly as shown in Fig. 15. and it can be conclude that 43L/h leakage increasing when injection temperature increase 5°C The secondary seal on two different locations influence the interior forces of the seal assembles. Compare the Fig. 16 with Fig. 17 and it can be conclude that the position variation of O-ring behind seal rings is more important than that of clamp ring's O-ring. Compare the Fig. 16 with Fig. 17 and it can be conclude that the radius of O-ring behind seal rings increase each 0.1mm, the leakage increase 60L/h. and the differential radius of clamp ring O-rings increase each 0.1mm, the leakage increase 2L/h. this conclusions above are obtained on conditions of 15.5MPa and 55°C injection temperature.
The MTEHM can couple the influence of thermal effective, multi-body contacting effective, hydrostatic effective and rigid-flexible body deformation effective very well. The MTEHM results are approximate to simulation data from Westinghouse and real operation data from Saint Alban nuclear plant 52PO, 54PO as shown in Fig. 18. Except for that, It can be found in Fig. 18, the testing result about pressure-leakage curve match the MTEHM simulating result very well. That means this model can be used to predict the leakage behavior of RCP No.1 seal in engineering applications. And it can be used to optimize the design too.

References

1. C J Ruger and J C Higgins. Reactor Coolant Pump Seal Issues and Their Applicability to New Reactor Designs[R]//NUREG/49115, Brookhaven National Laboratory, Upton, NY, 1993.
2. Robert J, Lutz Jr, J Gene Zottola, etc. Westinghouse Owners Group Reactor Coolant Pump Seal Performance for Appendix R Assessments[R]. Prepared for Westinghouse owners Group, Developed under WOG Program, PA-RMSC-0181, Jan, 2005: 11-14.
3. T boardman, N jeanmougin, et al. Leak Rate Analysis of The Westinghouse Reactor Coolant Pump[R]// Research Report Prepared for U.S Nuclear Regulatory Commission, Energy Technology Engineering Center, 1985.
4. Erwan Galenne, Isabelle Pierre-Danos. Thermo-Elastic-Hydro -Dynamic Modeling of Hydrostatic Seals in Reactor Coolant Pumps [J]. Tribology Transactions, 2007, 50:466-476.
5. Brunetière N, Tournerie B and Frêne J. TEHD Lubrication of Mechanical Face Seals in Stable Tracking Mode [J]//Part1: Numerical Model and Experiments, ASME Journal of Tribology, 2003, 125(5): 608-616.
6. Wang Heshun, Zhu weibing, Huang Zepei, Zhang Chening. Main Pump Seal's Characteristics Affected by Cone Angle and Clearance [J]//TELKOMNIKA Indonesian Journal of Electrical Engineering, October, 2014, 12(10): 7233-7241.
7. Yasuhiro Sasaki, Kazuyuki Obata, Osamu Hisai, Toshihiko Matsuo, Naotaka Komatsu, Hidekazu Uehara. Improved Shaft Seal for 93A Type Reactor Coolant Pump [J]//Mitsubishi Heavy Industries, Technical Review, Feb, 2005, 42(1).
8. Liao C J, Huang W F, Suo S F, et al. fluid-solid strong-interaction model of mechanical seals in reactor coolant pumps [J]. Science china-technology sciences, 2011, 54(9): 2339-2448.
9. Huang Weifeng, Liao Chuanjun, Liu Xiangfeng, Suo Shuangfu, Liu Ying, and Wang Yuming. Thermal Fluid-Solid Interaction Model and Experimental Validation for Hydrostatic Mechanical Face Seals [J]. Chinese Journal of Mechanical Engineering, 2014, 27(5): 949-957.
10. Lebeck. Principles and Design of Mechanical Face Seals [M]. 1991: 153-155.
11. Xinmin Z, X Yanqiu, Z Qing, F Xin. Analysis for Rotordynamic Coefficients of Partially Tapered Short Annular Seals [J]. Lubrication Engineering, 2004, 163(3): 71-75.
12. Etsion I, Pascovici M D. A thermodynamic analysis of a misaligned mechanical face seal [J]. Tribology Transactions, 1993, 36(3): 589-596.
13. Zhenmin H, W Qinhu, C Zhipeng, H Ye. Theory Research and Application of Using Temperature Control Method to Avoid the Nuclear Main Pump High Stops Due on the 1st Shaft Seal High Leakage [J]. Pump Technology. 2008, (6): 7-13.
14. Dongbo M, C Jining, Z Qiuxiang, L Shuangxi. Study on the Seal Property of the Second Seals of Reactor Coolant Pump in Nuclear Station from Contacting to Non-contacting Mechanical Seal [J]. Lubrication Engineering. 2009, 34(7): 33-37.
15. Doust T G, Parmar A. An Experimental and Theoretical Study of Pressure and Thermal Distortions in Mechanical Seals [J]//ASLE Trans, 1986, 29(2): 151-159.
16. Tournerie B, J Frêne. Influence of Fluid Flow Regime on Performances of Non-Contacting Liquid Face Seals [R]. 2002.

Analysis and Research on an on-line measuring device for machining accuracy of parts used in NC machine tools

Zhiqiang Wan, Qingqing Zhang

Anhui Technical College of Mechanical and Electrical Engineering, Wuhu, Anhui, China

Abstract: The utility model relates to an on-line detection device for the machining accuracy of parts of a NC machine tool, which comprises a storage flat box, a reciprocating driving mechanism, an on-line detection assembly and a trigger probe. Through the cooperation of the reciprocating driving mechanism and the storage flat box, the on-line detection assembly and the trigger probe can be switched between the storage and detection states. There is no need to disassemble the trigger probe and the cutter shaft of the NC machine tool during the detection, so as to improve the detection efficiency, the time required for the NC machine tool to recover from the detected state to the machining state is shortened.

Key words: Numerical control machine; testing; Machining accuracy

1. Introduction

High level machinery manufacturing is inseparable from high-precision machine tools and machining centers. CNC machine tools are the machining machine of manufacturing industry and an important foundation of the national economy. It provides equipment and means for all departments of the national economy.

Due to the inevitable errors of CNC machine tools, and the increase of errors will be caused by parts aging, mechanical vibration and other reasons, it is necessary to regularly detect the machining accuracy of CNC machine tools, especially when the parts have high requirements for machining accuracy, it is more necessary to shorten the detection cycle to ensure that the parts processed by CNC machine tools meet the requirements.

Generally, the machining accuracy detection device of NC machine tool is mainly the trigger probe replaced on the tool shaft, that is, when the machining accuracy detection is required, the tool on the tool shaft is removed and replaced by manual or automatic tool magazine, and the required tool is installed on the tool shaft after the detection is completed, which not only complicates the operation and reduces the detection efficiency, but also due to the inevitable error in each installation of the trigger probe, It will also increase the detection error.

The purpose of this paper is to provide an on-line detection device for the machining accuracy of parts of NC machine tools, so as to solve the technical problem of reducing the detection efficiency caused by the disassembly and assembly of the trigger probe during and after the machining accuracy detection in the existing technology[1].

2. Technical analysis and research

The utility model relates to an on-line detection device for machining accuracy of parts of a NC machine tool, which comprises a storage flat box, a reciprocating drive mechanism, an on-line detection assembly and a trigger probe. The storage flat box is installed on the side wall on one side of the operation window of the NC machine tool, a storage cavity penetrating one end is opened in the storage flat box, the on-line detection assembly is slidably installed in the storage cavity, and the reciprocating drive mechanism is installed on the storage flat box, The trigger probe is installed on the on-line detection assembly;

The online detection assembly comprises a lifting mechanism, a telescopic rocker arm, a steering mechanism and a pitch adjustment mechanism. The lifting mechanism is slidably arranged in the storage cavity and connected with the reciprocating drive mechanism. The outside of the lifting mechanism is rotationally connected with the tail end of the telescopic rocker arm through the rotating mechanism. The rotating mechanism is rotationally installed at the head end of the telescopic rocker arm, and the trigger probe is movably installed on the steering mechanism through the pitch adjustment mechanism, The rotary guide axis of the steering mechanism is perpendicular to the swing axis of the pitch adjustment mechanism. The rotary mechanism drives the telescopic rocker arm to switch between the horizontal detection state and the storage state aligned with the storage cavity. The trigger probe detects the machining accuracy of the parts processed by the NC machine tool from multiple directions under the adjustment of the steering mechanism and the pitch adjustment mechanism[2].

The lifting mechanism and the rotating mechanism are slidably installed in the storage cavity through the sliding shell, the outer wall of the sliding shell is fitted with the inner wall of the storage flat box, the reciprocating driving mechanism is connected with the sliding shell, the front of the sliding shell is provided with a vertical groove, and a guide shaft connecting the telescopic rocker arm and the rotating mechanism is installed in the vertical groove.

The reciprocating driving mechanism comprises a rack, a traveling stepping motor and a gear, the rack is arranged on the cavity wall of the storage cavity along the sliding direction of the sliding shell, the outer wall of the sliding shell is provided with a guide groove inserted and matched with the rack, and the traveling stepping motor is embedded and installed on the sliding shell and meshed with the rack.

The rack is provided with a plurality of, and the sliding shell is provided with a plurality of guide grooves and a plurality of progressive motors.

The telescopic rocker arm comprises a sleeve connecting the rotating mechanism, a hollow connecting rod sliding and mating with the sleeve, and a telescopic driving component for driving the hollow connecting rod to expand and contract along the sleeve. The length from the tail end of the hollow connecting rod in the sleeve to the trigger probe is less than the length of the sleeve, so that the trigger probe can be protected in the sleeve.

The telescopic drive assembly is a telescopic stepping motor and a lead screw, the central hole for installing the lead screw is opened at the axis of the hollow connecting rod, and the nut of the lead screw is connected with the hollow connecting rod.

The trigger probe is installed on the pitch adjustment mechanism through the plug-in end, and the trigger probe is electrically connected with the plug-in end.

The interior of the sleeve is a cylindrical cavity, the shape of the plug-in end is adapted to the internal contour of the sleeve, the inner wall of the open end of the sleeve is sliding and sealed with the side wall of the plug-in end, a plurality of guide holes through both ends are arranged in the wall of the plug-in end, and the intersection of the axes of the plurality of guide holes is located at the head of the reset trigger probe, so that the plug-in end is in the process of receiving the sleeve, The gas in the sleeve overflows from a plurality of guide holes to form an air flow for cleaning the head of the trigger probe.

A plurality of axially distributed annular grooves are arranged on the inner wall of the open end of the sleeve, a sealing ring is installed in the annular groove, and the side wall of the plug-in end is slidably sealed and matched with the inner wall of the sleeve through the sealing ring.

A baffle is installed on the outside of the telescopic rocker arm relative to the guide shaft, and the size of the baffle is larger than the size of the storage cavity[3].

3. Example given to illustrate

The utility model relates to an on-line detection device for machining accuracy of parts of a NC machine tool, which comprises a storage flat box 1, a reciprocating drive mechanism 2, an on-line detection assembly 3 and a trigger probe 4. The storage flat box 1 is installed on the side wall on one side of the operation window of the NC machine tool, a storage cavity 101 penetrating one end is opened in the storage flat box 1, the on-line detection assembly 3 is slidably installed in the storage cavity 101, and the reciprocating drive mechanism 2 is installed on the storage flat box 1, The trigger probe 4 is installed on the on-line detection assembly 3

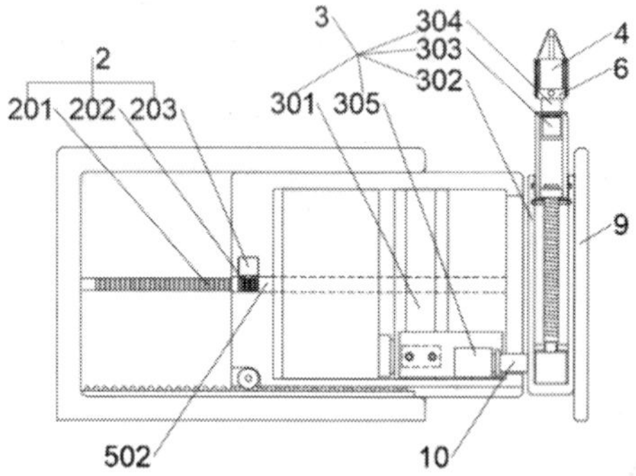

Fig. 1 Overall structure diagram

1 - storage of flat box; 2 - reciprocating drive mechanism; 3 - online detection assembly; 4 - trigger probe; 5 - sliding shell; 6 - plug end; 7 - annular groove; 8 - sealing ring; 9 - baffle; 10 - guide shaft;101 - storage chamber;201 rack; 202 - traveling stepping motor; 203 - gear;
301 - lifting mechanism; 302 - telescopic rocker arm; 303 steering mechanism; 304 - pitch adjustment mechanism; 305 - rotating mechanism;3021 sleeve; 3022 hollow connecting rod; 3023 telescopic stepping motor; 3024 lead screw;501 - vertical groove; 502 - guide groove; 601 - diversion hole.

The online detection assembly 3 includes a lifting mechanism 301, a telescopic rocker arm 302, a steering mechanism 303 and a pitch adjustment mechanism 304. The lifting mechanism 301 is slidably arranged in the storage cavity 101 and connected with the reciprocating drive mechanism 2. The outside of the lifting mechanism 301 is rotationally connected with the tail end of the telescopic rocker arm 302 through the rotating mechanism 305, and the steering mechanism 303 is rotationally installed at the head end of the telescopic rocker arm 302, The trigger probe 4 is movably installed on the steering mechanism 303 through the pitch adjustment mechanism 304. The rotation guide axis of the steering mechanism 303 is perpendicular to the swing axis of the pitch adjustment mechanism 304. The rotation mechanism 305 drives the telescopic rocker arm 302 to switch between the horizontal detection state and the storage state aligned with the storage cavity 101, The trigger probe 4 detects the machining accuracy of the parts processed by the NC machine tool from multiple directions under the adjustment of the steering mechanism 303 and the pitch adjustment mechanism 304.

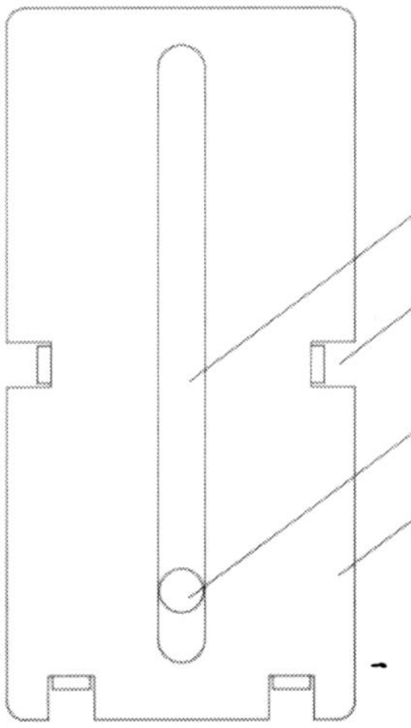

Fig. 2 Structural diagram of sliding shell

The baffle plate 11 is connected with the The function of the reciprocating driving mechanism 2 is to adjust the position of the trigger probe 4 in the horizontal length direction when the telescopic rocker arm 302 is in the detection state, and when the telescopic rocker arm 302 is in the storage state, the reciprocating driving mechanism 2 drives the online detection assembly 3 into the storage cavity 101 as a whole, so as to facilitate the cleaning and protection of the online detection assembly 3 and the trigger probe 4, so as to maintain the repeated detection accuracy of the online detection device, And avoid the negative impact of the exposed on-line detection assembly 3 on the operation of the NC machine tool, and control the trigger probe 4 through the on-line detection assembly 3 that can be stored to regularly detect the machining accuracy of the parts to be tested processed by the NC machine tool, and avoid installing the trigger probe 4 on the cutter shaft of the NC machine tool and driving the trigger probe 4 through the cutter shaft to detect the parts to be tested, There are disadvantages that the detection efficiency and the machining efficiency of NC machine tool are affected due to the disassembly and assembly of trigger probe 4, and the detection error is increased due to the installation error of trigger probe 4.

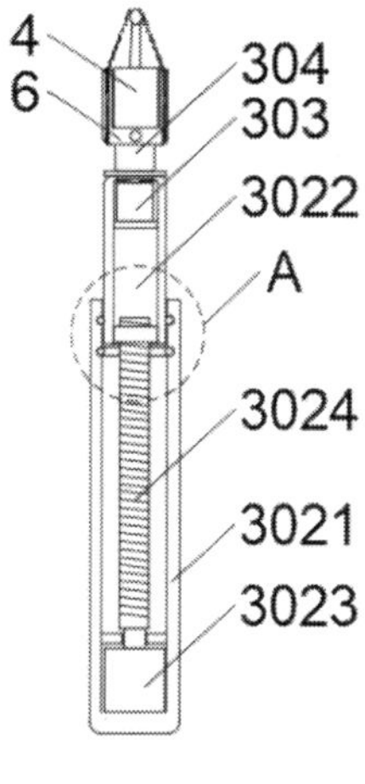

Fig. 3 Structural diagram of telescopic rocker arm

Among them, the function of the lifting mechanism 301 is to adjust the height of the trigger probe 4, the function of the telescopic rocker arm 302 is to adjust the distance between the trigger probe 4 and the surface of the part to be measured on the NC machine tool, so that the trigger probe 4 moves on the surface of multiple directions of the part to be measured and contacts the surface of the part to be measured for detection, and the function of the pitch adjustment mechanism 304 is to adjust the angle of the trigger probe 4 relative to the axis of the telescopic rocker arm 302, for example, The trigger probe 4, which is originally on the same straight line as the telescopic rocker arm 302, rotates 90 ° under the driving trend of the rocker arm, and detects the upper surface of the part to be measured under the adjustment of the lifting mechanism 301 and the telescopic rocker arm 302. The function of the steering mechanism 303 is to drive the trigger probe 4 to rotate, so that the trigger probe 4 can adjust the angle to any direction under the adjustment of the steering mechanism 303 and the pitch adjustment mechanism 304, To adapt to the detection of parts to be tested with complex geometric structure.

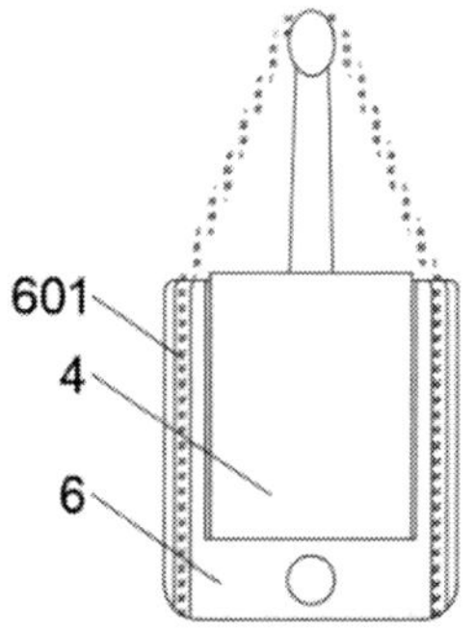

Fig. 4 Structural diagram of plug-in terminal

The device switches the on-line detection assembly 3 and the trigger probe 4 between the storage and detection states through the cooperation of the reciprocating drive mechanism 2 and the storage flat box 1. The lifting mechanism 301, the telescopic rocker arm 302, the steering mechanism 303 and the pitch adjustment mechanism 304 of the on-line detection assembly 3 cooperate with each other to realize the detection of the surface of the parts to be measured by the trigger probe 4 in multiple directions, There is no need to disassemble the trigger probe 4 and the cutter shaft of the NC machine tool during detection, which improves the detection efficiency, shortens the time required for the NC machine tool to recover from the detected to the processing state, and avoids the disadvantage of increasing the detection error due to the installation error of the trigger probe 4 every time. The on-line detection assembly 3 and the trigger probe 4 can be stored in the storage flat box 1 without disturbing the operation of the NC machine tool.

In the above implementation process, it is further optimized that the lifting mechanism 301 and the rotating mechanism 305 are slidably installed in the storage cavity 101 through the sliding shell 5, the outer wall of the sliding shell 5 is fitted with the inner wall of the storage flat box 1, the reciprocating driving mechanism 2 is connected with the sliding shell 5, and the setting of the sliding hole shell is conducive to improving the stability and positioning accuracy of the reciprocating movement of the lifting mechanism 301. In addition, the front of the sliding shell 5 is provided with a vertical groove 501, A guide shaft 10 connecting the telescopic rocker arm 302 and the rotating mechanism 305 is installed in the vertical groove 501.

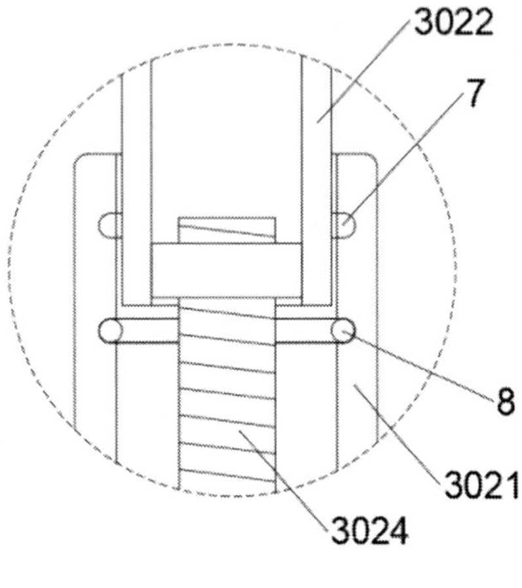

Fig. 5 Structural diagram of part a in Fig3

The reciprocating driving mechanism 2 includes a rack 201, a traveling stepping motor 202 and a gear 203. The rack 201 is arranged on the cavity wall of the storage cavity 101 along the sliding direction of the sliding housing 5. The outer wall of the sliding housing 5 is provided with a guide groove 502 inserted and matched with the rack 201. The traveling stepping motor 202 is embedded and installed on the sliding housing 5 and engaged with the rack 201.

The arrangement of the guide groove 502 enables the sliding housing 5 to move stably along the rack 201, and the rack 201 arranged on the side of the storage cavity 101 or on the ground cooperates with the travel stepping motor 202 arranged at the corresponding position on the sliding housing 5 to form a distributed structure of the reciprocating drive mechanism 2. While realizing the reciprocating drive function of the reciprocating drive mechanism 2, the disadvantage of increasing the length of the storage flat box 1 caused by the setting of the reciprocating drive mechanism 2 is avoided, In order to facilitate the good combination of on-line detection device and NC machine tool.

Moreover, in order to improve the stability of the reciprocating motion of the sliding housing 5, a plurality of racks 201 are arranged on the sliding housing 5, and a plurality of guide grooves 502 and a plurality of progressive motors 202 are arranged on the sliding housing 5. For example, racks 201 are arranged on both sides of the bottom cavity wall of the storage cavity 101 and on both sides of the cavity wall of the storage cavity 101, and the racks 201 on both sides of the bottom stably support the sliding housing 5, The two racks 201 on the side can support the upper half of the sliding housing 5. When the front half of the sliding housing 5 extends out of the storage cavity 101, the two racks 201 on the side can inhibit the rear half of the sliding housing 5 from tilting upward.

The telescopic rocker arm 302 includes a sleeve 3021 connecting the rotating mechanism 305, a hollow connecting rod 3022 sliding and mating with the sleeve 3021, and a telescopic driving component for driving the hollow connecting rod 3022 to expand and contract along the sleeve 3021. The length from the tail end of the hollow connecting rod 3022 in the sleeve 3021 to the trigger probe 4 is less than the length of the sleeve 3021, so that the trigger probe 4 can be protected in the sleeve 3021. The steering mechanism 303 and the reset pitch adjustment mechanism 304 meet the conditions of being included in the sleeve 3021. It is possible to adjust the structure of the steering mechanism 303 and the pitch adjustment mechanism 304, increase the internal cross-sectional area of the sleeve 3021 or change the internal shape of the sleeve 3021.

The telescopic drive assembly is any device or component with reciprocating drive function, which is conducive to the air pump and the piston installed on the hollow connecting rod 3022, that is, the piston drives the hollow connecting rod 3022 to telescopic action by adjusting the air pressure in the sleeve 3021 through the air pump, but the accuracy is affected by ambient temperature, piston aging, etc. Therefore, the telescopic drive assembly preferably adopts the telescopic stepping motor 3023 and the lead screw 3024. The central hole for installing the lead screw 3024 is opened at the axis of the hollow connecting rod 3022, and the nut of the lead screw 3024 is connected with the hollow connecting rod 3022, or compared with the matching mode of the telescopic stepping motor 3023 and the lead screw 3024

The trigger probe 4 is installed on the pitch adjustment mechanism 304 through the plug-in end 6. The trigger probe 4 is electrically connected with the plug-in end 6 to facilitate the disassembly and assembly of the trigger probe 4.

In addition, the interior of the sleeve 3021 is a cylindrical cavity, the shape of the plug-in end 6 is adapted to the internal contour of the sleeve 3021, and at least the inner wall of the open end of the sleeve 3021 is sliding and

sealed with the side wall of the plug-in end 6. A plurality of guide holes 601 penetrating both ends are arranged in the wall of the plug-in end 6, and the intersection of the axes of the plurality of guide holes 601 is located at the head of the reset trigger probe 4, so that the plug-in end 6 can be inserted into the sleeve 3021, The gas in the sleeve 3021 overflows from a plurality of guide holes 601 to form an air flow for cleaning the head of the trigger probe 4, so that the trigger probe 4 can clean the head by air injection after each detection, so as to avoid the debris on the surface of the part to be tested sticking to the head of the trigger probe 4 and affecting the accuracy of repeated detection of the trigger probe 4.
The inner wall of the open end of the sleeve 3021 is provided with a plurality of axially distributed annular grooves 7, in which a sealing ring 8 is installed. The side wall of the plug-in end 6 and the inner wall of the sleeve 3021 are slidably sealed and matched through the sealing ring 8. Compared with setting the sealing ring 8 on the side of the plug-in end 6, setting the sealing ring 8 on the inner wall of the sleeve 3021 is conducive to keeping the sealing ring 8 clean.
In addition, a baffle 9 is installed on the outside of the telescopic rocker arm 302 relative to the guide shaft 10, and the size of the baffle 9 is larger than the size of the storage cavity 101, so that when the telescopic rocker arm 302 is collected into the storage cavity 101, the baffle 9 on the outside can close the storage cavity 101, so as to further prevent the on-line detection assembly 3 and the trigger probe 4 from being polluted by dust, water vapor and oil in the environment, The service life and detection accuracy of the on-line detection device are negatively affected[4].

4. Conclusion

The device switches the on-line detection assembly and the trigger probe between the storage and detection states through the cooperation of the reciprocating drive mechanism and the storage flat box, while the lifting mechanism, telescopic rocker arm, steering mechanism and pitch adjustment mechanism of the on-line detection assembly cooperate with each other to realize the detection of the trigger probe from the surface of the parts to be measured in multiple directions, without disassembling the trigger probe and the cutter shaft of the NC machine tool during the detection, While improving the detection efficiency, the time required for the NC machine tool to recover from the detected state to the machining state is shortened.
The labels in the article figure are shown as follows:
1 - storage of flat box; 2 - reciprocating drive mechanism; 3 - online detection assembly; 4 - trigger probe; 5 - sliding shell; 6 - plug end; 7 - annular groove; 8 - sealing ring; 9 - baffle; 10 - guide shaft;101 - storage chamber;201 rack; 202 - traveling stepping motor; 203 - gear;
301 - lifting mechanism; 302 - telescopic rocker arm; 303 steering mechanism; 304 - pitch adjustment mechanism; 305 - rotating mechanism;3021 sleeve; 3022 hollow connecting rod; 3023 telescopic stepping motor; 3024 lead screw;501 - vertical groove; 502 - guide groove; 601 - diversion hole.

Acknowledgements

Supported by the scientific research project of Anhui Provincial Department of Education (KJ2021A1512, KJ2021A1515).

References

1. Research on process optimization and fixture design of thin-walled sleeve parts [J] Zhu Ridong, Peiyan, Gao Jun Papermaking equipment and materials 2021(05)
2. Design and implementation of automatic fixture of NC machine tool for connecting rod parts based on PMC [J] Wei Jianjun, Liu Keming, Li paixia, Huang Linsheng Machine tools and hydraulics 2020(02)
3. Application of multi part clamping in engine parts processing [J] Yang Feng, Zheng chaozhuan, Kang Zhaojun Engineering technology research 2018(05)
4. NC machining process analysis of rim and fixture design of machining center [J] Zhang Guozheng, Zhou Yuanzhi Machine tools and hydraulics 2014(14)

Design and research of a conveying device for metal parts processing

Zhiqiang Wan, Qingqing Zhang, Qianbin Zhang

Anhui Technical College of Mechanical and Electrical Engineering, Wuhu, Anhui, China

Abstract: The utility model relates to a conveying device for processing metal accessories with vibration type chip removal, which comprises an outer frame body, a conveying roller, a support plate, a conveying belt and a sprocket mechanism. The left and right sides of the bottom surface of the outer frame body are fixed with a fixed sleeve, and the internal screw of the fixed sleeve is installed with a second return spring, the top screw of the second return spring is fixed with a support rod, and the top screw of the support rod is fixed with an inner frame body, In addition, the left and right ends of the inner frame main body are internally connected with a conveying roller, the outer side of the conveying roller is fitted with a conveying belt, the inner wall of the upper surface of the conveying belt is fitted with a support plate, and the front and rear ends of the support plate are connected with the screws of the inner frame main body. The conveying device for processing metal parts with vibration type chip removal is provided with a cross bar and an eccentric wheel, so that the metal parts on the conveyor belt can be transported while vibrating, so that the debris and impurities on the surface of the metal parts can be well vibrated off.

Keywords: Vibration; Chip removal; Metal fittings; Conveyor

1. Introduction

Metal accessories refer to parts made of metal. Metal accessories need to be processed in such processes as chip cutting and grinding. In this process, conveying devices are required for conveying. Although there are many kinds of conveying devices on the market, there are still some shortcomings, such as:

The traditional conveying device only conveys the metal parts. After a series of processing, some debris and impurities will adhere to the surface of the metal parts. During the conveying process, the debris and impurities on the surface of the metal parts cannot be removed well, affecting the later processing process. At the same time, the conveying device has a single function and cannot be used well.

2. Technical Design Scheme

The utility model relates to a conveying device for processing metal accessories with vibration type chip removal, which comprises an outer frame body, a conveying roller, a support plate, a conveying belt and a sprocket mechanism. The left and right sides of the bottom surface of the outer frame body are fixed with a fixed sleeve, the internal screw of the fixed sleeve is installed with a second return spring, the top screw of the second return spring is fixed with a support rod, and the top screw of the support rod is fixed with an inner frame body, The left and right ends of the inner frame main body are internally connected with a conveying roller, and the outer side of the conveying roller is fitted with a conveying belt, the inner wall of the upper surface of the conveying belt is fitted with a support plate, and the front and rear ends of the support plate are connected with the screws of the inner frame main body, and the upper surface of the conveying belt is fitted with a brush, the upper screw of the brush is installed with a connecting plate, and the rear side of the connecting plate is provided with a rack, And the upper part of the rack is meshed and connected with a half gear, the internal key of the half gear is connected with a fixed rod, and the rear end of the fixed rod is connected with the outer frame main body bearing, and the outer side of the rear end of the fixed rod is connected with the cross bar through a sprocket mechanism. The inner side wall of the rear side of the outer frame main body is provided with a groove, and the interior of the groove is provided with a fixed rod and a rack, and the front and rear ends of the cross bar are connected with the outer frame main body bearing, The rear end of the cross bar penetrates the outer frame body, the middle outer key of the cross bar is connected with an eccentric wheel, and the top end of the eccentric wheel fits with the bottom surface of the inner frame body.

Preferably, the front side inner wall of the outer frame body is provided with a clamping groove, the rear side inner wall of the outer frame body is provided with a chute,

https://doi.org/10.1051/matecconf/202236301021

and the internal screw of the chute is fixed with a first return spring.

Preferably, the support rod is arranged in an inverted "t" shape structure, the support rod and the fixed sleeve are arranged in one-to-one correspondence, and the support rod forms a lifting structure with the fixed sleeve through the second return spring.

Preferably, the eccentric wheel is in intermittent contact with the inner frame body through the cross bar, and the inner frame body forms a lifting structure with the outer frame body through the eccentric wheel.

Preferably, the bottom screw of the rack is fixed with a sliding block, the lower outer side of the sliding block is connected with a sliding groove, and the left screw of the sliding block is fixed with a first return spring.

Preferably, the front-end screw of the connecting plate is fixed with a clamp block, the outside of the clamp block is clamped and sliding connected with a clamp slot, the rear-end screw of the connecting plate is fixed with a connecting block, and the rear end of the connecting block is connected with the rack through screws.

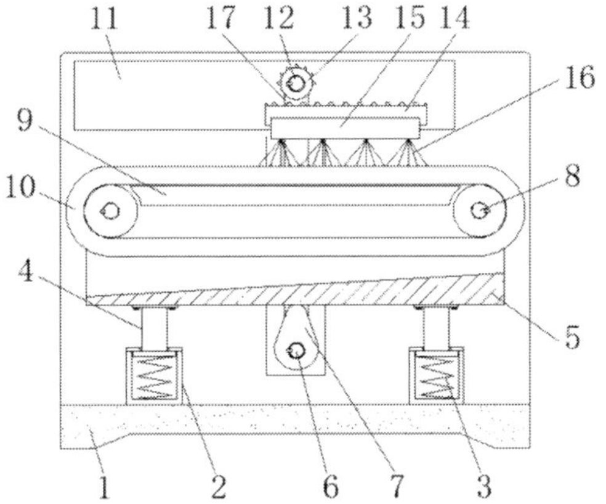

Fig. 1 Schematic diagram of main sectional structure of the device

1. Outer frame body; 101. Card slot; 102. Chute; 103. The first return spring; 2. Fixing sleeve; 3. A second return spring; 4. Support rod; 5. Inner frame body; 6. Cross bar; 7. Eccentric wheel; 8. Conveying roller; 9. Support plate; 10. Conveyor belt; 11. Grooves; 12. Fixed rod; 13. Half gear; 14. Rack; 141. Sliding block; 15. Connecting plate; 151. Clamping block; 152. Connecting block; 16. Brush; 17. Sprocket mechanism.

3. Example given to illustrate

The utility model relates to a conveying device for processing metal accessories with vibratory chip removal, which comprises an outer frame body 1, a clamping groove 101, a sliding groove 102, a first return spring 103, a fixed sleeve 2, a second return spring 3, a support rod 4, an inner frame body 5, a cross bar 6, an eccentric wheel 7, a conveying roller 8, a support plate 9, a conveying belt 10, a groove 11, a fixed rod 12, a half gear 13, a rack 14, a sliding block 141, a connecting plate 15, a clamping block 151, a connecting block 152, a brush 16 and a sprocket mechanism 17, The left and right sides of the bottom surface of the outer frame main body 1 are fixed with a fixed sleeve 2 by screws, the internal screws of the fixed sleeve 2 are installed with a second return spring 3, the top screws of the second return spring 3 are fixed with a support rod 4, the top screws of the support rod 4 are fixed with an inner frame main body 5, the left and right ends of the inner frame main body 5 are connected with a conveying roller 8, and the outside of the conveying roller 8 is fitted with a conveying belt 10, The inner wall of the upper surface of the conveyor belt 10 is fitted with a support plate 9, and the front and rear ends of the support plate 9 are connected with the screws of the inner frame body 5, and the upper surface of the conveyor belt 10 is fitted with a brush 16, the upper screw of the brush 16 is installed with a connecting plate 15, and the rear side of the connecting plate 15 is provided with a rack 14, the upper part of the rack 14 is meshed with a half gear 13, and the internal key of the half gear 13 is connected with a fixed rod 12, The rear end of the fixed rod 12 is connected with the bearing of the outer frame main body 1, and the outside of the rear end of the fixed rod 12 is connected with the cross bar 6 through the sprocket mechanism 17. The inner side wall of the rear side of the outer frame main body 1 is provided with a groove 11, and the inside of the groove 11 is provided with a fixed rod 12 and a rack 14. The front and rear ends of the cross bar 6 are connected with the bearing of the outer frame main body 1, and the rear end of the cross bar 6 passes through the outer frame main body 1, and the middle outer key of the cross bar 6 is connected with an eccentric wheel 7, The top end of the eccentric wheel 7 fits with the bottom surface of the inner frame body 5;

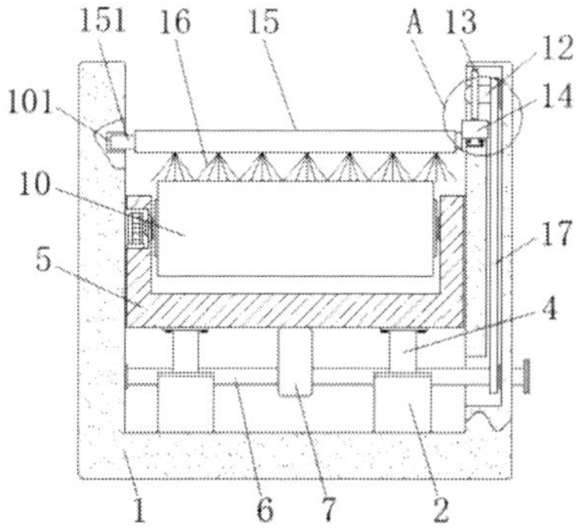

Fig. 2. Right view structure diagram

The inner wall of the front side of the outer frame main body 1 is provided with a clamping groove 101, and the inner wall of the rear side of the outer frame main body 1 is provided with a sliding groove 102, and the internal screw of the sliding groove 102 is fixed with a first return spring 103, so as to facilitate the clamping and sliding of the clamping block 151 through the clamping groove 101, so as to facilitate the stable movement of the connecting plate 15;

https://doi.org/10.1051/matecconf/202236301021

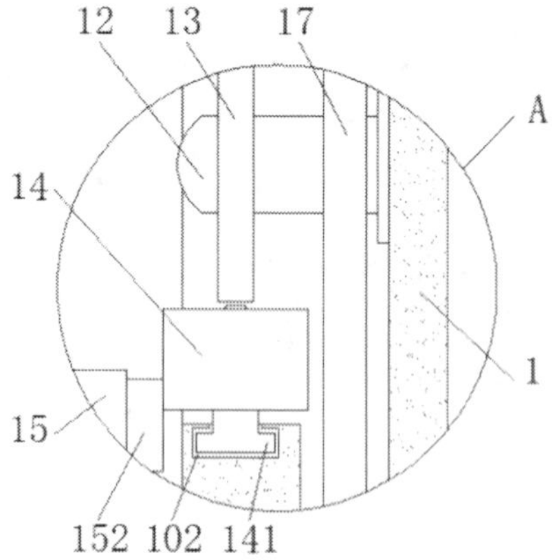

Fig. 3. Figure 2 enlarged structure diagram at a

The support rod 4 is arranged in an inverted "t" shape structure, and the support rod 4 is arranged in one-to-one correspondence with the fixed sleeve 2, and the support rod 4 forms a lifting structure with the fixed sleeve 2 through the second return spring 3. The support rod 4 is arranged in an inverted "t" shape structure, so that the support rod 4 can stably carry out lifting work in the fixed sleeve 2;

The eccentric wheel 7 is in intermittent contact with the inner frame main body 5 through the cross bar 6, and the inner frame main body 5 forms a lifting structure with the outer frame main body 1 through the eccentric wheel 7. The eccentric wheel 7 is in intermittent contact with the inner frame main body 5 through the cross bar 6, so as to facilitate the intermittent lifting of the inner frame main body 5;

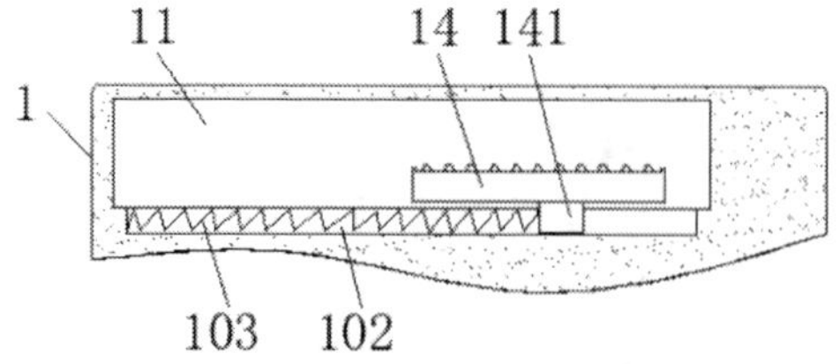

Fig. 4. Main sectional structure diagram of rack and outer frame main body connection

The bottom screw of the rack 14 is fixed with a sliding block 141, the lower outer side of the sliding block 141 is connected with a sliding groove 102, and the left screw of the sliding block 141 is fixed with a first return spring 103, which facilitates the stable movement of the rack 14 through the sliding of the sliding block 141 in the sliding groove 102;

The front-end screw of the connecting plate 15 is fixed with a clamp block 151, and the outside of the clamp block 151 is connected with a clamp slot 101 by clamping and sliding, and the rear-end screw of the connecting plate 15 is fixed with a connecting block 152. At the same time, the rear-end of the connecting block 152 is connected with the rack 14 through screws, so that the connecting plate 15 can well drive the brush 16 to move back and forth.

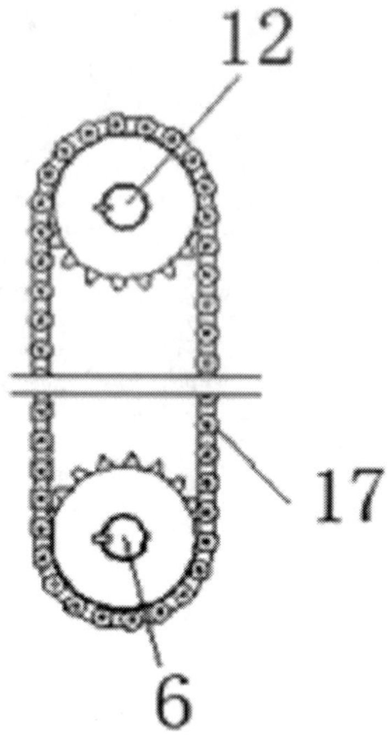

Fig. 5. Main view structure diagram of sprocket mechanism

4. Conclusion

Compared with the prior art, the innovation of a conveying device for metal parts processing with vibration chip removal is as follows:

(1) A cross bar and an eccentric wheel are arranged, and the rotation of the cross bar drives the eccentric wheel to rotate, so that the eccentric wheel drives the inner frame main body to rise and fall back and forth, and then the inner frame main body drives the inner conveyor belt to rise and fall back and forth, so that the metal accessories on the conveyor belt can be transported while vibrating, so that the debris and impurities on the surface of the metal accessories can be well shaken off;

(2) A fixed rod is installed. The rotation of the cross bar drives the fixed rod to rotate through the sprocket mechanism, so that the fixed rod drives the outer half gear to rotate, so that the half gear intermittently drives the rack to move. At the same time, with the use of the first return spring, the tooth strip drives the connecting plate and brush to move left and right, so as to facilitate the left and right pushing of the metal accessories on the conveyor belt, The utility model is convenient to brush away the debris and impurities on the surface of metal fittings, and improves the efficiency of removing debris.

Acknowledgements

Supported by the scientific research project of Anhui Provincial Department of Education (KJ2021A1512, KJ2021A1515).

References

1. A galvanizing production process [p] Wangyi, Zhang Jian, Chengmin, tuxiaoqing, sunhuajun Chinese patent: cn111945091a, November 17, 2020
2. A workpiece surface grinding device with intermittent rotation for machining [p] Zhudongdong Chinese patent: cn211966953u, November 20, 2020
3. A fully automatic surface sandblasting device [p] Chenyuefu Chinese patent: cn111958499a, November 20, 2020
4. A method of coating zinc protective layer on workpiece surface [p] Chenzhiwei, Yu Jinzhang Chinese patent: cn111996483a, November 27, 2020
5. Deburring machine for workpiece surface treatment [p] Xuejiabin Chinese patent: cn212286994u, 2021-01-05
6. A surface leveling device for machining workpiece [p] Jiangtingdao Chinese patent: cn112192340a, 2021-01-08

Silicon/Needle Coke Composites as Efficient Anodes for Lithium Ion Batteries

Jing Xia [1,*], Shuxiang Li [1], Yuan Gao [1], Dakui Zhang [2], Qiongguang Li [3], Yanhong Wang [3]

[1]Ansteel Beijing Research Institute CO., LTD., Beijing 102200, China
[2]Angang Chemical Technology CO., LTD., Anshan 114000, China
[3] Institute of Process Engineering, Chinese Academy of Sciences, Beijing 100190, China

Abstract. In this research, a facile method was reported to prepare silicon/carbon composites by heating Si nanosheets and coal-based needle cokes in the assistance of binder glucopyranose. The microstructures and electrochemical performances of samples were analyzed. It was found that Si nanosheets adhered to needle cokes forming silicon/carbon composites. Compared with needle cokes, the composites showed higher capacity and initial coulombic efficiency. Also, they improved the cycle stability of silicon materials. The silicon/carbon anode had a reversible capacity of 381.8 mAh/g at a current density of 100 mA/g after 170 cycles. In our work, relatively inexpensive Si nanosheets and coal-based needle cokes with low price were employed as the silicon and carbon sources respectively. Therefore, this method provides a possible strategy to reduce costs of silicon/carbon anodes, accelerating their commercial applications.

1. Introduction

Lithium ion battery, one of the major rechargeable energy storage device, has dominated wide applications in electric vehicles and portable electronic devices.[1-5] To improve the energy density of lithium ion battery, the anode material with high specific capacity has always been a hot topic.[6, 7] Silicon with high specific capacity of 4200 mAh/g and low working potential is the most promising candidate for the next generation lithium ion battery anode.[8, 9] However, the huge volume change of silicon during the Li insertion/extraction processes hinders its practical application as the anode material.[10, 11] In order to suppress the volume expansion of silicon, preparation of silicon/carbon composites has been developed.[12, 13] Such composites are able to improve the electronic conductivity of electrodes and alleviate the volume expansion of silicon.[14, 15] To fabricate silicon/carbon anodes, various strategies have been reported, for example, synthesis of silicon/carbon microspheres through chemical vapor deposition, composite of silicon nanowire, graphene and carbon nanotube, coating of carbon on silicon nanoparticles, and preparation of silicon/carbon particles through chemical etching.[16, 17] However, the current methods are relatively costly[18-20] (such as expensive raw materials and instruments) and complicated[21-23] (such as magnesiothermic reduction and electrospray), which severely limits the commercial application of silicon/carbon anodes in lithium ion batteries. Thus, developing a low-cost and facile method to synthesize silicon/carbon anodes is critical. In this work, relatively inexpensive Si nanosheets and coal-based needle cokes were chosen as the raw materials. The silicon/carbon composite was prepared through simply heating of a mixture of Si nanosheet, needle coke and binder glucopyranose. Furthermore, the microstructures and electrochemical performances of the obtained composites were explored.

2. Experimental

2.1 Materials

The Glucopyranose and ethanol were purchased from Sinopharm Chemical Reagent Co., Ltd.. The coal-based needle coke was bought from Ansteel Group CO., LTD.. Its volatile, ash content, density and sulfur residue value were 5.2%, 0.01%, 1.41 g/cm^3 and 0.45%, respectively.

2.2 Preparation of Silicon/Carbon Composites

The needle coke (0.475 g), silicon nanosheet (0.025 g), and glucopyranose (0.05 g) were added into ethanol (80 mL), followed by stirring of this mixed solution at room temperature for 3 h. Then the mixed solution was dried at 100 °C for 12 h. Finally, the mixture was heated at a rate of 5 °C/min to 600 °C, and kept at this temperature for 3 h under the protection of nitrogen. The product was denoted as Si/Nc-600. The sample Si/Nc-900 was prepared at the same condition but the heating temperature was changed to 900 °C. For comparison, samples Nc-600 and Nc-900 were produced through heating of pristine needle cokes at 600 °C and 900 °C respectively, without addition of Si nanosheets.

* Corresponding author: xiaj0001@126.com

https://doi.org/10.1051/matecconf/202236301022

2.3 Characterization

The X-ray diffraction (XRD) and atomic force microscopy (AFM) were conducted by Bruker D8 Advance with a Cu K radiation source of 1.5405 Å and Park NX20, respectively. The scanning electron microscopy (SEM) and element mapping were conducted by Gemini 560. The coin cells (CR2025) were used to analyze the electrochemical performances of samples. The active material, Super P and sodium alginate were mixed with a weight ratio of 7:2:1 in water to prepare a slurry. The working electrode was prepared by spreading the above slurry onto copper foil and dried at 120 °C for 12 h. A solution of 1 M $LiPF_6$ in ethylene (EC) and dimethyl carbonate (DMC) (v:v=1:1) was used as the electrolyte, and the counter electrode was the lithium foil. The charge/discharge tests were carried out in a voltage range from 0.01 to 2 V at room temperature.

3. Results and Discussion

Figure 1a shows the SEM image of the coal-based needle coke. Its size ranges from 2-10 μm. The Si nanosheet has a flaky morphology with the length of several micrometers and the height of 2-4 nm (figure 1b,c). As shown in the XRD pattern of the Si nanosheet, strong peaks of $d_{(111)}$, $d_{(220)}$ and $d_{(311)}$ were observed at 28.48°, 47.32° and 56.14°, respectively, demonstrating the crystalline structure of Si nanosheet (figure 1d).

As shown in the initial discharge-charge curve of Si nanosheet, a long voltage plateau is observed below 0.45 V during the charging process, which is the extraction of lithium ions from Si (figure 2a). The charge capacity and coulombic efficiency of Si nanosheet are 1537 mAh/g and 72.4% in the first cycle at a current density of 100 mA/g, respectively (figure 2b). However, the reversible capacity of Si nanosheet fades dramatically to 282.1 mAh/g in the 50th cycle (figure 2b). After 170 cycles, it is only 56 mAh/g (figure 2b). Si nanosheet has high capacity but low cycle stability. The capacity fading may be caused by the volume expansion and pulverization of Si nanosheets.[13]

To improve the electrochemical performances of carbon materials and increase the cycle stability of silicon materials, the silicon/carbon anodes were prepared through heating of the mixtures of needle cokes and Si nanosheets at 600 °C and 900 °C, respectively. The obtained silicon/carbon composites were denoted as Si/Nc-600 and Si/Nc-900. Figure 3 presents the SEM images and elemental analysis of Si/Nc-600 and Si/Nc-900. Compared with the pristine needle coke, some small particles (about several hundred nanometers in size) are observed on the surfaces of Si/Nc-600 (figure 1a and 3a). The elemental analysis shows these small particles are Si (figure 3c,e). Similar results are observed in the sample Si/Nc-900 (figure 3b, d and f). The above results indicate that Si nanosheets are crushed to small particles in the stirring, and adhere on the surfaces of needle cokes by glucopyranose, forming the silicon/carbon composites.

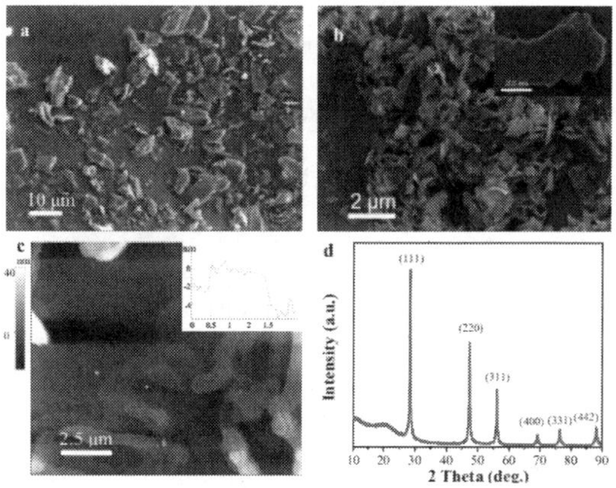

Figure 1. Typical SEM images of a) needle cokes, and b) Si nanosheets (inset: enlarged SEM image of Si nanosheet). c) The AFM image (inset: the corresponding height profile) and d) XRD pattern of Si nanosheets.

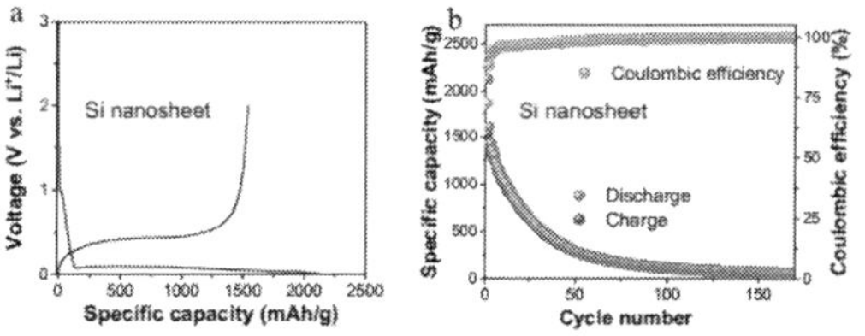

Figure 2. a) The initial discharge-charge curve and b) cycle performance and the corresponding coulombic efficiency of Si nanosheet. (Current density: 100 mA/g)

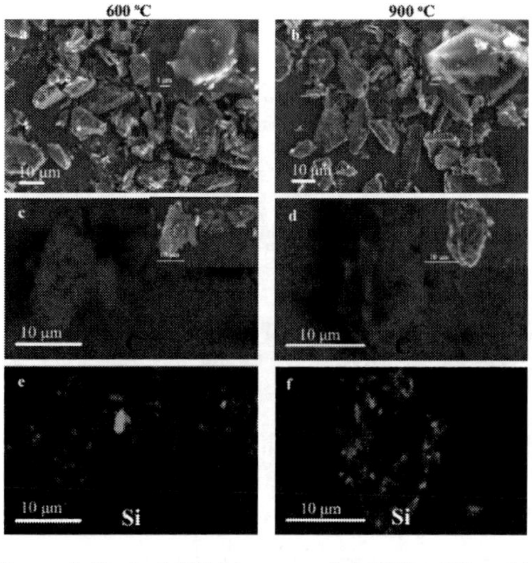

Figure 3. Typical SEM images of a) Si/Nc-600 and b) Si/Nc-900 (inset: enlarged SEM images). EDS elemental mapping images of c,e) Si/Nc-600 and d,f) Si/Nc-900 (inset: the corresponding SEM images).

Figure 4a shows the initial discharge-charge curve of the Nc-600, which was obtained through heating of needle cokes without addition of Si nanosheets at 600 °C. Its charge capacity and coulombic efficiency are 352.9

mAh/g and 46.2% in the first cycle at a current density of 100 mA/g, respectively (figure 4a). Compared with Nc-600, the initial charge capacity and initial coulombic efficiency of silicon/carbon composite Si/Nc-600 are increased to 411.9 mAh/g and 54.7%, respectively (figure 4c). After 170 cycles, the reversible capacity of Si/Nc-600 declines to 336.2 mAh/g, but it is higher than that of Si nanosheet (figure 2b and 4e). When the heating temperature was increased to 900 °C, similar phenomenon was observed. Both the initial charge capacity and initial coulombic efficiency of Si/Nc-900 (345.6 mAh/g, 66.6%) are higher than those of Nc-900 (302.5 mAh/g, 54.8%) (figur 4b,d). The reversible capacity of Si/Nc-900, which is 381.8 mAh/g, is also higher than that of Si nanosheet after 170 cycles (figure 2b and 4f). These results suggest that the addition of Si nanosheets could improve the initial charge capacity and initial coulombic efficiency of needle cokes, because of the high initial charge capacity of Si nanosheets. The composites of needle cokes and Si nanosheets may restrict the volume expansion of Si, leading to the improvement of the cycle stability.[13]

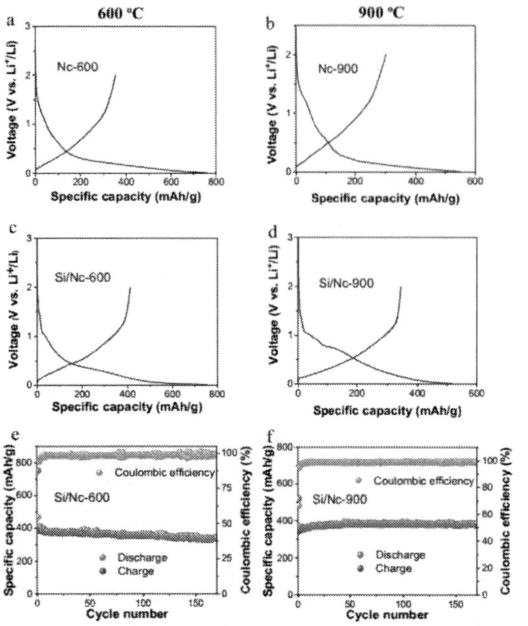

Figure 4. The initial discharge-charge curves of a) Nc-600, b) Nc-900, c) Si/Nc-600 and d) Si/Nc-900. Cycle performance and the corresponding coulombic efficiency of e) Si/Nc-600 and f) Si/Nc-900. (Current density: 100 mA/g)

In comparison with the Si/Nc-600, the Si/Nc-900 shows a little lower initial charge capacity but higher initial coulombic efficiency (table 1). This is because a higher heating temperature could cause the decrease of initial charge capacity and increase of initial coulombic efficiency of needle cokes, as reported in our previous work (table 1). After 170 cycles, the Si/Nc-900 (381.8 mAh/g) presents a better cycle stability than Si/Nc-600 (336.2 mAh/g, figure 4e,f). Maybe higher heating temperature benefits the combination of carbon and silicon, resulting in improvement of cycle stability. But the exact reason is still under research.

Table 1. The electrochemical performances of Nc-600, Nc-900, Si/Nc-600 and Si/Nc-900.

Sample	Initial charge capacity (mAh/g)	Initial coulombic efficiency (%)
Nc-600	352.9	46.2%
Nc-900	302.5	54.8%
Si/Nc-600	411.9	54.7%
Si/Nc-900	345.6	66.6%

4. Conclusion

In this work, a low-cost method was provided to prepare the silicon/carbon anode by using the Si nanosheet with relatively low price and cheap coal-based needle coke as the raw materials. After the stirring and heat treatment, crushed Si nanosheets adheres to the surfaces of needle cokes, forming the silicon/carbon composites Si/Nc-900. This composite structure helps to improve the capacity and initial coulombic efficiency of needle cokes. Meanwhile, it benefits the cycle stability of silicon/carbon anode. After 170 cycles, the reversible capacity of Si/Nc-900 keeps at 381.8 mAh/g. This simple and low-cost fabrication process makes Si nanosheet/needle coke anode a candidate for the practical application in the Li-ion battery.

Acknowledgments

The authors are very grateful for the financial support from the Ansteel Beijing Research Institute CO., LTD..

References

1. An, W., et al., Scalable Synthesis of Pore-Rich Si/C@C Core–Shell-Structured Microspheres for Practical Long-Life Lithium-Ion Battery Anodes. ACS Applied Materials & Interfaces, 2022. 14(8): p. 10308-10318.
2. Cao, X., et al., Preparation of Novel 3D Graphene Networks for Supercapacitor Applications. 2011. 7(22): p. 3163-3168.
3. Dayou, S., et al., Direct growth of graphene on MgO by chemical vapor deposition for thermal conductivity enhancement of phase change material. Materials Chemistry and Physics, 2017. 202: p. 352-357.
4. Chen, W., et al., Self-Assembly and Embedding of Nanoparticles by In Situ Reduced Graphene for Preparation of a 3D Graphene/Nanoparticle Aerogel. 2011. 23(47): p. 5679-5683.
5. Han, Q., et al., Durable, flexible self-standing hydrogel electrolytes enabling high-safety

rechargeable solid-state zinc metal batteries. Journal of Materials Chemistry A, 2018. 6(45): p. 23046-23054.

6. Jiang, C., et al., Flexible Interface Design for Stress Regulation of a Silicon Anode toward Highly Stable Dual-Ion Batteries. 2020. 32(17): p. 1908470.
7. Li, X., et al., Review on comprehending and enhancing the initial Coulombic efficiency of anode materials in lithium-ion/sodium-ion batteries. Nano Energy, 2020. 77: p. 105143.
8. Nulu, A., V. Nulu, and K.Y. Sohn, Si/SiOx Nanoparticles Embedded in a Conductive and Durable Carbon Nanoflake Matrix as an Efficient Anode for Lithium-Ion Batteries. 2020. 7(19): p. 4055-4065.
9. Zhou, X., et al., Synthesis of nano-sized silicon from natural halloysite clay and its high performance as anode for lithium-ion batteries. Journal of Power Sources, 2016. 324: p. 33-40.
10. Kasavajjula, U., C. Wang, and A.J. Appleby, Nano- and bulk-silicon-based insertion anodes for lithium-ion secondary cells. Journal of Power Sources, 2007. 163(2): p. 1003-1039.
11. McDowell, M.T., et al., 25th Anniversary Article: Understanding the Lithiation of Silicon and Other Alloying Anodes for Lithium-Ion Batteries. 2013. 25(36): p. 4966-4985.
12. Manthiram, A., A reflection on lithium-ion battery cathode chemistry. Nature Communications, 2020. 11(1): p. 1550.
13. Ren, W., et al., Preparation of porous silicon/carbon microspheres as high performance anode materials for lithium ion batteries. 2015. 3(11): p. 5859-5865.
14. Rümmeli, M.H., et al., Direct Low-Temperature Nanographene CVD Synthesis over a Dielectric Insulator. ACS Nano, 2010. 4(7): p. 4206-4210.
15. Salah, M., et al., Pure silicon thin-film anodes for lithium-ion batteries: A review. Journal of Power Sources, 2019. 414: p. 48-67.
16. Magasinski, A., et al., Erratum: High-performance lithium-ion anodes using a hierarchical bottom-up approach. Nature Materials, 2010. 9(5): p. 461-461.
17. Wang, B., et al., Intertwined Network of Si/C Nanocables and Carbon Nanotubes as Lithium-Ion Battery Anodes. ACS Applied Materials & Interfaces, 2013. 5(14): p. 6467-6472.
18. Chae, C., et al., A High-Energy Li-Ion Battery Using a Silicon-Based Anode and a Nano-Structured Layered Composite Cathode. 2014. 24(20): p. 3036-3042.
19. Nguyen, B.P.N., et al., Nanosilicon-Based Thick Negative Composite Electrodes for Lithium Batteries with Graphene as Conductive Additive. 2013. 3(10): p. 1351-1357.
20. Ren, J.-G., et al., Silicon–Graphene Composite Anodes for High-Energy Lithium Batteries. 2013. 1(1): p. 77-84.
21. Yu, Y., et al., Reversible Storage of Lithium in Silver-Coated Three-Dimensional Macroporous Silicon. 2010. 22(20): p. 2247-2250.
22. Yin, Y.-X., et al., Electrospray Synthesis of Silicon/Carbon Nanoporous Microspheres as Improved Anode Materials for Lithium-Ion Batteries. The Journal of Physical Chemistry C, 2011. 115(29): p. 14148-14154.
23. Wen, Z., et al., Rational design of carbon network cross-linked Si–SiC hollow nanosphere as anode of lithium-ion batteries. Nanoscale, 2014. 6(1): p. 342-351.

Influence of coal mining on the lower Yellow River region and related management countermeasures-- Taking Heze mining area as an example

Xinghua Li [1], Qi Zhang [1,*], Zhaoping Liu [2], Jianchi Yan [2], kai Zhang [2]

[1] Shandong University of Science and Technology, Qingdao, China
[2] Shandong Ding'an Testing Technology Co., Ltd, Jinan, China

Abstract: To realize the ecological protection and high-quality development of the Yellow River basin is related to the centenary project of the Chinese nation. The middle and lower reaches of the Yellow River are rich in coal resources and have a long history of mining. Large-scale coal mining activities lead to the destruction of underground aquifer, surface subsidence and the reduction of cultivated land, which seriously threatens the safety of local ecological environment, land resources and water resources. Luxi coal base Heze mining area, for example, the analysis of coal mining activities and land resources, ecological environment, the contradiction between water resources protection and its causes, and put forward preventive measures, to help alleviate the lower Yellow River coal mining and land resources, ecological environment, water resources protection, to realize luxi mining area and the lower Yellow River ecological environment protection and high quality development is of great significance.

Key words: lower Yellow River area; coal mining activities; soil and water protection; ecological protection; high-quality development

1. Heze mining area overview

Heze mining area is rich in coal resources, is one of the national key planning and construction of 14 large coal production base — an important part of coal base, the main distribution of giant wild coalfield, cao county coalfield, shan county coal field, accumulated proved coal reserves of 10.058 billion tons, accounting for about 21% of the proved coal reserves in Shandong province.Mining area mainly mining 3 coal seam, coal seam in average buried depth between 600m and 1200m, the mining area is a total of 7 mine production, a pair of the mine under construction, mining area in 2021, coal resources output of 18.1 million tons, accounting for 19.5% of coal production in Shandong province, Heze mining area safety production for the whole luxi coal base energy supply has an important impact.

2. Surface moving deformation mechanism

After coal extraction, the original stress balance of the goaf rock layer destroyed, the goaf rock layer in the surrounding rock pressure and its gravity, falling, fracture, bending, through the rocks, eventually to the surface, forming a nearly oval much larger than the goaf area sinking basin[1] .Affected by the fluctuation, destruction and bending of the overlying rock formation in the goaf, the hydrological environment and the surface of the underground rock formation covered by the goaf have also changed significantly. There are two main functions:
(1) Under the perennial alluvial of the Yellow River in the lower reaches of the Yellow River, the geological conditions characterized by huge thick loose layer and high diving level have been formed.Under such geological conditions, after coal extraction, the surface is obviously deformed by [2]. Affected by the high diving level in the quaternary loose layer, after the surface subsidence reaches a certain value, the water flows out, forming a contiguous coal mining subsidence water area, which seriously affects the use function of the land.
(2) Before the coal is extracted, the whole geological environment is in a relatively stable state, and the relative change range of the groundwater level is small.After coal extraction, the water in the upper aquifer under gravity, will penetrate down through the microcrack, with the erosion effect of the water flow, the original microcrack in water pressure, gravity, and water erosion, further expand, expand the water, destroy the original stability of the aquifer.See the details shown in Figure 1.

* Corresponding author: 3466692521@qq.com

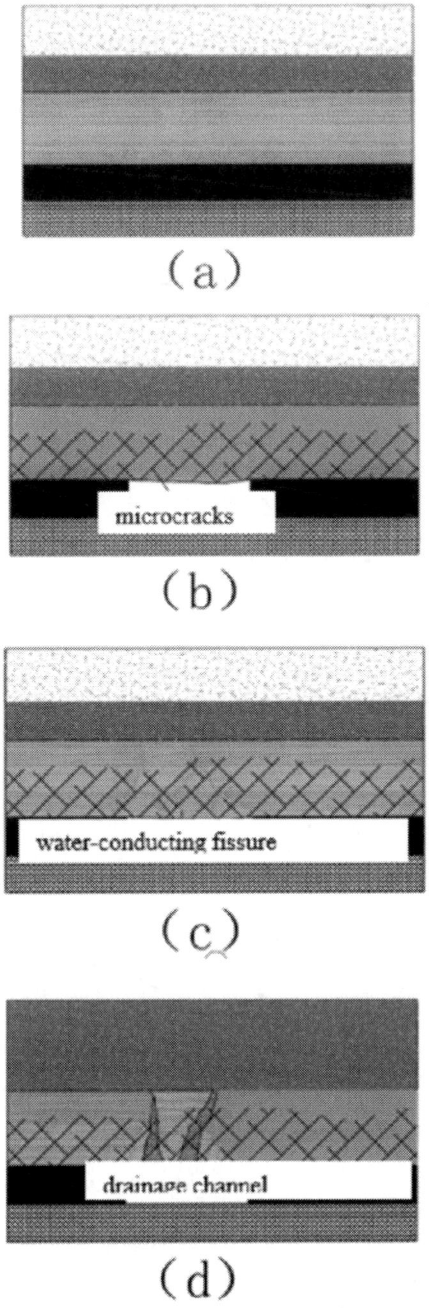

Figure 1 Schematic diagram of the conduction process of overcovering aquifer after coal seam mining

3. Problems caused by coal mining

3.1 Damage to cultivated land

According to relevant statistics, there are existing coal mining subsidence areas of 4301.7511 hectares of Heze mining area, and the Heze mining area is drawn according to the survey data.See the details shown in Figure 2.

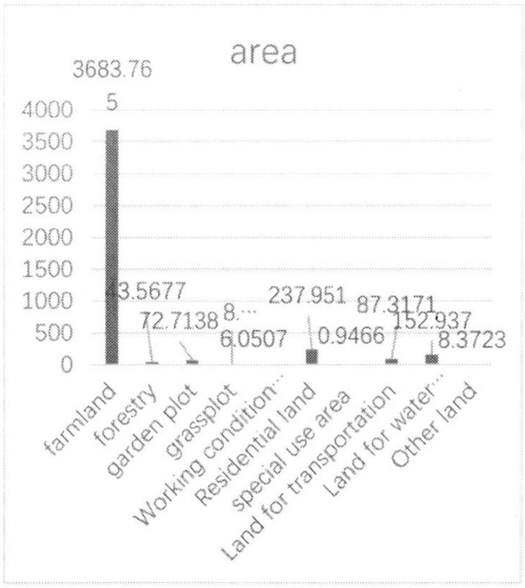

Figure 2 Map of various land damage in Heze mining area

According to FIG. 2, there are 10 land types of damaged cultivated land, residential land, water area and water conservancy facilities land, garden land, woodland, grassland, industrial and mining storage land, special land, transportation land, and other land.The maximum cultivated land area is 3683.7650 hectares, accounting for 85.63% of the total subsidence area; second, residential land area 237.9510 hectares, 5.53% of the subsidence area; water conservancy area 152.9370 hectares, 3.56% of the subsidence area, and 94.72% of the total subsidence area.

3.2 Ecological destruction

After coal extraction, large seasonal water areas and perennial water areas will be formed, which will not only destroy the original agricultural ecological balance, and seriously damage the original relatively stable soil structure and geological environment, but also threaten the biodiversity.

According to relevant statistics, at the end of 2021, the coal mining subsidence area was 810.5372 hectares, accounting for 18.84% of the total subsidence area, 302.4563 hectares, 7.03% of the total subsidence area, and 3188.7576 hectares, accounting for 74.13%.

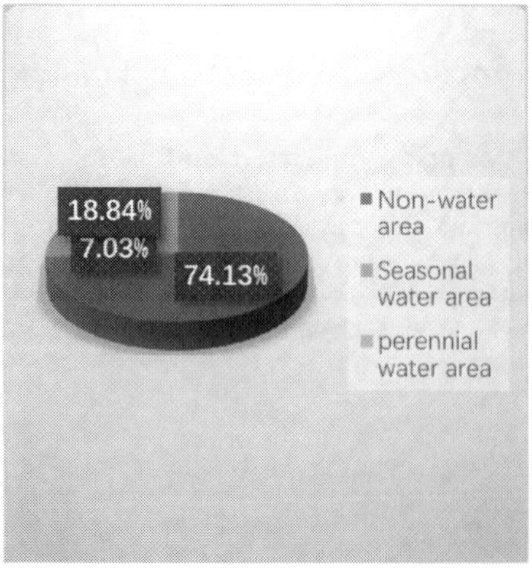

Figure 3 Distribution of coal mining subsidence land in Heze mining area

3.3 Waste of water resources

Heze mining area is mainly agricultural production, dense population, many towns, the demand for water resources is very huge.Large-scale coal mining activities have caused a series of water resources protection problems, and coal mining activities caused by water resources damage is mainly reflected in two aspects:

(1) Coal mining activities have seriously damaged the underground aquifer, leading to the reduction of the water level of the groundwater aquifer, and aggravated the dependence of the lower Yellow River region on surface water.

(2) Coal production activities will produce a large amount of waste water [3], dust and harmful gases, coal gangue, fly ash and other accumulated waste will be dissolved by rainwater will also pollute the channels, rivers and lakes.In addition, due to the decline of underground aquifer water level, agricultural irrigation and other production activities are very dependent on the irrigation of the Yellow River water resources[4-5], resulting in the utilization of water resources of the Yellow River is tight .

4. Prevention and governance measures

4.1 Land resource protection

"Very cherish and reasonable use of every inch of land, to protect cultivated land" is a basic state policy of our country, protect land resources must be based on the present, the "source control" and "process management" as the new concept of mining ecological environment management [2], based on accurate mining damage prediction, mining and ecological restoration, the organic combination of land reclamation .Drainage drainage, shallow excavation and filling method, and dynamic pre-reclamation method and active prevention method shall be adopted in the predicted areas.

4.2 Ecological and environmental protection

The ecological security of the Yellow River basin is not only related to the ecological security of the north, but also related to the ecological security of the whole country.Lucid waters and lush mountains are mountains of gold and silver, and a sound ecological environment is the most beneficial to people's livelihood.Coal mining activities destroy the ecological environment of the surface and lead to the deterioration of the lower reaches of the Yellow River.

4.3 Water resources protection

The advantages and wealth of water resources affect the social stability, economic development and people's happiness index of a region.While coal mining, it is necessary to avoid the pollution and damage of the underground aquifer, reduce blocking and cut water loss and pollution; adopt green coal mining process to reduce the possibility of cracks caused by surface deformation and prevent the surface water through the cracks of surface water leakage.

Water resources should be rationally distributed and utilized, rationally allocate water consumption for industrial, domestic and agricultural purposes, and ensure that the principle of "water, land, people and production by water" should be adhered to, and the overall distribution of water use in cities, towns and villages should be reasonably planned.

5. Conclusion and Outlook

The Yellow River is the mother river of China. Protecting the ecological environment of the Yellow River will benefit for generations to come.According to local conditions, reasonable remediation, adhere to the concept that clear water and green mountains are gold and silver mountains, broaden the concept of ecological governance; adhere to the source control, process governance, solve the problem of land resource damage in the development process; choose reasonable mining methods to reduce the impact of underground mining on the underground aquifer, reasonable distribution of water resources, to solve the problem of unreasonable distribution of water resources in the Yellow River basin.The coordinated development of coal mining activities and ecological protection, water resources protection and land resources protection plays a positive role in promoting the ecological protection and high-quality development in the lower Yellow River basin.

References

1. Chinese President Xi Jinping. Speech at the Yellow River Basin ecological protection and high-quality development symposium]J].China water conservancy, 2019(20) : 1 – 3 .
2. Hu Zhenqi, Yuan Dongzhu. Key technology and Strategic Thinking of ecological environment Restoration in Yellow River Basin]J] . Journal of Coal, 2020, 45(4) : 1211 – 1221 .

3. Wang Xiaotong, Hu Zhenqi, Liang Yusheng. Optimization of soil structure based on Hydrus-1D[J].Journal of Agricultural Engineering,2022,38(02):76-86.
4. Wang Shihu. Environmental Problems of Historical Mine under the Background of Ecological Civilization Construction[J]Mining safety and environmental protection.2018,45;No.242(06):88-91+96.
5. Sun Guang, Li Chengzhi, Dong Qinghuan, Li Xueliang. Practice and Thinking of comprehensive coal mining subsidence area in Rencheng District of Jining City[J].Mining safety and environmental protection, 2020,v.47;No.254(06):113-117+121.

Advances in water-resistant modification of aqueous acrylic resins: modification methods, mechanism of action

Jiajun Wei, Chunfa Ouyang, Xiaoqian Shan, Qun Gao, Kangsheng Zheng, Wei Luo

School of Shanghai Institute of Technology University, Shanghai, China

Abstract. Waterborne acrylic resin is widely available in architectural coatings, road adhesives, etc. Its non-toxicity, non-irritating odour, and low cost are characteristics that have received increasing attention in recent years. Concerning the water-resistance of the resin, this review introduces the effect of inorganic materials like silane coupling agents, fluorine silicone, and copolymers on the hydrophobic properties of aqueous acrylic resins, analyses and discusses the mechanism of action of various methods and their factors affecting hydrophobic properties, such as the formation of cross-linked structures and micro-convex structures on the surface, also points out the problems and challenges in the water-resistance modification of aqueous acrylic resins.

1. Introduction

Water-based resin is a resin system with water as a dispersion medium. Low volatile organic compounds (VOC) and good film-forming properties are outstanding advantages as functional materials cite coatings, adhesives, and sealants. [1-3]. Many water-based resins serve as epoxy resin, polyurethane, alkyd resin, acrylic resin, etc. [4-8]. In specific industrial applications, the cost needs to be comprehensively considered. And performance advantages to choose from. Waterborne acrylic resin is one of the fastest-growing resins. It is synthesized with acrylate or methacrylate as the main monomer [9]. It has intense light, thermal and chemical stability [10]. It is used as coatings for interior and exterior walls in the construction industry, adhesives for colored pavements in the road industry, etc. However, we found that water-based acrylic resins have defects from specific applications, including low water resistance, high film-forming temperature, and inadequate corrosion resistance. [9, 11, 12], which are limited in specific applications. Water-resistance refers to the resistance of the paint film to water. However, whether in the construction or road industry, water damage is a significant problem for water-based acrylic resins when used as coatings and adhesives. It is determined by the chemical structure of the resin, as proof of weak molecular polarity in the resin structure, the high surface energy of the film after film formation, insufficient self-crosslinking, etc. These factors will make it easier for water molecules to attack the resin coating. It increases its water absorption rate, thereby affecting its water resistance.

The water resistance of water-based acrylic resins is improved by two commonly used modification methods: physical modification and chemical modification. Physical transformation is to mechanically and physically blend two or more inorganic/organic substances with acrylic resin. It is described as convenient and straightforward, but the modification effect of resin is generally inadequate. Due to the system's poor compatibility to single physical blending, remarkable substances undergo nucleation and growth phase separation, showing an "island structure, " which affects the modification effect. However, the chemical method achieves molecular level compatibility between various substances through copolymerization or cross-linking between the modified importance and the acrylic resin by illustrating the formation of covalent bonds, grafting points, etc. The resin has outstanding advantages in performance, thus showing ethical application value in the follow-up.

This paper reviews different modifiers for water resistance modification of waterborne acrylic resins, namely silane coupling agents, fluorine and silicone substances, resin copolymerization, and analyses. It discusses the effects and characteristics of the different modification methods.

2. Water-based resin modification methods

2.1 Physical modification

Modification of aqueous resins is a standard modification method achieved by simple physical blending. To obtain a matte surface for acrylic resins, Ou et al. [13] prepared acrylic resin/micro-SiO_2/polymethylurea (AC/ SiO_2/PMU) composites by physically blending acrylic resin emulsions (AC), polymethylurea resins (PMU) and modified silica (SiO_2). This blended composite has increased interfacial tension between polymer and inorganic, committing phase separation in the AC/ SiO_2/PMU composite system,

* Corresponding author: ouyoung_0916@163.com

e.g., nucleation of SiO_2 particles on account of aggregation of particles through weak compatibility with the polymer. Aqueous acrylic coatings with a matte surface are obtained.

It is possible to see that physical modification has its unique advantages in obtaining certain specific properties of aqueous resins and can be used in synergy with chemical modification methods to achieve maximum transformation of the properties of aqueous resins.

2.2 Chemical modification

The chemical approach is to achieve molecular compatibility between the various substances utilizing copolymerization between the modifier and the aqueous resin, etc. Pablo J. Peruzzo et al. [14] We have synthesized polyurethane/acrylic resin composites with a cross-linked structure by emulsion polymerization of acrylic monomers with preformed polyurethane chains with polymerizable terminal vinyl—a polyurethane-acrylic resin blend dose preparation. The polyurethane/acrylic resin composite with a cross-linked structure performed significantly better than the polyurethane-acrylic resin blend. Physical or chemical modifications of aqueous resins have benefits and shortcomings.

Physical blending is convenient and straightforward. Chemical changes are stable and long-lasting and provide more significant performance improvements. In addition, waterborne resins have high water resistance requirements for coating applications, so it is crucial to prepare waterborne resins with high hydrophobicity by chemical modification. At the same time, it has excellent application value.

3. Water-resistant modification of water-based acrylic resins

3.1 Water-resistance modification of aqueous acrylic resins by silane coupling agents

Silane coupling agents can cross-link with aqueous acrylic resins or other substances to amend the chemical structure of the resin, thereby improving its water resistance. Jiao et al. [15] operate 2-(3,4-epoxy)ethyltriethoxysilane (KH567) to modify aqueous acrylic resins, and KH567 underwent a grafting reaction with the acrylic resin, changing the molecular chain structure of the resin. The migration of the KH567 side-chain structure in the aqueous acrylic resin to the film surface resulted in a reduction in the surface energy of the resin film [16], consequently improving its water resistance. However, excessive amounts of silane coupling agent can introduce more -OH groups, which worsens the water resistance, and the dose should generally control between 10-15 wt%. Zheng et al. [17] utilized 2-butoxyethanol, cellulose acetate butyrate, and acrylic acid monomer as raw materials, N, N - dimethylethanolamine as a neutralizing agent, and γ-aminopropyltriethoxysilane (KH-550) as curing agent to prepare a new waterborne acrylic resin modified with cellulose acetate by homogeneous solution polymerization. A three-dimensional network structure was formed between KH-550 and the resin. Increased crosslink density and the addition of the curing agent reduced the surface tension, lowered the surface energy, and enlarged the contact angle [18].

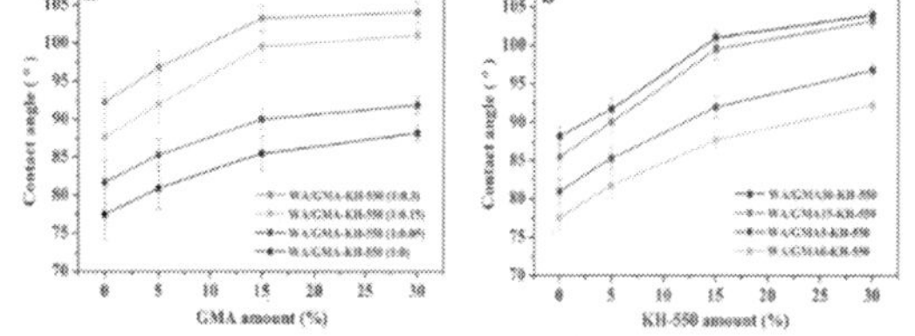

Figure 1. (a) Influence of the GMA amount on the contact angle for different amount of KH-550. (b) Influence of the KH-550 amount on the contact angle for different amount of GMA.

Guo et al. [19] applied aminopropyltriethoxysilane (KH-550) as a curing agent to cure the GMA-modified resin and then modified the aqueous acrylic resin with the GMA-modified resin. The data in Fig. 1 showed that the increase in both GMA and KH550 content brought forth a more extensive water contact angle of the resin film, which resulted from the cross-linking reaction between KH550 and the system, altering the physicochemical properties of the coating [20, 21]. Wu et al. [22] synthesized multi-walled carbon nanotubes (CNTs)/waterborne acrylic resin nanocomposite coatings using 3-aminopropyltriethoxysilane (KH-550) coupling agent as a curing agent by a solution mixing method. The synthesis mechanism in Fig. 2 and the contact angles between the CNTs/WA nanocomposite coatings with different CNTs loadings and water droplets in Fig. 3. As a result of the small particle size and large specific surface area of CNTs, they are not compatible with water droplets. The resin changes its internal structure after mixing [23]. And the surface energy of the coating of the water-based wax decreases. Hence, the contact angle of the coating shows an increasing trend with the increase of the CNTs content, but when the carbon nanotube content increases from 6wt% to 10wt%, The ordered dispersion state results in a slow rise in the contact angle. It can be deduced that when the carbon nanotube content continues to increase, the effect on the contact angle will tend to a stable value.

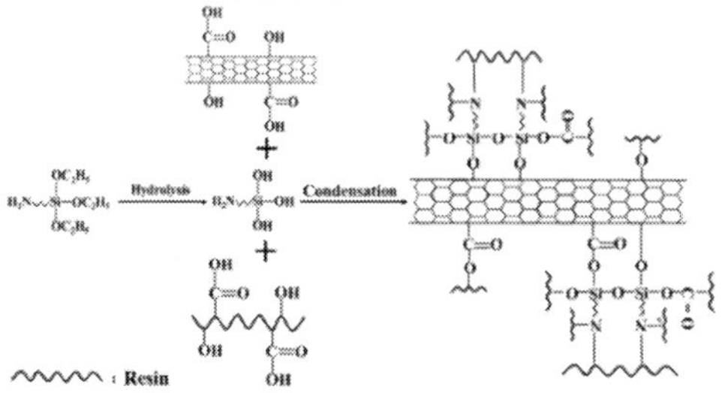

Figure 2. Mechanism of CNTs/WA nanocomposite synthesis.

The relationship between the contact angle and the equilibrium of the three-phase boundary forces does investigate by utilizing Young's equation [24]. Where γ_{sv}、γ_{sl}、γ_{lv} and θ represent the surface free energy, solid-liquid interfacial tension, liquid-vapor interfacial tension, and contact angle of the solid-vapor interface, respectively.

$$\gamma_{sv} - \gamma_{sl} = \gamma_{lv} cos\theta \quad (1)$$

The Newman equation of state can yield the surface energyn [25]:

$$\gamma_{sl} = \gamma_{lv} + \gamma_{sv} - 2\sqrt{\gamma_{lv}\gamma_{sv}} e^{-\beta(\gamma_{lv}-\gamma_{sv})^2} \quad (2)$$

Simultaneous (1) and (2):

$$\cos\theta = -1 + 2\sqrt{\gamma_{sv}/\gamma_{lv}} e^{-\beta(\gamma_{lv}-\gamma_{sv})^2} \quad (3)$$

Combined with the equations and the analysis of the test results, it is clear that the surface energy tends to decrease, and the hydrophobicity improves as the loading of carbon nanotubes increases. Nanocomposites containing 10wt% carbon nanotubes have the lowest surface energy (28 $mN \cdot m^{-1}$), much lower than some other resins [26-28]. Low surface energy coatings are widely used for antifouling based on their hydrophobicity and have potential applications for non-toxic antifouling coatings.

Aiming at the water resistance modification of water-based acrylic resin by different silane coupling agents and other modifiers, the change law of its water contact angle makes the comparison, as shown in Table 1. By comparing the three groups of KH567, KH550/GMA, and KH550/CNTs, It can be seen from the water contact angle data of the pure WA resin coating that the contact angle of the pure WA resin coating is above 80 before the modification and the contact angle increases after the conversion. The KH550/GMA and KH550/CNTs modified WA resins. The rise in the contact angle of the coating film is above 10, while that of KH567 is relatively weak. In addition, the KH550/CAB modified system has a relatively significant change in the contact angle of the WA resin coating film, with a growth rate of 21.4%. Still, the water contact angle itself is relatively low, 65.1° and 78.1° before and after modification, respectively, indicating that the system Compared with the other three scenarios, the hydrophobicity of the WA resin coating film is relatively weak. The hydrophobic type of the KH550/GMA modified system is the greatest.

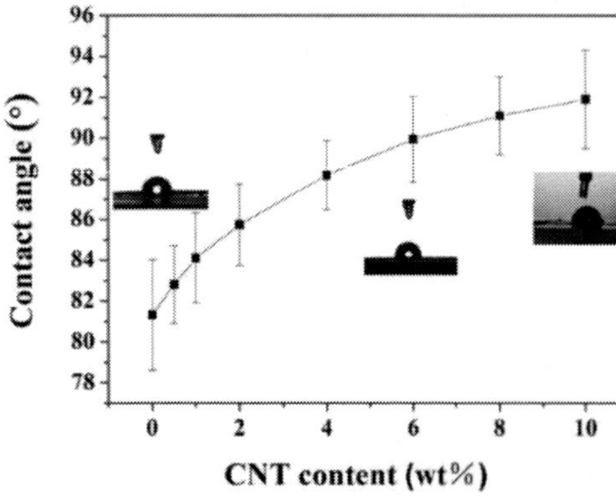

Figure 3. Influence of the CNT amount on the contact angle.

Table 1. The change of WA water contact angle

	KH567	KH550/CAB	KH550/GMA	KH550/CNTs
Contact angle before modification(°)	84.7	65.1	87.7	81.32
Contact angle after modification(°)	90.4	79.1	99.5	90.0
Rate of change(%)	6.7	21.4	13.4	10.7

In summary, silane coupling agents ameliorate the water-resistance of waterborne acrylic resins can be summarized. Silane coupling agents can cross-link with the resin to form a stable three-dimensional network structure, thereby increasing the cross-link density and reducing the surface energy of the coating [29-32]. There are some differences in the effect of silane coupling agents for different modification systems, However, all have improved the water-resistance of the waterborne acrylic resin.

3.2 The effect of fluorine and silicones on the water-resistance of aqueous acrylates

Inorganic materials mixed with polymers to form composites will to some extent, improve the performance of polymer-based composites, e.g., fluorine and silicones in aqueous acrylates can have a positive effect on the water-resistance of resin coatings. Yang, Du, Li et al. [33-35] all used silane-based substances to pretreat SiO_2 particles, and then the modified SiO_2 particles were then mixed with WA. the water-resistance properties were characterized. Yang et al. modified SiO_2particles with 3-(methacrylate)propyltrimethoxysilane and prepared fluorinated acrylic acid/ SiO_2 composite emulsions (F1) by emulsion polymerization, as shown in Fig. 4. The results showed a significant increase in the water contact angle of the resin film, e.g., S2b (modified SiO_2) compared to S2 (unmodified SiO_2), the water contact angle of the resin film increased from 93° to 100°. With increasing fluorine content, as in S4-S7, the water contact angle remained steady at nearly 110°. In addition, the surface roughness (Rq) of the resin film was explored by characterizing its AFM/nm. Du et al. modified the nano-TiO_2/ SiO_2composite particles with a silane coupling agent (KH-570) and a titanate coupling agent (NDZ-101) and then embedded them into the acrylic resin by a co-blending method to obtain a nano-TiO_2/ SiO_2 composite modified acrylic resin (F2), probably due to the addition of nano-TiO_2/ SiO_2 composite particles and the hydrophobic effect of the coupling agent, thus forming a nano-convex structure on the surface of the acrylic resin coating. As a result, the water absorption of the nano-

TiO_2/ SiO_2 acrylic resin was less than 5%, and the water-resistance of the resin material was improved.

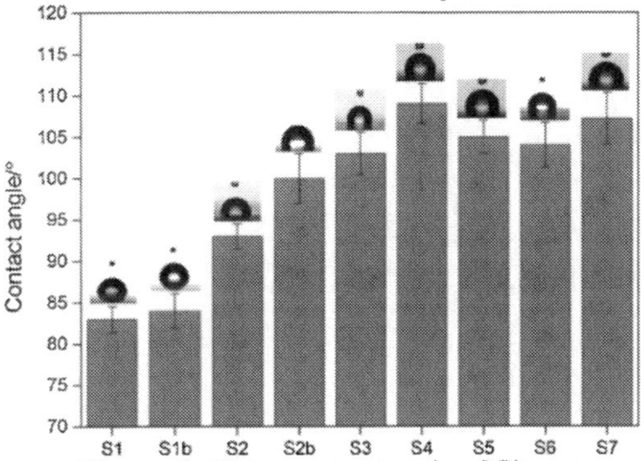

Figure 4. Water contact angle of films.

From the analysis in F2, it can speculate that the reason for the change of the water contact angle in F1 is that the functionalized SiO_2forms a nano-convex surface in the resin film, and the hydrophobic effect of the fluorine substance itself also has a certain degree of influence on the system. It increases the hydrophobicity of the system. Similarly, Li et al. prepared functionalized SiO_2-modified aqueous fluorinated acrylate emulsions (F3) using KH-570 for surface treatment of SiO_2 nanoparticles, and the water absorption and water contact angle of the resin films in Fig. 5. Overall, the water absorption of the films decreased gradually with increasing SiO_2content, with the SiO_2content at 3 %. To further investigate the intrinsic reason for the effect of SiO_2content on water absorption, the surface morphology of the resin film was also characterized by AFM, as shown in Fig. 6. From the 3D image in Fig. 6(a), there is no doubt that the unmodified SiO_2corresponds to The surface morphology of the resin film is relatively flat. As the SiO_2content increases, the raised structure on the film surface gradually increases, with the most significant number of micro-convex facilities appearing at 3% SiO_2 content. When the SiO_2content continues to grow, as shown in Fig. 6(e), due to aggregation and combined with the above data on water absorption and contact angle, this reduces its water resistance, as shown in Fig. 6(b). The water absorption, water contact angle, and AFM analysis show that the amount of functionalized SiO_2 affects the surface microconvex structure of the resin film and thus the hydrophobic ability of the film.

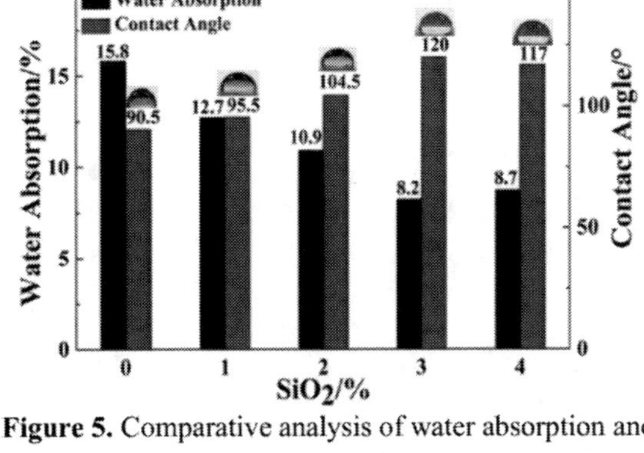

Figure 5. Comparative analysis of water absorption and contact angle.

A common rule was found by analyzing F1, F2, and F3. Through the treatment of SiO_2functionalization, the internal reason for changing the water-resistance of WA film is the degree of influence on the micro-convex structure on the surface of the film, and the micro-convex network is dense. And the larger the number, the rougher the surface of the film and the better the water resistance. The surface roughness (Ra, Rq) of F1 and F3 in Table 2 goes without saying that F3 has a tremendous advantage in water resistance over the F1 system.

Table 2. F1、F3 Surface roughness

F1	Rq(nm)	F3	Ra(nm)
S1	1.45	0%SiO_2	1.92
S1b	1.75	1%SiO_2	3.13
S2	1.86	2%SiO_2	4.0
S2b	3.33	3%SiO_2	6.02
S4	0.426	4%SiO_2	6.91

Organic or inorganic substances containing fluorine, silicon, etc., which are inherently hydrophobic, strengthen the water-resistance of aqueous resins to a certain extent. Li et al. [36] prepared aqueous organic/inorganic hybrids by dispersing superhydrophobic SiO_2 nanoparticles in acrylate copolymers, and the water contact angle of the original hybrid coating reached 164°. As shown in Fig. 6(a), the SiO_2 nanoparticles were more uniformly dispersed in the acrylate copolymer. The nanoparticles showed a specific raised structure, which was closely related to the superhydrophobic nature of the coating, Fig. 6(d-f) shows the surface of the layer after a certain number of cycles of wear. Hao et al. [37] prepared a fluorosilicone propylene emulsion using 2,2,3,4,4-hexafluorobutyl acrylate (HFBA) and vinyltriethoxysilane (VTES) for emulsion polymerization with acrylic acid monomers liquid and the magnitude of water absorption of resin coatings with 10%F, 10%Si, and 10%F/Si additions were measured, respectively, as shown in Fig. Fig. 7. It can see that the water absorption of 10%F films grew more rapidly with time. The water absorption of 10%Si films was significantly lower than that of others. In addition, the water absorption of the 10% F/Si film is 8% due to the cross-linking reaction of silicon in the system, forming a three-dimensional cross-linked network, which is similar to the role of silane coupling agents in the resin modification above, and the synergistic

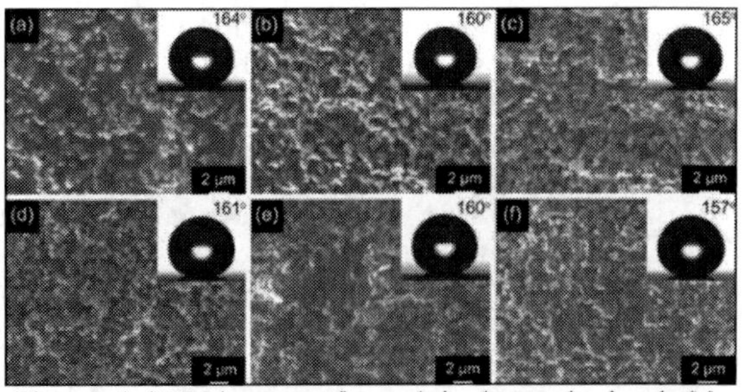

Figure 6. SEM images of an original superhydrophobic hybrid coating surface (a), and the

surfaces of superhydrophobic hybrid coating after 50(b), 100(c), 200(d), 250(e) and 300(f) cycles of abrasion, the insets are the corresponding CAs of the corresponding surfaces.
effect of fluorine. It is identical to the position of silane coupling agents in the resin modification above. The synergistic effect of fluorine contributes to the outstanding advantages of the films in terms of water resistance over single fluorine or silicon modifications. Xiong et al. [38] synthesized an acrylic copolymer emulsion containing fluorine, silicone, and epoxy resin. The surface tension decreased due to the aggregation of fluorine chains on the film's surface, resulting in a more hydrophobic modified film. F. Sbardella et al. [39] synthesized a waterborne acrylic resin coating of silica-coated polymer particles. They coated it on Carrara marble and Lecce stone, where the water penetration rate was slowest at 10% SiO_2 content. Due to differences in the internal pore size of Carrara marble and Lecce stone, the water penetration rate was also inconsistent.

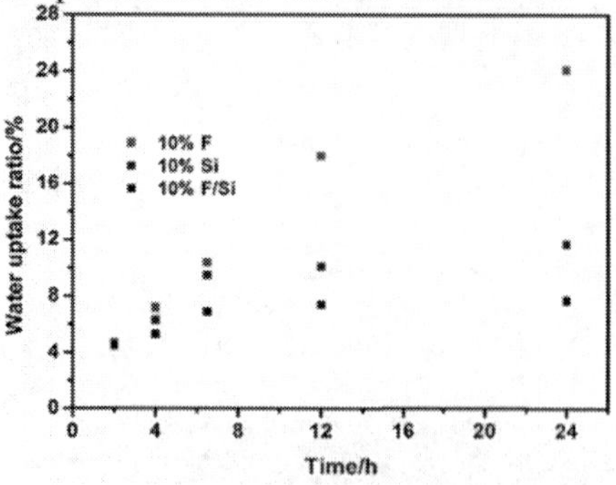

Figure 7. Water absorption of the films.

By analyzing the effect of substances such as fluorine and silicon on the water-resistance of waterborne acrylic resin, it is pretty clear that the principle of its modification of waterborne acrylic resin is forcing the roughness of the film surface. When such substances are mixed with the resin by physical or chemical, the fluorine substances will act on the resin surface, making the raised structure of the coating surface more evident than before the modification. In addition, the cross-linking effect of silicon will physically build upon the coating surface, which will also affect the roughness of the coating surface, making the water absorption of waterborne acrylic resin decrease and water resistance increase.

3.3 Resin copolymer modified aqueous acrylate.

The resin copolymerization method can give full play to the advantages of each polymer in specific properties to achieve the purpose of modification. Ji et al. [40] synthesized acrylic acid-alkyd copolymer by free radical polymerization and studied its water resistance and alkyd acid. Resin content. As shown in Fig. 8(a), when the alkyd content is 10%, the water contact angle of the copolymer reaches more than 90°, and when the content is 25%, the water contact angle keeps at about 95°. From the point of view of contact angle alone, the range of alkyd will have a specific influence on the hydrophilicity of acrylic acid-alkyd copolymer. When the content of alkyd is 10%, the acrylic acid-alkyd copolymer has a specific hydrophobic type. It is due to hydrophobic alkyd species [41, 42]. The change of the contact angle with time was also qualitatively studied. It can see from Fig. 8(b) that with the increase of the alkyd content, the decreasing trend of the water contact angle is more and more gentle. When the alkyd content

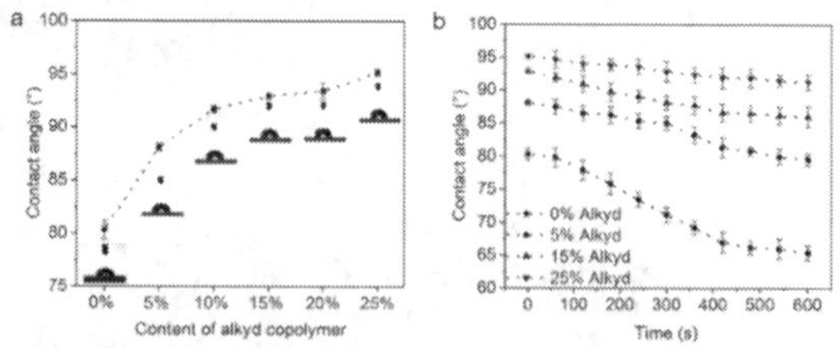

Figure 8. (a) Water contact angles of the acrylic-alkyd copolymer films at 0 s, and (b) the change of water contact angles of the acrylic-alkyd copolymer films.

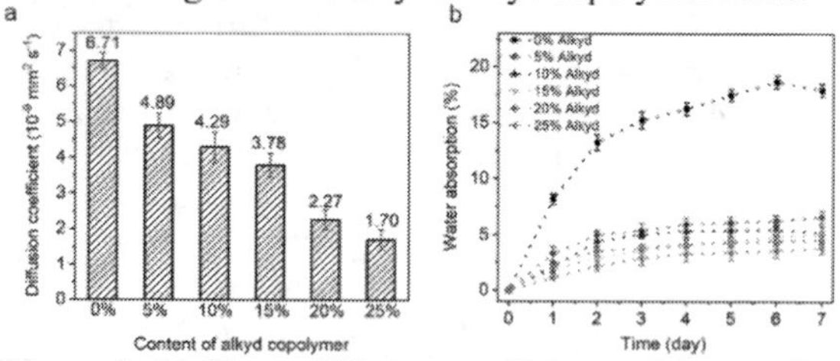

Figure 9. (a) Water diffusion coefficients of the acrylic-alkyd copolymer films, and (b) water uptake of the copolymer films at room temperature.

is 25 %, the decreasing trend tends to the level, indicating that the hydrophobicity of the copolymer film at this time has a higher level.
In addition, the pattern of variation of the water diffusion coefficient and water absorption of the acrylic-alcoholic acid copolymer is also consistent with the results of the above analysis, as shown in Fig. 9(a), where the water diffusion coefficient changes from 6.71×10-9 $mm^2 \cdot s^{-1}$ to $1.70\times10^{-9} mm^2 \cdot s^{-1}$ as the alkyd content increases. And the water absorption is also significantly lower, as shown by the results in Fig. 9(b). This lower water diffusion coefficient and water absorption rate will help improve the copolymer film's corrosion resistance [43].
The crosslinking density of the acrylic copolymer can be changed by the esterification synthesis method that can enhance the water resistance of the coating. Esterified the epoxy resin with methacrylate in the first step by Duan et al. [44]. Then added acrylamide to the esterification reaction product, obtaining the bisphenol A epoxy resin modified acrylic resin in

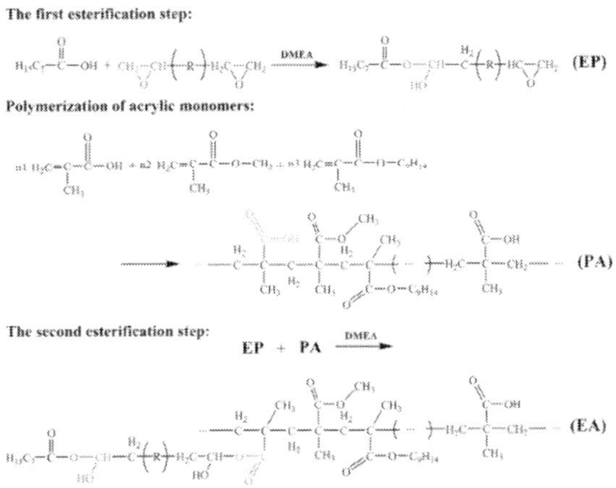

Figure 10. The schematic diagram for the two-step esterification process.

the end. Liu et al. [45] synthesized the waterborne epoxy-acrylic graft copolymer resin by a two-step esterification method. The synthesis method in Fig. 10.

Both of the above synthesis methods are used to change the internal structure of aqueous acrylic resins, forming a cross-linked network with a specific strength that can hinder the attack of water molecules. As in the case of acrylic resins modified with bisphenol epoxy resins, a cross-linked structure is formed between the O-H and amide groups in acrylamide and polar groups such as epoxy and carboxyl groups; Concerning waterborne epoxy-acrylic graft copolymer resins, the molecular weight of the resin and the carboxyl group content affect the degree of cross-linking of the coating, thus affecting the water-resistance of the layer.

Similarly, crosslinking agents can also chemically modify water-based acrylic resins. Shen et al. [46] added aziridine crosslinking agents to crosslink with carboxyl groups in acrylic resins, thereby reducing the content of carboxyl groups and making them hydrophilic The groups are facilitated, and the water-resistance of the water-based polish is improved. Pi et al. [47] used adipic acid dihydrazide (ADH) as the crosslinking agent to synthesize a room temperature self-crosslinking acrylic resin by solution polymerization. The content of the diacetone acrylamide (DAAM) monomer affects the crosslinking density. The water absorption of the film was reduced from 37.14% to 9.2%. Sumit Parvate et al. [48] used butyl acrylate (BA), styrene, acrylic acid (AA), acrylamide,

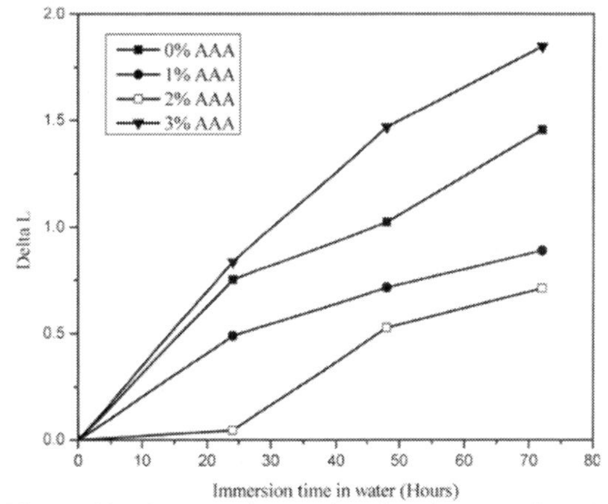

Figure 11. ΔL versus immersion time in water (hours).

and allyl acetoacetate (AAA) as monomers and hexamethylenediamine (HDA) as the crosslinking agent. A stable one-component room temperature-curing copolymer was synthesized by seed emulsion polymerization. The degree of crosslinking between AAA and HAD affects the water absorption of the film. By testing the degree of whiteness of binder film as a function of the time of the resin, it is clear that when the content of AAA is 2%, the ΔL value is relatively low, and the film has the best water resistance at this time, as shown in Fig. 11. It is because too much AAA will introduce more amide-based hydrophilic groups. It can see from the comparison with 0% AAA that this effect is more dominant than the promotion effect of cross-linking on the water resistance. The above modification of the resin utilizing cross-linking is similar to the modification principle of the silane coupling agent above, both of which form a specific cross-linked network structure, thereby improving the hydrophobicity of the resin film.

4. Conclusion

Waterborne acrylic resins are widely used in architectural coatings, road adhesives, and other applications as environmentally friendly materials. To prolong the life of the material, different modifiers play a role in waterborne acrylic resins for the water-resistance aspect of the wax, and waterborne materials with excellent water resistance are obtained.

Silane coupling agents can crosslink with resin to form a stable three-dimensional network structure, thereby increasing the crosslinking density, reducing the surface energy of the coating, and achieving the purpose of hydrophobic modification of the resin. The resin copolymerization modification method also has a similar effect. The principle of improvement of water-based acrylic resin with fluorine, silicon, and other substances is mainly focused on the impact of surface roughness. When such substances are physically or chemically mixed with the resin, the fluorine substances will make the surface of the resin convex. The degree of modification is changed. In addition, the chemical crosslinking reaction of silicon substances or the physical accumulation on the surface of the coating also affects the surface roughness of the layer, which reduces the water absorption rate of water-based acrylic resin and increases water resistance. These modification methods have played a positive role in the water-resistance modification of water-based acrylic resins. However, in specific industrial applications, such as when used as road adhesives, the cost will directly affect the possibility of the resin application. Therefore, based on the above-mentioned various research methods, it is necessary to consider the dosage ratio of a specific modifier or use a variety of modifiers to modify the resin to achieve low cost while maintaining performance: development and application. Additionally, there are still some problems and challenges in the water-resistant modification of water-based acrylic resins, such as the influence of the crosslinked structure of the modified resin on the resin film-forming temperature, the relationship between the solid content of the water-based polish and its water resistance, etc. are questions that need to be researched. It is hoped that the water-based acrylic resin will continue to develop in a green and sustainable direction while maintaining excellent application performance.

Acknowledgments

This work was financially supported by Research and Demonstration of Key Technologies for Ecological Road Construction, 19DZ1204205, Shanghai Science and Technology Commission fund.

References

1. Padget J C. Polymers for water-based coatings-a systematic overview[J]. Journal of coatings technology, 1994, 66(839): 89-105.
2. Annable T, Brown R A, Padget J C, et al. Improvements in the application properties of water-based low VOC coatings[J]. Surface coatings international, 1998, 81(7): 321-329.
3. Jackson K. Recent advances in water-borne protective coatings[J]. Surface coatings international, 1999, 82(7): 340-343.
4. Ifijen I H, Maliki M, Odiachi I J, et al. Review on Solvents Based Alkyd Resins and Water Borne Alkyd Resins: Impacts of Modification on Their Coating Properties[J]. Chemistry Africa, 2022: 1-15.
5. MENG X, GAO Y. Development of Studies on Water-based Epoxy Resin[J]. Materials Review, 2006, 11(20): 384-386.
6. Si P, Zhao B. Water-based polyurethanes for sustainable advanced manufacture[J]. The Canadian Journal of Chemical Engineering, 2021, 99(9): 1851-1869.
7. Guo K Y, Wu Q, Mao M, et al. Water-based hybrid coatings toward mechanically flexible, super-hydrophobic and flame-retardant polyurethane foam nanocomposites with high-efficiency and reliable fire alarm response[J]. Composites Part B: Engineering, 2020, 193: 108017.
8. Jiao C, Sun L, Shao Q, et al. Advances in waterborne acrylic resins: Synthesis principle, modification strategies, and their applications[J]. ACS omega, 2021, 6(4): 2443-2449.
9. Sainz M F, Souto J A, Regentova D, et al. A facile and green route to terpene derived acrylate and methacrylate monomers and simple free radical polymerisation to yield new renewable polymers and coatings[J]. Polymer Chemistry, 2016, 7(16): 2882-2887.
10. Karami S, Motahari S, Pishvaei M, et al. Improvement of thermal properties of pigmented acrylic resin using silica aerogel[J]. Journal of Applied Polymer Science, 2018, 135(1): 45640.
11. Li M, Li Y, Xue F, et al. Water-based acrylate copolymer/silica hybrids for facile preparation of robust and durable superhydrophobic coatings[J]. Applied Surface Science, 2018, 447: 489-499.
12. Zhong S, Li J, Cai Y, et al. Novel surfactant-free waterborne acrylic-silicone modified alkyd hybrid resin coatings containing nano-silica for the corrosion protection of carbon steel[J]. Polymer-Plastics Technology and Materials, 2019, 58(8): 866-878.
13. Ou J, Zhang M, Liu H, et al. Matting films prepared from waterborne acrylic/micro-SiO2 blends[J]. Journal of Applied Polymer Science, 2015, 132(13).
14. Peruzzo P J, Anbinder P S, Pardini O R, et al. Waterborne polyurethane/acrylate: Comparison of hybrid and blend systems[J]. Progress in Organic Coatings, 2011, 72(3): 429-437.
15. Jiao C, Shao Q, Wu M, et al. 2-(3, 4-Epoxy) ethyltriethoxysilane-modified waterborne acrylic resin: preparation and property analysis[J]. Polymer, 2020, 190: 122196.
16. Wang Y, Wu J, Zhang D, et al. Design of salt-responsive and regenerative antibacterial polymer brushes with integrated bacterial resistance, killing, and release properties[J]. Journal of Materials Chemistry B, 2019, 7(38): 5762-5774.
17. Zheng B, Ge S, Wang S, et al. Effect of γ-aminopropyltriethoxysilane on the properties of cellulose acetate butyrate modified acrylic waterborne coatings[J]. Reactive and Functional Polymers, 2020, 154: 104657.
18. Shukla S, Rai J S P. Synthesis and characterizations of waterborne poly (acrylate) latexes[J]. High Performance Polymers, 2014, 26(6): 653-659.
19. Guo X, Ge S, Wang J, et al. Waterborne acrylic resin modified with glycidyl methacrylate (GMA): Formula optimization and property analysis[J]. Polymer, 2018, 143: 155-163.
20. Paul S. Water-borne acrylic emulsion paints[J]. Progress in Organic Coatings, 1977, 5(1): 79-96.
21. Zhai L, Berg M C, Cebeci F C, et al. Patterned superhydrophobic surfaces: toward a synthetic mimic of the Namib Desert beetle[J]. Nano Letters, 2006, 6(6): 1213-1217.

22. Mingyang Wu,Shengsong Ge,Cuiyan Jiao,Zhangyin Yan,Heyun Jiang,Yunfeng Zhu,Binbin Dong,Mengyao Dong,Zhanhu Guo. Improving electrical, mechanical, thermal and hydrophobic properties of waterborne acrylic resin-glycidyl methacrylate (GMA) by adding multi-walled carbon nanotubes[J]. Polymer,2020,200(prepublish).
23. Li S, Li H, Wang X, et al. Super-hydrophobicity of large-area honeycomb-like aligned carbon nanotubes[J]. The journal of physical chemistry B, 2002, 106(36): 9274-9276.
24. Mahadik D B, Rao A V, Rao A P, et al. Effect of concentration of trimethylchlorosilane (TMCS) and hexamethyldisilazane (HMDZ) silylating agents on surface free energy of silica aerogels[J]. Journal of colloid and interface science, 2011, 356(1): 298-302.
25. Liu J, Huang J, Wujcik E K, et al. Hydrophobic electrospun polyimide nanofibers for self-cleaning materials[J]. Macromolecular Materials and Engineering, 2015, 300(3): 358-368.
26. Brostow W, Dutta M, Rusek P. Modified epoxy coatings on mild steel: tribology and surface energy[J]. European Polymer Journal, 2010, 46(11): 2181-2189.
27. Jia S, Zhu Y, Wang Z, et al. Influences of PP-g-MA on the surface free energy, morphologies and mechanical properties of thermoplastic polyurethane/polypropylene blends[J]. Journal of Polymer Research, 2015, 22(8): 1-10.
28. Chen J S, Sun Z, Guo P S, et al. Effect of ion implantation on surface energy of ultrahigh molecular weight polyethylene[J]. Journal of applied physics, 2003, 93(9): 5103-5108.
29. Xie Q, Xu J, Feng L, et al. Facile creation of a super-amphiphobic coating surface with bionic microstructure[J]. Advanced materials, 2004, 16(4): 302-305.
30. Yang X, Zhu L, Zhang Y, et al. Surface properties and self-cleaning ability of the fluorinated acrylate coatings modified with dodecafluoroheptyl methacrylate through two adding ways[J]. Applied surface science, 2014, 295: 44-49.
31. Shin M S, Lee Y H, Rahman M M, et al. Synthesis and properties of waterborne fluorinated polyurethane-acrylate using a solvent-/emulsifier-free method[J]. Polymer, 2013, 54(18): 4873-4882.
32. Xu H, Qiu F, Wang Y, et al. Preparation, mechanical properties of waterborne polyurethane and crosslinked polyurethane-acrylate composite[J]. Journal of applied polymer science, 2012, 124(2): 958-968.
33. Yang W, Zhu L, Chen Y, et al. Surface topography and hydrophobicity of waterborne fluorinated acrylic/silica hybrid coatings[J]. Colloids and Surfaces A: Physicochemical and Engineering Aspects, 2015, 484: 62-69.
34. Du B, Chen F, Luo R, et al. Synthesis and characterization of nano-TiO2/SiO2-acrylic composite resin[J]. Advances in Materials Science and Engineering, 2019, 2019.
35. Li L, Li X, Shen Y, et al. Hydrophobicity and corrosion resistance of waterborne fluorinated acrylate/silica nanocomposite coatings[J]. e-Polymers, 2021, 21(1): 779-792.
36. Meng Li,Yu Li,Fang Xue,Xinli Jing. Water-based acrylate copolymer/silica hybrids for facile preparation of robust and durable superhydrophobic coatings[J]. Applied Surface Science,2018,447.
37. Hao G, Zhu L, Yang W, et al. Investigation on the film surface and bulk properties of fluorine and silicon contained polyacrylate[J]. Progress in Organic Coatings, 2015, 85: 8-14.
38. Xiong P, Guan R, Lu D, et al. Characterization of acrylic copolymer latex modified by fluorine, silicon, and epoxy resin[J]. e-Polymers, 2008, 8(1).
39. Sbardella F, Bracciale M P, Santarelli M L, et al. Waterborne modified-silica/acrylates hybrid nanocomposites as surface protective coatings for stone monuments[J]. Progress in Organic Coatings, 2020, 149: 105897.
40. Ji S, Gui H, Guan G, et al. Molecular design and copolymerization to enhance the anti-corrosion performance of waterborne acrylic coatings[J]. Progress in Organic Coatings, 2021, 153: 106140.
41. Wang S T, Schork F J, Poehlein G W, et al. Emulsion and miniemulsion copolymerization of acrylic monomers in the presence of alkyd resin[J]. Journal of Applied Polymer Science, 1996, 60(12): 2069-2076.
42. Wu X Q, Schork F J, Gooch J W. Hybrid miniemulsion polymerization of acrylic/alkyd systems and characterization of the resulting polymers[J]. Journal of Polymer Science Part A: Polymer Chemistry, 1999, 37(22): 4159-4168.
43. Zhong S, Li J, Yi L, et al. Cross-linked waterborne alkyd hybrid resin coatings modified by fluorinated acrylate-siloxane with high waterproof and anticorrosive performance[J]. Polymers for Advanced Technologies, 2019, 30(2): 292-303.
44. Duan Y, Huo Y, Duan L. Preparation of acrylic resins modified with epoxy resins and their behaviors as binders of waterborne printing ink on plastic film[J]. Colloids and surfaces a: physicochemical and engineering aspects, 2017, 535: 225-231.
45. Liu M, Mao X, Zhu H, et al. Water and corrosion resistance of epoxy–acrylic–amine waterborne coatings: Effects of resin molecular weight, polar group and hydrophobic segment[J]. Corrosion Science, 2013, 75: 106-113.
46. Shen Y, Zhao C, Zhou J, et al. Application of waterborne acrylic emulsions in coated controlled release fertilizer using reacted layer technology[J]. Chinese Journal of Chemical Engineering, 2015, 23(1): 309-314.
47. Pi P, Chen X, Wen X, et al. Preparation and characterization of ambient-temperature self-

crosslinkable water-soluble acrylic resin for PE film ink[J]. Journal of Coatings Technology and Research, 2016, 13(1): 73-80.

48. Parvate S, Mahanwar P. Insights into the preparation of water-based acrylic interior decorative paint: tuning binder's properties by self-crosslinking of allyl acetoacetate-hexamethylenediamine[J]. Progress in Organic Coatings, 2019, 126: 142-149.

A Review of Research on Recyclable Polymer Materials

Liyun Yuan and Yong Shen*

State Key Laboratory Base of Eco-Chemical Engineering; College of Chemical Engineering, Qingdao University of Science and Technology, Qingdao 266042, China.

Abstract. Polymer materials have been widely used in applications ranging from aerospace, automobile transportation, medical and health care due to their excellent properties. The current linear production and disposal model of polymeric materials has raised concerns about the continuous consumption of limited fossil fuels and the severe environmental crises. To address the dual challenges of the environment and resources, it is necessary to develop sustainable polymers and more promising recycling strategies. This contribution summarizes the recent research on the preparation of sustainable polymers and their chemical recycling, including polyesters, polycarbonates, polythioesters and polyurethanes.

1. Introduction

Polymer materials play an important role in people's lives. With the development of the economy and the improvement of living standards, people have put forward higher requirements for polymer materials. Polymer materials have been widely used in the fields of clothing, packaging, agriculture, electronics, biotechnology and aerospace due to their low cost, light weight, high strength, and facile processing. The production and varieties of polymer materials are increasing year by year, which have occupied a very important position in the material industry. The vast majority of polymer materials come from petrochemical products, such as rubber, plastics and synthetic fibers, which are difficult to degrade in nature, and will cause serious social, economic and environmental problems after they are discarded.[1] In order to meet the dual challenges of environment and resources, it is necessary to develop sustainable polymers and implement more promising recycling strategies, and many researchers have focused their attention on bio-based polymer materials.[2] Another advantage of these materials is that their raw materials are all derived from biomass, completely getting rid of the current situation that polymer materials are completely dependent on petrochemical resources, realizing carbon cycle, greatly reducing white pollution, improving the earth's atmospheric environment, and having an alternative to traditional the potential of polymer materials.[3, 4] However, most bio-based polymer materials have good durability and are difficult to decompose in the natural environment, causing white pollution. In order to solve the problem of environmental pollution, recyclable bio-based polymer materials have become a research hotspot.[5]

2. Recyclable polyester, polycarbonate, polythioester

With the wide application of bio-based polymer materials, the bio-based polymer material industry has been listed as one of the seven strategic emerging industries in China. In recent years, bio-based polymer material industry is growing at a rate of more than 20% per year, with a total output of more than 6 million tons per year in China. At the same time, it is also facing serious energy and environmental crises. There are two solutions: one is to develop new polymers that avoid the use of plasticizers; the other is to find a polymer monomer that can be chemically recycled.[6, 7] The ideal sustainable polymers should include but not limited to the following criteria: (1) they are completely biologically derived and make maximum use of industrial waste. (2) They improve the properties of existing bio-derived plastics, especially reducing the use of plasticizers. (3) Allowing selective use of catalytic chemistry for recovery. At present, great achievements have been made in recyclable polymer materials. For example, there have been outstanding achievements in the research of polyesters, polycarbonates, polythioesters, polyacetals and so on.

Polyester is a large class of biodegradable polymer materials, which are widely used in aerospace, biomedicine, and textiles.[8] Aliphatic polyesters can be prepared by ring-opening polymerization of cyclic lactones or polycondensation of diols and diacids. Currently the most studied are poly(ε-caprolactone) (PCL), poly(glycolide) (PGA) and poly(L-lactide) (PLLA) due to their good biocompatibility and degradable, which can be degraded into carbon dioxide and water in nature. It is a kind of environmental friendly polymer material with great market potential. Its general structural formula is shown in Figure 1.

* Corresponding author: shenyong@qust.edu.cn

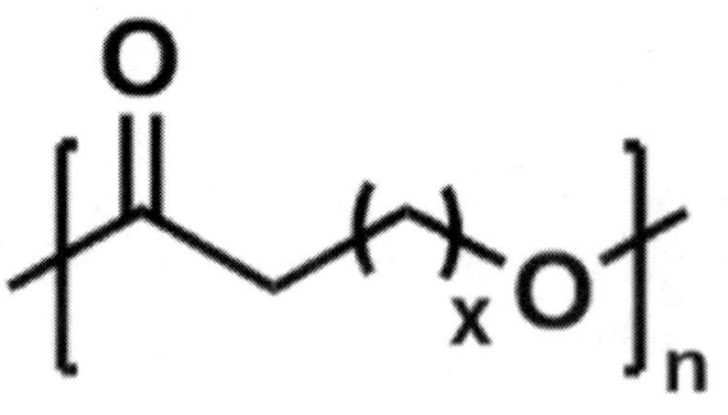

Figure 1. The structure of aliphatic polyester

Alpha-methylene-gamma-butyrolactone (MBL) occurs naturally in tulips and can be obtained from biological sources. MBL has been intensively investigated as a sustainable alternative to petroleum-based methyl methacrylate (MMA) monomers due to the higher glass transition temperature of MBL-derived polymers compared to polymethyl methacrylate (PMMA) and better solvent durability, so MBL is a functionalizable monomer with potential chemical recyclability. In 2020, our group achieved the selectively ring-opening polymerization of MBL by using an organophosphazene base/urea binary catalyst. The prepared polymer P(MBL)ROP can be completely depolymerized to recover MBL monomer.[9]

Aliphatic polycarbonate (APC) and polythioester (PTE) have also made great breakthroughs due to their potential degradability and biocompatibility. In 2022, Zhu and coworkers prepared a class of seven-membered ring carbonates containing trans-cyclohexyl fused rings. This kind of monomer has high activity for ring-opening polymerization, and the molecular weight of the prepared polycarbonate is as high as 673 kg·mol-1. Depolymerization studies were carried out on the prepared polycarbonate, and it was found that it can be completely depolymerized into its corresponding cyclic dimer, which can be repolymerized to obtain the same polymer as the original monomer polymerization, indicating that the chemical cycle is realized potential pathways (Figure 2).[10]

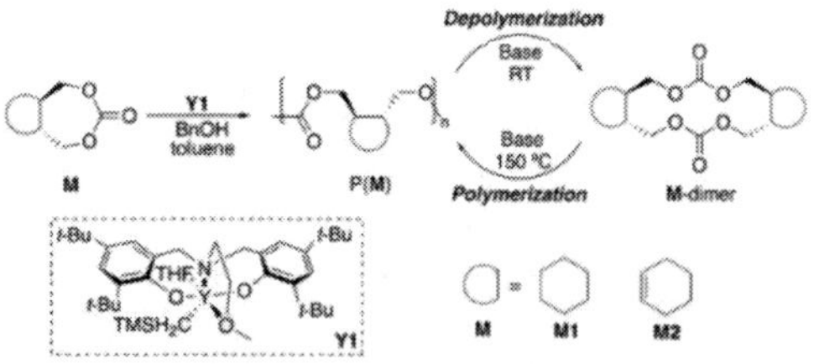

Figure 2. Accessing Chemically Recyclable Polycarbonates through Ring-Opening Polymerization of M1 and M2.[10] Reprinted with permission from ref. 10. Copyright (2022) American Chemical Society.

Lu et al from Peking University is committed to using renewable amino acids as raw materials to construct recyclable functional polymer materials. They synthesized three proline derivatives with different substituents, NBoc- PTL, NCbz-PTL and Nene-PTL. Polythioesters with high molecular weights and narrow molecular weight distributions were then prepared using bases such as triethylamine, 1,8-diazabicyclo[5.4.0] undec-7-ene (DBU) as catalysts and benzyl mercaptan as initiator. The polythioester PNBoc-PTL can be depolymerized and recycled back to monomers under mild conditions. When the more basic DBU was used, complete depolymerization of obtained polythioesters was achieved within 2 min at 50 °C (Figures 3).[11]

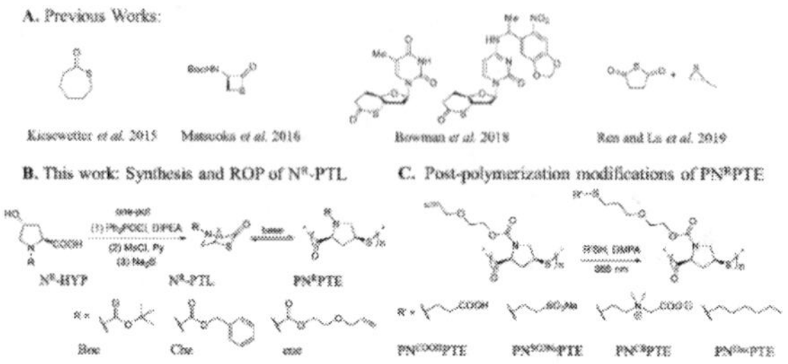

Figure 3. Preparation of Recyclable Polythioesters from Proline Derivatives.[11] Reprinted with permission from ref. 11. Copyright (2019) American Chemical Society.

3. Recyclable Polyurethane

As one of the important members of the polymer material family, polyurethane is the only organic synthetic material that has application value in the fields of plastics, rubber, fibers, coatings, adhesives and functional polymers because of its superior properties. However, the recyclability of polyurethane has not been fully studied. To date, a common strategy to achieve chemical recovery of polyurethanes is to recover polyols by alcoholysis or hydrolysis. For example, in 2017, Chen and coworkers explored an optimal process condition for recycling and reusing polyurethane flexible foam waste by alcoholysis. The optimum alcoholysis agent, reaction temperature, reaction time and material ratio were studied to achieve the maximum utilization of resources.[12]

The use of alcoholysis or hydrolysis has disadvantages such as large use of solvent, complicated purification involved in the recovery process, high recovery cost, and low yield. In order to overcome these deficiencies, it is attractive to develop biodegradable, recyclable and recyclable bio-based polymer materials that can be completely depolymerized into monomers under mild and energy-saving conditions. In 2016, Hillmyer and colleagues used bio-degradable poly(β-methyl-delta-valerolactone) as polyol precursor to synthesize thermoplastic polyurethane and flexible foam. The chemical recycling of prepared polyurethanes to recover β-methyl-delta-valerolactone (MVL) was achieved by the tandem dissociation of carbamate groups and the pyrolysis of poly(β-methyl-delta-valerolactone) polyols. This strategy bypasses many technical challenges and realizes the recycling of building blocks with excellent purity under mild conditions for the first time.[13]

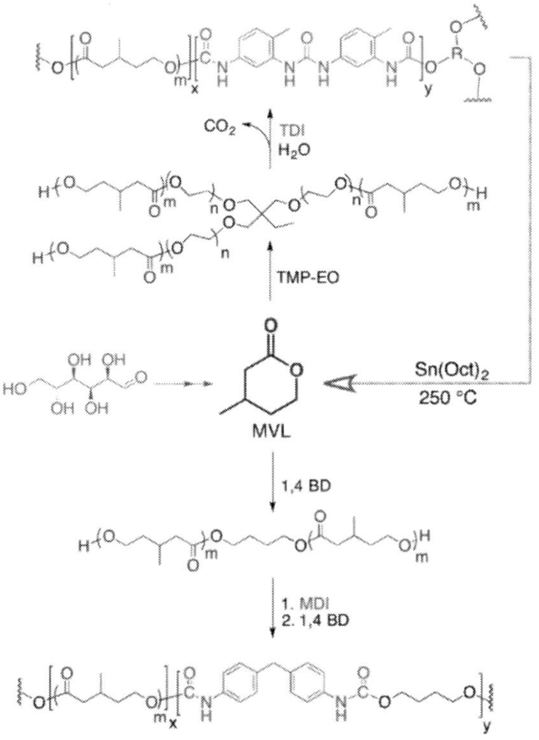

Figure 4. Synthesis mechanism and degradation of polyurethane and foam.[13] Reprinted with permission from ref. 13. Copyright (2016) American Chemical Society.

4. Conclusion

With the advent of the material age, countries around the world are looking for solutions to address the challenge of energy and the environment. It is attractive to develop more sustainable polymers with capability to chemical recycling back to pristine monomers from bio-derived resources, thus successfully establish a circular closed-loop life cycle. Despite the recent advancements, the polymers with closed-loop life cycle are still in their infancy. More efforts should be devoted to solve the current challenges including tedious monomer syntheses, harsh conditions for polymer preparation as well as poor thermal and mechanical properties of obtained polymer materials.

References

1. Moore T, Adhikari R, Gunatillake P, Chemosynthesis of bioresorbable poly(γ-butyrolactone) by ring-opening polymerisation: a review. Biomaterials, 2005, 26 (18): 3771-3782.
2. O'Keefe B J, Hillmyer M A, Tolman W B, Polymerization of lactide and related cyclic esters by discrete metal complexes. Journal of the Chemical Society, Dalton Transactions, 2001, (15): 2215-2224.
3. Zhu Y, Romain C, Williams C K, Sustainable polymers from renewable resources. Nature, 2016, 540 (7633): 354-362.
4. Gu L, Wu Q-Y, Recyclable bio-based crosslinked polyurethanes with self-healing ability. Journal of Applied Polymer Science, 2018, 135 (21).
5. Morales-Cerrada R, Tavernier R, Caillol S, Fully Bio-Based Thermosetting Polyurethanes from Bio-Based Polyols and Isocyanates. Polymers (Basel), 2021, 13 (8).
6. Zhu Y, Radlauer M R, Schneiderman D K, et al, Multiblock Polyesters Demonstrating High Elasticity and Shape Memory Effects. Macromolecules, 2018, 51 (7): 2466-2475.
7. Xu T Q, Yu Z Q,Zhang X M, Recyclable Vinyl-Functionalized Polyesters via Chemoselective Organopolymerization of Bifunctional α-Methylene-δ-Valerolactone. Macromolecular Chemistry and Physics, 2019, 220 (12).
8. Mija A, Louisy E, Lachegur S, et al, Limonene dioxide as a building block for 100% bio-based thermosets. Green Chemistry, 2021, 23 (24): 9855-9859.
9. Shen Y, Xiong W, Li Y, et al, Chemoselective Polymerization of Fully Biorenewable α-Methylene-γ-Butyrolactone Using Organophosphazene/Urea Binary Catalysts Toward Sustainable Polyesters. CCS Chemistry, 2021, 3 (1): 620-630.
10. Zhang W, Dai J, Wu Y-C, et al, Highly Reactive Cyclic Carbonates with a Fused Ring toward Functionalizable and Recyclable Polycarbonates. ACS Macro Letters, 2022, 11 (2): 173-178.

11. Yuan J, Xiong W, Zhou X, et al, 4-Hydroxyproline-Derived Sustainable Polythioesters: Controlled Ring-Opening Polymerization, Complete Recyclability, and Facile Functionalization. Journal of the American Chemical Society, 2019, 141 (12): 4928-4935.
12. Chen Jiabin, Ma Meng, Zhu Jiajie, Alcoholysis recovery and reuse of waste polyether-based polyurethane flexible foam. 2017 Innovation-driven and accelerated development of Lishui ecological industry—Proceedings of the New Materials Academic Forum, 2017.
13. Schneiderman D K, Vanderlaan M E, Mannion A M, et al, Chemically Recyclable Biobased Polyurethanes. ACS Macro Letters, 2016, 5 (4): 515-518.

Discussion on Causes and Solutions of the Biting Mechanism of Stainless Steel Bolts

Jianjun Zheng[1,*], Hao Chen[2], Yanfei Zhang [1]

[1] Inner Mongolia Power (Group) Co., Ltd., Inner Mongolia Power Research Institute Branch, Hohhot 010020, China
[2] Inner Mongolia Enterprise Key Laboratory of High Voltage and Insulation Technology, Hohhot 010020, China

Abstract. Due to the influence of the peculiar material characteristics, austenitic stainless steel bolts are always easy to be locked during disassembly. This is not only seriously affected the service performance of austenitic stainless steel bolts, but also affects the reuse of bolts. In this paper, the main causes and mechanism of the locking of austenitic stainless steel blots in the process of fastening is summarized, and the effective measures containing correct material selection, carefully clean before assembly, spraying loosening agent, thermal expansion method, raw material belt tape winding and development of anti-locking austenitic stainless steel fastening bolt to prevent locking through theoretical analysis and experimental verification are also provided, which is hoped that certain guiding significance can be provided to prevent the phenomenon of bolt locking from the source as far as possible.

Keywords: Austenitic stainless steel; bolt; locked; mechanism; measurement

1. Introduction

Bolt is indispensable standard parts in daily life and industrial production, which is always used from mechanical products to electronic products, from ships, vehicles, hydraulic engineering to chemical experiments, from engineering, construction, showing that the bolts are widely used and convenient. Austenitic stainless steel is a common bolt material with stable mechanical properties, beautiful appearance and corrosion resistance[1-4]. However, due to its peculiar material characteristics, austenitic stainless steel bolts (Fig.1) are usually prone to lock the thread connection during installation[5-6]. Locking is a phenomenon of bolts in the process of assembly, which shows that the bolts stick together after several assemblies and are difficult to disassemble and tighten. Bolt locking will directly affect the maintenance and repair of the equipment, and may even lead to the damage and scrapping of the equipment. When the austenitic stainless steel bolt is locked, the pressure and heat generated between the teeth will destroy the chromium oxide layer between them, making the metal teeth block or shear, and then adhesion. As the continue of this phenomenon, the austenitic stainless steel bolt will be completely locked and can no longer be removed or locked[7].

Fig.1 Physical drawing of austenitic stainless steel bolt

2. Locking cause and mechanism of austenitic stainless steel bolt

The locking of austenitic stainless steel bolts is similar to a cold melting process. The main reason is that the instantaneous high temperature caused by friction during bolt tightening leads to metal fusion and bonding[8], as shown in Fig.2. After being damaged, the surface of austenitic stainless steel material will produce a thin oxide layer to prevent further corrosion. When the austenitic stainless steel bolt is locked, the pressure and heat generated between the teeth will destroy this oxide layer, resulting in blockage or shear between the metal threads, and then adhesion. The stress increases with the bolt locked, and finally the austenitic stainless steel fastener is completely bitten and cannot be removed. Usually, the locking process of the austenitic stainless steel bolt includes blocking, shearing, sticking, and locking takes place in just a few seconds. At the same time, the scraps produced during the stickiness of the austenitic stainless

* Corresponding author: dkyzjj@163.com

steel are easily to stick to the nut profile, thus increasing the friction. With the gradual increase of the tightening force to overcome the friction, the locking degree will be further aggravated, especially between the similar austenitic stainless steel fasteners. In addition, if the tightening force is too large in the process of material assembly, it will also produce greater pressure on the tooth pattern, and then lock up will occur. Besides, when the tightening direction is not perpendicular to the axial direction of the bolt resulting in a wrong angle, the bolt will also be locked.

Fig.2 Picture showing the thread locking of austenitic stainless steel bolt

3. Introduction of improvement measures

3.1 Correct material selection

Products of different materials can be assembled under the condition of design permission, such as 304 austenitic stainless steel bolts with relatively low price and 316 austenitic stainless steel nuts with relatively high price. In addition, the length of austenitic stainless steel bolts shall be selected appropriately, subject to one or two tooth pitches exposed after tightening. Practice shows that this method is feasible in product assembly [9].

3.2 Carefully clean before assembly

Due to the good ductility of austenitic stainless steel, rough tooth lines or foreign matters sticking, it is easy to increase the friction force, which often leads to the locking of austenitic stainless steel bolts. Therefore, clean the austenitic stainless steel bolts before assembly, improving the working environment of austenitic stainless steel bolts and the locking condition.

3.3 Spraying loosening agent

Using surface coating can greatly reduce the friction coefficient and effectively reduce the probability of austenitic stainless steel thread biting. Experience shows that reducing the mutual matching friction of austenitic stainless steel thread pairs can better avoid the locking problem. Based on this, the design principle of anti locking material formula is put forward, which is reducing the friction and lubrication effect is good, preventing the direct friction of matching thread pairs, and improving the matching surface between threads. In addition, the coating material design should also consider adapting to various working conditions of the connector as much as possible, such as high temperature, cleanliness and other factors. Coating materials commonly used in industry include butter, molybdenum disulfide, etc.

3.4 Thermal expansion method

For rusted fasteners or other fasteners that are ineffective in spraying thread loosening agent,the thermal expansion method is advised to be adopted, which is mainly heat the nut to make it easy to take out after thermal expansion. Heat is an important factor of thread biting. Friction in the assembly process leads to heat accumulation, and the heat increases with the acceleration of speed. Reducing the assembly speed can also reduce the risk of biting. Before using this method, first consider whether there is hot work space between bolts and nuts and the heat resistance of fasteners. This disassembly method can be adopted only if there is hot work space between the bolt and nut and the fastener has good heat resistance. Jiang et al. found in the theoretical study of locking failure at elevated temperature that the high temperature of the connecting part will lead to locking, this is because the metal is fusion welded under the influence of local high temperature. Therefore, reducing the contact temperature of the metal surface can improve the sliding friction and anti locking ability of the bolt. However, the above two methods are all required a certain amount of loosening agent to penetrate into the gap and hot work space at the connection between bolt and nut.

3.5 Raw material belt tape winding

Winding the raw material belt on the austenitic stainless steel bolt thread with appropriate thickness along the locking direction, so as to avoid direct contact between thread and nut when the austenitic stainless steel bolt is locked, thus preventing the thread surface being damaged. At the same time, due to the self-lubricating and non stick surface of raw material belt, the process of blockage→ shear→adhesion→locking in the fastening of austenitic stainless steel bolts can be effectively avoided. Besides, the raw materials also have cold fluidity, using which can generate the pressure between threads during the initial tightening and make the raw material belt fill the gap between the thread surfaces of austenitic stainless steel bolts, achieving the effect similar to film coating and improve the smoothness of thread surface. Even if the wrapped raw material belt is fell off during disassembly, it can still play a certain role in lubrication, so that the austenitic stainless steel bolts can be reused. In addition, the corrosion resistance of the raw material belt also meets the anti-corrosion application requirements of austenitic stainless steel bolts. And its non water absorption characteristics can also protect the bolt threads from the influence of humid environment with no environment pollution.

3.6 Development of anti-locking austenitic stainless steel fastening bolt

Biting or locking is also a common phenomenon in austenitic stainless steel fasteners. The root cause is the lack of sufficient hardness difference between bolts and screw holes, which makes metal threads blocked or sheared, resulting in adhesion and biting. In order to prevent the thread of bolt and screw hole from locking, they usually use different materials, so there must be differences in physical properties. However, if the physical properties between the bolt and the screw hole do not match, it is easy to loosen due to heating and other reasons, resulting in fatigue failure and fracture of the bolt. However, if the materials are exactly the same, there will be no hardness difference on the surface of bolts and screw holes, and it is easy to lock. If different hardness can be obtained through different preparation processes in the same material, the problem of fastener biting can be solved and the problem of bolt loosening can be avoided [10]. Therefore, to sum up, starting from the overall material selection idea, this project uses the nitriding technology to form a dense nitriding layer on the screw thread surface to improve its wear resistance and hardness [11], so as to form a hardness difference between the screw and the nut, and then solve the technical problem that the austenitic stainless steel fastening bolt is easy to lock.

4. Conclusion

This paper analyzes and discusses the causes, treatment methods and preventive measures of austenitic stainless steel bolt locking. The locking of austenitic stainless steel bolts is mainly resulted from the serious adhesive wear occurred on the surface of steel screw teeth due to the high temperature between screw teeth caused by rapid preloading and disassembly. Taking the measures such as correct material selection, carefully clean before assembly, spraying loosening agent, thermal expansion method, raw material belt tape winding and development of anti-locking austenitic stainless steel fastening bolt is considered to be effectively improve the locking problem of austenitic stainless steel bolts. It is hoped that through the discussion of this paper, certain help and guidance can be provided to prevent the phenomenon of bolt biting from the source as far as possible, and improve the on-site work efficiency.

Acknowledgments

This research was financially supported by 2022 Employee Innovation Studio innovation project of Inner Mongolia Power (Group) Co.,Ltd (Grant No. 2022-51).

References

1. N. Mingolo, A.P. Tschiptschin, C.E. Pinedo. On the formation of expanded austenite during plasma nitriding of an AISI 316L austenitic stainless steel. Surface & Coatings Technology, 2006, 201(7): 4215-4218.
2. A.D. Schino, I. Salvatori, J.M. Kenny. Effects of martensite formation and austenite reversion on grain refining of AISI 304 stainless steel. Journal of Materials Science, 2002, 37(21): 4561-4565.
3. Y. Lv, H. Luo, J. Tang, et al. Corrosion properties of phase reversion induced nano/ultrafine grained AISI 304 metastable austenite stainless steel[J]. Materials Research Bulletin, 2018, 107: 421-429.
4. J. Nomani, A. Pramanik, T. Hilditch, et al. Machinability study of first generation duplex (2205), second generation duplex (2507) and austenite stainless steel during drilling process[J]. Wear, 2013, 304(1-2): 20-28.
5. C.B. Tang, X.H. Yuan . Preparation of scuffing resistance friction-reducing material for bolt in high temperature[J]. Lubrication Engineering, 2005, (1): 99-1114.
6. W.D. Li, M.J. Kong, W.D. Jiang. Analysis of screw pairs galling of stainless stell[J]. Mechanical & Electrical Engineering Technology, 2013, 42(1): 97-99.
7. L. Wan, L. Sun, Y.D. Yang. Failure analysis for screw thread of joint[J]. Aerospace Materials & Technology, 2015, 45(1): 77-81.
8. G.F. Zhang, S.U. Wei, J. Zhang, et al. Effects of shoulder on interfacial bonding during friction stir lap welding of aluminum thin sheets using tool without pin[J]. Trans. Nonferrous Met. Soc. China, 2010, 20(12): 2223-2228.
9. X.Z. Xiao, S.H. Liang. The discussion of deadlock about stainless steel bolt[J]. Development &

Innovation of Machinery & Electrical Products, 2010, 23(3): 65-66.

10. J. Liu, F. Sun, H. Yu, et al. Monte carlo simulation of nitriding in iron by a mixing technology with laser and plasma beams[J]. Current Applied Physics, 2007, 7(6): 683-689.
11. Y.Y. Jiang, M. Zhang, C.H. Lee. A study of early stage self-loosening of bolted joints[J]. Journal of Mechanical Design, 2003, 125(3): 518-526.

Digital realization method of FFT and its application in adjusting velocity damping coefficient of inertial navigation system

Yujie Wang [1], Yang Zhang [2,*], Xintai Ma [1], Minghao Gao [1] and Zhipeng Zheng[1]

[1] Tianjin Institute of navigation instruments No.1, dingzigu Road, Hongqiao District, Tianjin, China
[2] Naval Research Academy, Beijing, 100161, China

Abstract. Aiming at the velocity damping feedback system commonly used in inertial navigation, this paper proposes the principle of decomposing the output attitude of inertial system based on FFT algorithm and gives the principle of spectrum adjustment ratio. The effectiveness of the algorithm is verified from two aspects: data simulation and actual system solution. In addition, in the design and implementation of FFT algorithm, the butterfly computation of operator symmetry is applied to effectively reduce the complexity of the algorithm, and the dual core chip technology is applied to reduce the impact of the complex algorithm on the system timing. The engineering applicability of the algorithm is demonstrated.

1. Introduction

The fast Fourier transform algorithm (FFT) was proposed in 1965. Based on the discrete Fourier statistical method, it effectively reduces the complexity of the solution cycle, making it possible to solve it in real time in the low-frequency processor system. The inertial system usually relies on the external velocity damping feedback to suppress the velocity oscillation trend. The damping coefficient absolutely determines the damping effect. When the external reference velocity error is small, the strong damping coefficient has better inhibition ability, making the velocity error converge quickly. When the external reference velocity error is large, the high damping coefficient will introduce the external reference velocity error. Therefore, real-time analysis of inertial navigation system attitude data spectrum to adjust the damping ratio can be used to improve the navigation accuracy of inertial navigation system.

2. Analysis of FFT algorithm principle

2.1 DFT transformation

For any continuous function *f(t)*, the frequency ω at any frequency can be obtained by Fourier transform Spectral density function $F(\omega)$ as follows.

$$f(t) = \int_{-\infty}^{+\infty} \frac{1}{2\pi} \left(\int_{-\infty}^{+\infty} f(t)\, e^{-j\omega t} dt\right) e^{j\omega t} d\omega \tag{1}$$

$$F(\omega) = \int_{-\infty}^{+\infty} f(t)\, e^{-j\omega t} dt \tag{2}$$

For any discrete function, Dirac sampling function should be introduced to analyze its Fourier decomposition. Dirac function has the following definitions.

$$\int_{-\infty}^{+\infty} \delta\,(t) dt = 1 \text{ and if } t \neq 0 \;\; \delta(t) = 0 \tag{3}$$

According to its definition, we can get the following formula.

$$\int_{-\infty}^{+\infty} \delta\,(t - t_0) f(t) dt = f(t_0) \tag{4}$$

According to this formula, Dirac sampling function can assist in sampling single time value of continuous function The discrete function of any continuous function *x(t)* can be obtained by continuous sampling with period T_s.

$$x_s(t) = \sum_{n=-\infty}^{+\infty} x(t)\, \delta(t - nT_s) \tag{5}$$

The discrete spectral density of equation $x_s(t)$ can be obtained by Fourier transform as follows.

$$x_s(\omega) = \int_{-\infty}^{+\infty} \left(\sum_{n=-\infty}^{+\infty} x(t)\, \delta(t - nT_s)\right) e^{-j\omega t} dt \tag{6}$$

$$x_s(\omega) = \sum_{n=-\infty}^{+\infty} \int_{-\infty}^{+\infty} \left(x(t)\delta(t - nT_s)\right) e^{-j\omega t} dt \tag{7}$$

$$x_s(\omega) == \sum_{n=-\infty}^{+\infty} x(nT_s)\, e^{-j\omega nT_s} \tag{8}$$

For the actual system, the analysis of spectral density should consider the completion of sampling point data in a limited time, so the sampling function $x_s(t)$ and spectral density $x_s(\omega)$.

$$x_s(t) = \sum_{n=0}^{N-1} x(t)\, \delta(t - nT_s) x_s(\omega) = \sum_{n=0}^{N-1} x(nT_s)\, e^{-j\omega nT_s} \tag{9}$$

In order to make the expression of spectral density more concise, it is simplified to the following formula.

* Corresponding author: 18945028038@163.com

$$X[k] = \sum_{n=0}^{N-1} x[n]\, e^{-j\frac{2\pi}{N}kn} \quad \text{where } X[k] = x_s(\omega), x[n] = x(nT_s) \tag{10}$$

It can be seen from this formula that if you want to extract the spectral density of all frequency points within a sampling time, $2n^2$ operations are required. Systems with high real-time requirements such as inertial systems The operation of $O(N^2)$ complexity class may interfere with the solution speed and the real sequence set by the system Therefore, consider using $e^{j\theta}$Operator has periodicity to reduce the amount of computation.

2.2 FFT transform

Euler formula is as following.

$$e^{j\theta} = cos\theta + jsin\theta \tag{11}$$

According to this, $e^{j\theta}$ Operator has periodicity.

$$e^{-j\frac{2\pi}{N}2nk} = e^{-j\frac{2\pi n}{N/2}k} = e^{-j\frac{2\pi n}{N/2}(k+N/2)} \tag{12}$$

According to this characteristic, $X[k]$ obtained by DFT algorithm can be decomposed into two parts: n is odd and n is even.

$$X[k] = \sum_{n=0}^{N/2-1} x[2n]\, e^{-j\frac{2\pi}{N}2nk} + \sum_{n=0}^{N/2-1} x[2n+1]\, e^{-j\frac{2\pi}{N}(2n+1)k} \tag{13}$$

$$X[k] = \sum_{n=0}^{N/2-1} x[2n]\, W[2n,k] + e^{-j\frac{2\pi}{N}k} \sum_{n=0}^{\frac{N}{2}-1} x[2n+1] W[2n,k] \tag{14}$$

$$\text{where } W[n,k] = e^{-j\frac{2\pi}{N}nk}$$

It can be seen from this formula that by extracting the common factor, the calculation of the odd part can be transformed into the calculation of the even factor $W[2n,k]$ For the convenience of analysis, we can make the even part $X[k]$ and the odd part $O[k]$. Since $W[1,k]$ has periodicity, the following formula can be obtained.

$$X[k+N/2] = E[k] - W[1,k]O[k] \tag{15}$$

Taking n equals 8 as an example, we can get the following formula. It can be seen from this formula that in the actual calculation of the system, only the even part operator $e[k]$ of $x[1]$, $x[2]$, $x[3]$, $x[4]$ and the odd part operator $O[k]$ of $x[5]$, $x[6]$, $x[7]$, $x[8]$ need to be calculated. The operator $e[k]$ to be calculated can be further divided into odd and even parts according to the factor until only one term remains in one part According to this butterfly decomposition operation, the calculation diagram is as follows.

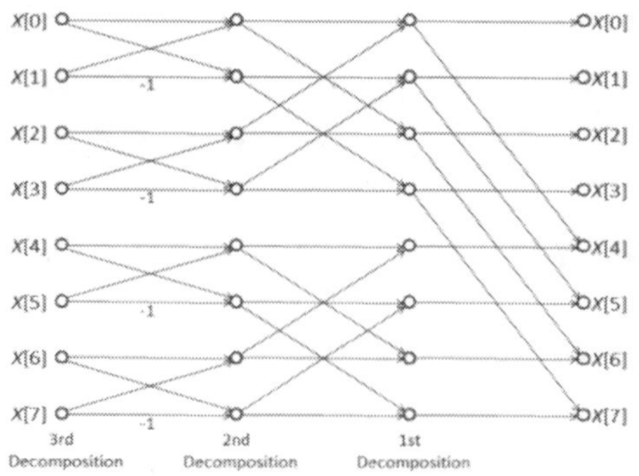

Figure 1. Schematic diagram of FFT iterative butterfly decomposition.

It can be seen from this formula that by extracting the common factor, the calculation of the odd part can be transformed into the calculation of the even factor $W[2n,k]$ For the convenience of analysis, we can make the even part $X[k]$ and the odd part $O[k]$. Since $W[1,k]$ has periodicity, the following formula can be obtained.

3. Application of FFT algorithm in adjusting the damping coefficient of inertial navigation system

The pure inertial navigation system only calculates the speed and position information according to the system data. This undamped system without external reference information input usually leads to 84.4 minutes Schuler oscillation in various error sources and initial condition errors, which is undesirable in many cases. Therefore, the inertial system usually introduces the external reference speed information for feedback, dampens the system speed, and then suppresses the Shula oscillation, so as to obtain higher precision navigation results.

3.1 Introduction to speed damping of inertial navigation system

The mechanical arrangement equation of geographic system (n-system) navigation system with horizontal velocity damping can be expressed by the following formula.

$$\dot{r}^n = v^n - [\rho^n]r^n \tag{16}$$

$$\dot{v}^n = f^n - [\omega_{ie}{}^n + \omega_{in}{}^n]r^n - Cv_d + g^n \tag{17}$$

Where $v_d = v^n - v_r{}^n$ represents the difference between the system inertia indication speed v^n and the external reference speed $v_r{}^n$. C refers to damping feedback network. Typical damping network forms include constant velocity error feedback network, phase lead series network and phase lag lead series network. The error transfer function diagram of the typical geographical system velocity Schuler damping loop is shown in the figure below.

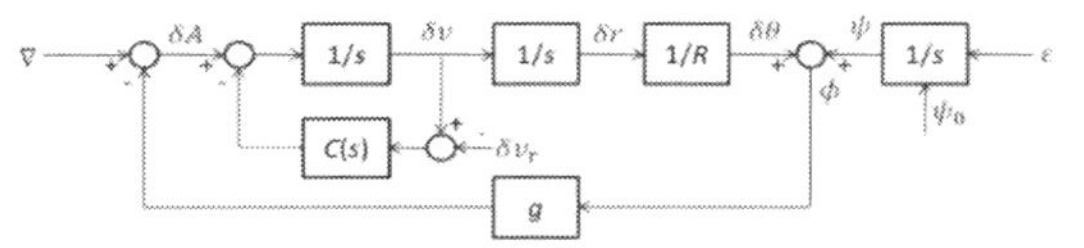

Figure 2. Block diagram of error transfer function of Schuler damping loop.

Let $Q(s) = s/(s + C(s))$, the following series Schuler loop error and error transfer formula and error transfer function block diagram can be obtained.

$$v(s) = \frac{1}{s^2+Q(s)\omega_s{}^2}\left[\left(1 - Q(s)\right)s^2\delta v_r(s) + Q(s)\left(s\nabla(s) - g\varepsilon(s)\right)\right] \quad (18)$$

$$\delta r(s) = \frac{1}{s^2+Q(s)\omega_s{}^2}\left[\left(1 - Q(s)\right)s^2\delta v_r(s) + Q(s)\left(\nabla(s) - g\psi(s)\right)\right] \quad (19)$$

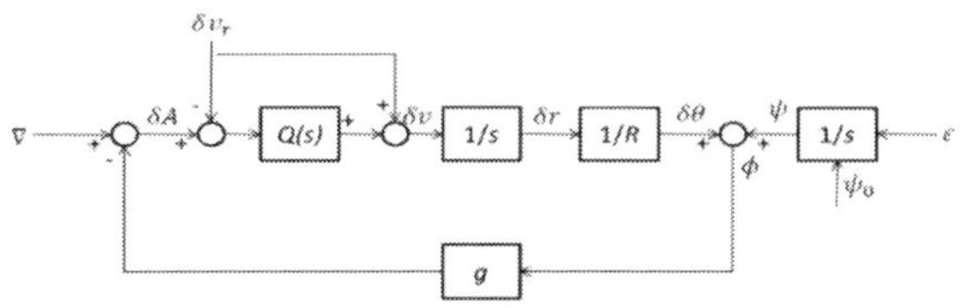

Figure 3. Error transfer function block diagram of series Schuler damping circuit.

Generally, the phase lag lead series damping network has the best performance. This paper will analyze this network. This type of damping network has the following transfer function.

$$Q(s) = \left((1 + 4\xi\eta)s^2 + 2(\xi + \eta)\omega_s + \omega_s{}^2\right)/\left(s^2 + 2(\xi + \eta) + \omega_s{}^2\right) \quad (20)$$

Consider these two conditions, $\omega \ll \omega_s$ and $\omega_s \ll \omega$, the following table can be obtained through the derivation and simplification of the error formula.

Table 1. Error transfer function block diagram of series Schuler damping circuit.

Error source	$\omega \ll \omega_s$ Reference speed error δv_r	$\omega \ll \omega_s$ Accelerometer error ∇	$\omega \ll \omega_s$ Reference speed error δv_r	$\omega_s \ll \omega$ Accelerometer error ∇
$\delta v(j\omega)$	$\frac{2\xi\omega^3}{\omega_s{}^3}$	$\frac{\omega}{\omega_s{}^2}$	$\frac{2\xi\omega_s}{\omega}(1 - 4\mu\eta)$	$\frac{1}{\omega}$
$\delta r(j\omega)$	$\frac{2\xi\omega^2}{\omega_s{}^3}$	$\frac{1}{\omega_s{}^2}$	$\frac{2\xi\omega_s}{\omega^2}(1 - 4\mu\eta)$	$\frac{1}{\omega^2}$

Therefore, when the external reference velocity error is large, the strong damping coefficient ξ will bring higher speed position error. Common external reference velocity sources include electromagnetic log, Doppler sonar log and acoustic correlation velocity log. The electromagnetic log measures the velocity of the carrier passing through the water, and its measured value may be polluted by the distorted water flow around the carrier. The error source of Doppler sonar is the same as that of the electromagnetic log, and excessive ocean current velocity will bring high measurement error. Doppler sonar log can measure relative sea bottom velocity, but it only has effective data within 200m depth. Using this velocity source has a high limit on the bottom contact depth of underwater precision navigation. Therefore, the error of current velocity will be analyzed below.

3.2 Damping coefficient adjustment principle

In the following cases, the speed of current measurement is large. When the ship makes a large maneuvering turn at a large angle, the relative ground speed of the ship is close to zero. However, due to the large current movement caused by the turning of the ship and the installation position of the speed measuring device is not close to the turning center, the speed measurement is large. Therefore, the error of external reference speed measurement during turning is large. In addition, when encountering rough and severe sea conditions (above level 3), the current velocity is larger than the ground speed of the ship itself, and the swing error of the ship itself is also large, so it brings higher velocity measurement error. It can be seen from the above that when the ocean current velocity is large, it usually affects the transformation of the large attitude angle of the ship body, and also affects the large transformation of the output attitude angle of the ship body fixed inertial navigation. Therefore, this paper considers extracting the frequency spectrum of the inertial navigation attitude angle through the FFT algorithm, and inversely separating the external reference velocity measurement error through the frequency spectrum. When the velocity measurement error is large, a smaller damping ratio is selected to bring higher measurement accuracy. When the velocity measurement error is small, A large damping ratio is selected to suppress the stretch oscillation of the inertial system.

Due to the complex characteristics of the system to be analyzed, there is no mathematical model reference in related fields, and different ship models and different inertial navigation models are also different. Therefore, this paper gradually optimizes the damping ratio adjustment rules through off-line data simulation and physical test retest of an inertial navigation system. The adjustment principle model obtained from the final set damping ratio is as follows.

$$\text{if} \sum_{n=0}^{n_{max}} k_n F(nw_0) < \xi_0$$

$$\xi = \xi_{max} - \sum_{n=0}^{n_{max}}\left(k_{Rn}F(nw, \theta_R) + k_{Pn}F(nw, \theta_P) + k_{Hn}F(nw, \theta_H)\right) \quad (21)$$

$$\text{if} \sum_{n=0}^{n_{max}} k_n F(nw_0) \geq \xi_0$$

$$\xi = 0 \tag{22}$$

In the formula, $\theta_R, \theta_P, \theta_H$ refers to the output attitude angle where $F(nw_0, \theta)$ Represents the spectral density of a specific frequency for a certain attitude angle after fast Fourier transform And k_{Rn}, k_{Pn}, k_{Hn} represents the adjustment weight of the spectral density. It is worth noting that K near the resonance frequency of the system is applied in the parameter adjustment process givek_na small weight to avoid the damping ratio caused by system resonance n_{max} represents the highest frequency band for FFT data analysis to limit the impact of high-frequency noise on damping ratio, ξ_{max} represents the maximum damping coefficient, indicating that the external information reference value at this time is fully trusted, and the system shall be damped with the highest damping ratio.

3.3 Inertial platform construction

In order to further verify the effectiveness of the algorithm, it needs to be verified in the actual inertial system. At the same time, in order to further improve the processing capacity of the system for FFT algorithm and separate the timing impact of FFT algorithm on the main program of the system, the system uses dual core DSP28379 chip to analyze the test data. The communication between dual cores adopts the form of IPC message and 28379 shared memory. The operation timing diagram of the whole system is shown in Figure 4 below.

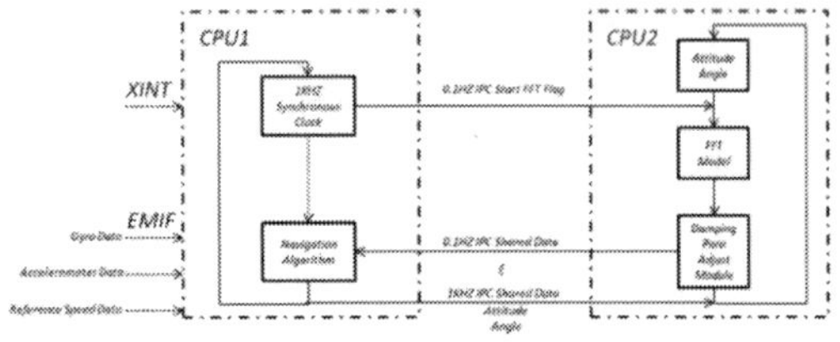

Figure 4. Dual core operation sequence of inertial system.

CPU1 will perform single-step calculation after receiving the external synchronization signal of 1000 frequency. After calculating the attitude information, CPU2 will send the attitude information to CPU2. At the same time, CPU2 will divide the frequency of 0.1 Hz in the 1000Hz frequency and send it to CPU2 as an FFT flag. After receiving the flag, CPU2 will perform FFT decomposition on the attitude data in this period and output it to the damping coefficient adjustment module. After calculating the damping ratio, CPU1 will send the real-time damping ratio back to CPU1. The navigation algorithm uses the updated damping ratio to solve the velocity and position.

4. Test data analysis

The navigation speed calculation results of a certain type of inertial equipment through dual channel FFT damping ratio adjustment algorithm and constant damping coefficient are shown in the following figure.

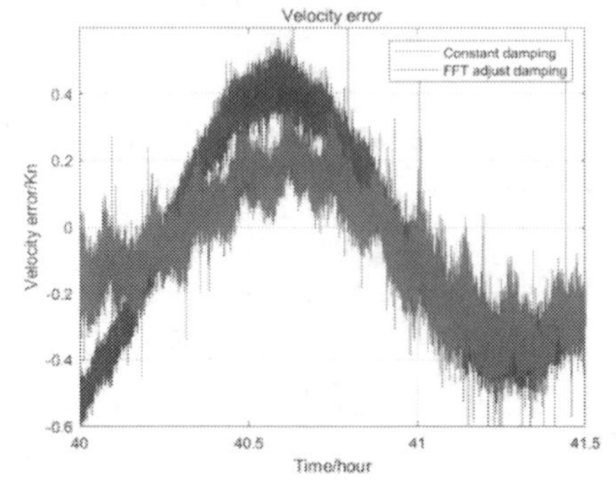

Figure 5. Navigation accuracy results under different damping conditions.

According to the test data curve, the velocity accuracy of the inertial system using FFT algorithm to adjust the damping ratio is improved by 5%, and the feasibility of using FFT algorithm to adjust the damping ratio is demonstrated.

5. Conclusion

Firstly, this paper analyzes the principle and digital implementation method of FFT algorithm, and gives the butterfly decomposition calculation diagram of FFT algorithm, which effectively reduces the computational complexity of the algorithm. On this basis, for the external reference velocity damping feedback system in inertial navigation, the quantitative expression between damping coefficient and external reference velocity is analyzed, and the adjustment principle of damping ratio is summarized through data simulation. In order to further verify the effectiveness of the algorithm, the dual core technology is used to separate the main program of the system and the FFT spectrum analysis algorithm to complete the verification on the inertial system, and higher speed accuracy is achieved. Through the above research, the conclusions are as follows.

1. The FFT algorithm will effectively reduce the complexity of spectrum analysis and calculation, and lay a foundation for its real-time operation on low dominant frequency systems.
2. Using dual core system to separate main program and FFT algorithm program segment can further reduce system pressure and reduce the possibility of complex algorithm affecting system timeliness.
3. By using FFT algorithm to analyze the output attitude of the inertial system in real time, and then analyze the complex situation of the current sea state, and reasonably planning the damping coefficient of the inertial system, the inertial system can obtain better navigation accuracy.

References

1. Qi Yuan,Kang Yang,Parallel implementation of large number FFT algorithm c6678 multi-core DSP, J. Electronic measurement technology. 38 (2015) 2.
2. FY Chen, DS Zhang, Z Wang, Research of the heterogeneous multi-core processor architecture design, J.Computer Engineering &Science, 33 (2011) 12.
3. Ning Hou,Yapeng Lu, Communication solution of multi-core chipset, J. Computer Era. 10 (2014) 17.
4. Haiming Wang,Weixia Zhang, Error compensation of gyro north seeking based on FFT,J. Sensors and Microsystems, 33 (2014) 2.
5. Fei Liu,Chao Liu, Horizontal damping algorithm of strapdown inertial navigation system based on Kalman filtering technology, J. Chinese Journal of inertial technology. 21 (2013) 3.
6. Yan Lu, Research on Celestial / inertial integrated navigation technology based on velocity damping, J. Optics and Optoelectronics Technology, 10 (2012) 4.

Experimental Study on Laser Ablation of C/C Composites

Jian Zheng [1,*], Yulong Yang [2], Yuhang Gao [1]

[1] School of Mechanic Engineering, NJUST, Nanjing, China
[2] Nanjing Chenguang Group Co., Ltd, Nanjing, China

Abstract. Aiming at the application of C/C composite in nozzle throat lining of solid rocket motor, the thermochemical ablation of C/C composites under CO2 or N2 different confining pressures is studied by using a continuous laser ablation test device, and some useful conclusions are obtained. In CO2 confining pressure environment, with the increase of pressure, the heat absorption effect of thermochemical reaction becomes more obvious, and the surface temperature of the sample decreases. The temperature at 1MPa and 2MPa decreases by 12% and 23% respectively, compared with the maximum temperature in N2 environment because there is no chemical reaction between N2 and C/C composites. The sample surface is carbonized after endothermic, and forms a large number of holes.

1. Introduction

The heat transfer and ablation of ablation resistant materials is a complex physical and chemical change process. Its heat transfer and ablation characteristics will be affected by many factors. C/C composite is widely used as throat lining material of rocket nozzle because of its good ablation resistance. The thermal environment of the nozzle throat is bad during the working process of the rocket motor. Due to the pyrolysis reaction of C/C composite at high temperature, it absorbs heat and produces pyrolysis gas. Then the size of its inner surface will change, which will affect the internal ballistic performance of the rocket motor[1].

In 2011 and 2013, K.Z. Li and B.L. Zha studied the actual ablative behavior of C/C composite material in solid rocket engine through ground ignition experiment, and the results showed that the fiber head was conical in high fiber density area, and the fiber was stripped in low fiber density area, and the surrounding matrix was ablated in shell shape. The oxidation of C/C composites results in the loss of the matrix material between the conical fiber and the shell and the formation of a large number of holes on the surface of the material[2,3]. F. Huang et al. used the DSMC method to reveal the ablation mechanism of C/C composites in air by studying the diffusion behavior of oxygen molecules in the ablation gap and the collision law with the wall at the micron level[4].

In 2015, J.Y. Zhi explained the thermal conductivity mechanism of C/C composites and the influence of graphitization degree, graphite scale and other factors on thermal conductivity from the micro scale. The results show that there is a linear relationship between the thermal conductivity of carbon matrix around the carbon fiber reinforcement and the thermal conductivity of C/C composites along the fiber direction[5].

In 2017, J. Lachaud et al. described the ablative mechanism of 3D C/C composites from the microscopic level. Based on the calculation of different ablative rates of carbon fiber and carbon matrix, the microscopic ablative morphology of the materials was obtained, which was consistent with the experimental results, proving the reliability of the simulation algorithm[6]. Y.Liu et al. studied the ablation characteristics of a 4D carbon/carbon composite under a high flux of combustion products with a high content of particulate alumina in a solid rocket motor[7] .

In 2018, J.M. Su et al. explored the factors affecting the ablative rate of C/C composite throat lining through several ground ignition experiments, and the results showed that pressure is the first factor affecting the ablative rate, the concentration of oxidizing components of propellant is the second factor affecting the ablative rate, and the combustion temperature of propellant has no obvious effect on the ablative rate[8].

Based on the fact that most of the previous scholars used numerical simulation analysis to study C/C composites, even if the ground ignition test of solid rocket motor was used which has a high cost. In this paper, the laser ignition test is proposed to study the ablation characteristics of C/C composites. It is more economical and convenient to operate.

* Corresponding author: zhengjian@njust.edu.cn

2. Test Equipment and Sample Preparation

During the working process of solid rocket motor, the C/C composite in the nozzle reacts with CO_2 and H_2O in the fuel gas. This chemical reaction results in the change of the inner surface of the nozzle. In the commonly used reaction model, similar heterogeneous reaction control equations are used for the reaction rates of $C+CO_2$ and $C+H_2O$. In this paper, the thermochemical ablation of C/C composites in CO_2 atmosphere at different pressures is studied by using a continuous laser ablation experimental device.

The continuous laser ablation experimental device generates laser from CO_2 laser with wavelength of 10.6 μm. The maximum power is 300W. The power and ablation time are controlled by the controller card. The sample is placed in a closed square cavity with a size of 110mm×120mm×260mm. In the experiment, different CO_2 atmosphere gas pressure conditions are realized by connecting the gas tank. After the experiment, the gas is discharged through the exhaust valve. In order to make the laser hit directly above the sample, a lens is set on the upper part of the square cavity, the focal length of the lens is 127mm, and finally the diameter of the light spot on the sample is 3mm. The experimental system is shown in Fig. 1. Calibrate the laser power to make the heat flux density of the heating surface of the sample the same before each experiment. The sample size of C/C composite is ϕ 5mm × 6mm cylindrical sample. The test sample is made by water cutting.

In the experiment, the infrared thermometer is used to measure the temperature of the heated surface. The model of the infrared thermometer is Ircon Modline 5. Its temperature measurement range is from 1000 °C to 3000 °C. The sample under ablation is shown in Fig. 2.

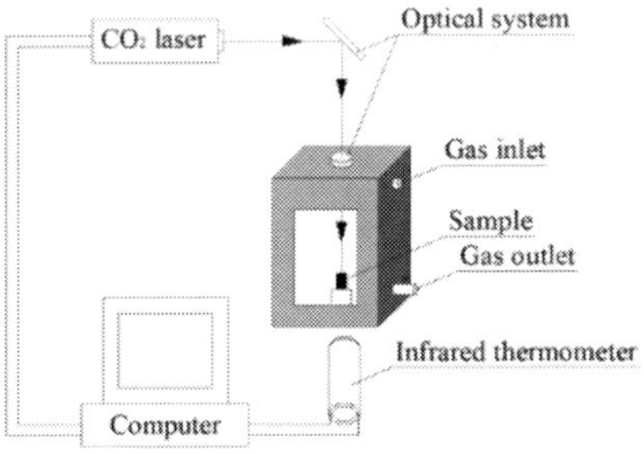

Figure 1. Schematic diagram of laser ablation experimental system

Figure 2. Sample during ablation

3. Experimental Results and Analysis

In the experiment, CO_2 or N_2 atmosphere gas with different pressure is introduced into the metal square cavity. The pressure is 0.1MPa, 1MPa, 2MPa and 3MPa respectively. The loading laser power are 120W for 60s continuously. The temperature of heating surface under different atmosphere gas conditions is measured by infrared thermometer, as shown in Fig. 3. The surface temperature is below 1273.15K, which cannot be measured due to the limitation of equipment range.

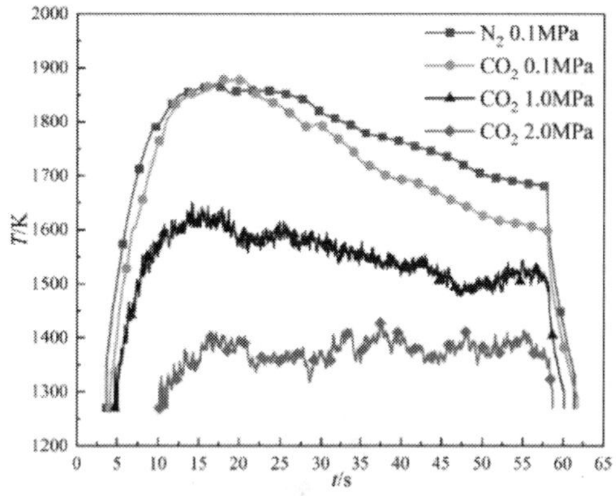

Figure 3. Variation of temperature on the heated surface of the sample with time

 https://doi.org/10.1051/matecconf/202236301028

In N_2 environment the sample temperature is higher, which results in reducing the surface heat flux of the sample due to the pore gas expansion flue gas to escape. And it affects the light path of infrared thermometer, so the temperature declines over time. Under the same 0.1 MPa CO2 gas environment, because of the thermal chemical ablation reaction of the sample, the temperature is slightly lower than the surface temperature in N_2,as shown in Fig. 3.

With the increase of pressure, the heat absorption effect of thermochemical reaction becomes more obvious in the CO_2 environment. The temperature at 1MPa and 2MPa decreases by 12% and 23% respectively, compared with the maximum temperature in N_2 environment. The temperature at 3MPa cannot be measured because it is lower than 1273.15K. But it can be inferred that the temperature decreases by more than 30%. The output temperature data of the infrared thermometer is the maximum temperature at the test area. Due to the uneven distribution of reaction rate on the surface, the temperature fluctuation is relatively large at 1MPa and 2MPa.

The morphology of C/C composites sample after the test was obtained by Hitachi SU3500 scanning electron microscope shown in Fig. 4. Sample heated surface microstructure is as shown in Fig. 5. Due to the thermal conductivity of carbon fiber higher than that of the pyrolytic graphite substrate, so the quantity of heat to the internal transfer faster through carbon fiber. It makes carbon fiber and CO2 react faster than the pyrolytic graphite substrates, so it forms a large amount of holes in the original position of carbon fiber after the experiment shown in Fig. 5.

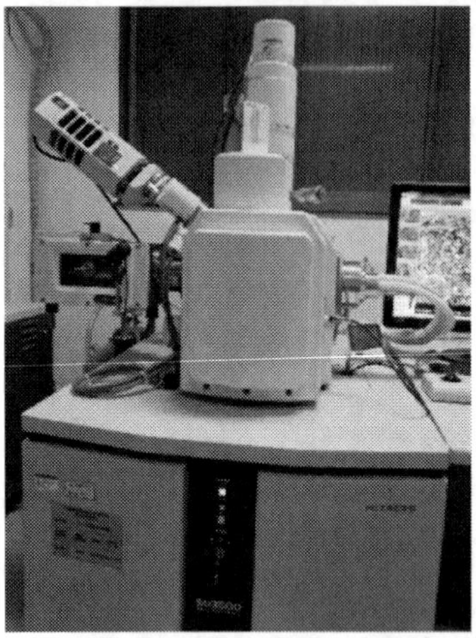

Figure 4. Scanning electron microscope(SEM)

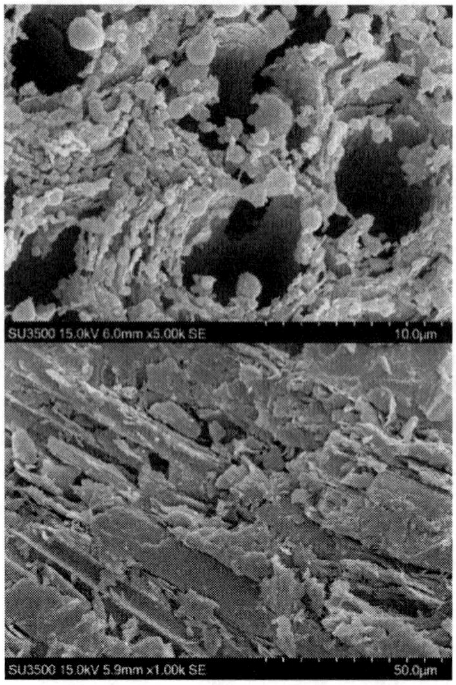

Figure 5. Microstructure of sample after ablation

4. Conclusion

In this paper, the thermochemical ablation of C/C composites under different CO_2 pressures is studied by using a continuous laser ablation test device, and the following conclusions are drawn.

(1) In CO_2 confining pressure environment, with the increase of pressure, the heat absorption effect of thermochemical reaction becomes more obvious, and the surface temperature of the sample decreases. The sample surface is carbonized after endothermic, forming a large number of holes.

(2) In N_2 confining pressure environment, the surface temperature of the sample is higher than that of CO_2 environment because there is no chemical reaction between N_2 and C/C composites.

Acknowledgements

This research was financially supported by the National Defense Foundation.

References

1. G.L.Vignoles, Y. Aspa, M. Quintard, Modelling of carbon–carbon composite ablation in rocket nozzles,Composites Science and Technology. 70(2010)1303-1311.
2. K.Z. Li, X.T. Shen, H.J. Li, Ablation of the carbon/carbon composite nozzle-throats in a small solidrocket motor, Carbon. 49(2011)1208-1215.

3. B.L. Zha, D.Y. Huang, S.L. Qiao, Ablation test and ablation mechanism of C/C composites, Solid Rocket Technology. 36(2013)692-696.
4. F. Huang, J.J. Yu, X.T. Li, Microscale oxygen diffusion characteristics and ablation analysis of carbon/carbon materials, Journal of Astronautics. 32(2011)1848-1853.
5. J.Y. Zhi. Theoretical study on thermal conductivity of pitch based C/C composite, North China Electric Power University, Beijing, 2015.
6. J. Lachaud , Y. Aspa , G.L. Vignoles, Analytical modeling of the steady state ablation of a 3D C/C composite, International Journal of Heat and Mass Transfer. 51(2008)2614-2627.
7. Y. Liu, J.Q. Pei, J. Li, Ablation characteristics of a 4D carbon/carbon composite under a high flux of combustion products with a high content of particulate alumina in a solid rocket motor, New Carbon Materials. 32(2017)143-150.
8. J.M. Su, S.J. Zhou, N.J. Xue, Effect of nozzle thermal environment on ablation rate of carbon-based material throat lining, New Carbon Materials. 33(2018)66-72.

 https://doi.org/10.1051/matecconf/202236301029

Research on the Extrusion Process of belt with Notch Descrition

Qi Liu, Yuwei Wang, Wenfang Zhu, Na Zhao

Northwest Institute of Mechanical and Electrical Engineering Xianyang Shaanxi China

Abstract. The extrusion process of the belt is a process in which the belt and the inner bore of the barrel undergo complex contact and collision, and the rifling gradually penetrates the belt, so that the material of the belt continues to fail and destroy. It has high instantaneous, strong impact, high temperature, high-speed change. There is a big gap between the existing theory and the actual extrusion process, and it is urgent to carry out research work on this problem. This paper analyzes the basic characteristics of elastic and plastic deformation during the extrusion process of the belt, and establishes the Johnson-Cook constitutive model of the deformation, flow and failure process of the belt material. Explicit dynamic calculations are carried out in the process of belt extrusion, and the contact problems, deformation of the belt and material flow problems during the extrusion process of the belt are discussed in depth

1. Introduction

When the artillery automatic weapon is fired, the projectile moves along the barrel under the action of high temperature and high pressure gunpowder and gas, and interacts with the barrel. The extrusion process of the belt has an important influence on the internal ballistic of the barrel weapon. However, due to the instantaneous nature of the extrusion process, it is difficult to accurately measure it by experiments. Therefore, the friction work and the rotational energy of the projectile are often considered as secondary work in the classical internal ballistics. Taking into account the energy balance equation of internal ballistics, and ignoring the extrusion process, it is considered that the projectile starts to move when the chamber pressure reaches the starting pressure, which is quite different from the actual situation [1].

The extrusion process of the belt is actually a process of contact and collision between the belt and the inner bore of the barrel, the rifling gradually penetrates the belt, and the material of the belt continues to fail and destroy. It has the characteristics of high instantaneous, strong impact, high temperature and high-speed change. The laws of advancing resistance and deformation are very complex [2-3]. Due to many difficulties, there is still a big gap between the existing research theories and the actual extrusion process, and it is urgent to carry out research work on the extrusion process.

With the development of computer level and the development of three-dimensional simulation calculation software, the research on the extrusion process has changed from the relatively simple two-dimensional calculation in the early stage to the complex three-dimensional calculation model; from the early ideal rigid-plastic material model to the current consideration of elastic belt a complex elastic-plastic constitutive model of material damage; from the early fully Lagrange finite element method to the current ALE adaptive technique [4]. With the development of technical means, the necessary means are provided for further research on the extremely complex mechanical process of belt extrusion. In order to deeply understand the essence of the extrusion process of the belt, it is necessary to establish a more complex mathematical model that can reflect the mechanics essence of the extrusion process to solve the highly nonlinear problem in the extrusion process. The existing large-scale finite element software and computer hardware level can already meet the needs of the extrusion process simulation research.

In this paper, relying on the National Natural Science Foundation of China project "Research on the Transient Extrusion Mechanism of Projectile Based on Different Gas Pressure Loading Rates", the method of nonlinear finite element numerical simulation is used to study the mechanical mechanism of the extrusion process of the belt, and the finite element software is used to establish the nonlinear dynamics model of the extrusion process, using the explicit numerical difference algorithm to numerically simulate the nonlinear dynamics of the extrusion process of the belt, to study the formation process of the belt groove, the extrusion movement law of the belt, and the dynamic extrusion resistance and the extrusion pressure, to study the effect of friction properties on the extrusion process of the belt. The above research can provide technical support for the design and use of projectiles and barrels, provide a necessary basis for further research on the movement of projectiles in the chamber, and provide a certain useful reference for the research on improving shooting accuracy.

2. Basic assumption

The force and deformation of the belt in the process of extrusion are very complex, and it has the characteristics of high instantaneous, strong impact, large deformation, and high-speed deformation[5]. In order to simplify the model and ignore the secondary influencing factors, the following basic assumptions are made for the process of the band being squeezed into the bore [6-7]:

(1) The jamming process is not considered, that is, the initial position of the projectile is the contact position between the elastic belt and the initial section of the slope bore, and the initial stress and deformation of the elastic belt are ignored;

(2) After the caliper is in place, it is considered that the belt and the inclination of the slope are closely matched, and this position is used as the beginning of the projectile extrusion process;

(3) Ignoring the imbalance problem, it is assumed that the projectile and the barrel axis are concentric, and the friction force and motion resistance are both along the axis direction;

(4) The influence of air resistance and shock wave at the front of the projectile is ignored;

(5) The stress and deformation of the barrel and projectile under gravity are not considered;

(6) The recoil movement of the body tube is not considered.

3. Model establishment

3.1 Johnson-Cook constitutive model

In the artillery structure, most structural materials adopt the elastic material constitutive relationship, and the constitutive relationship of the material in the elastic stage can be expressed as [8]:

$$\sigma = E\varepsilon \tag{1}$$

Where E is the elastic modulus.

The relationship between the stress and strain of the elastic band material, when the stress is less than the yield stress of the material, shows a linear relationship between the stress and the strain, and the deformation of the material is elastic [9]. As the material stress continues to increase beyond the yield stress, the relationship between stress and strain is no longer linear, as shown in Figure 1.

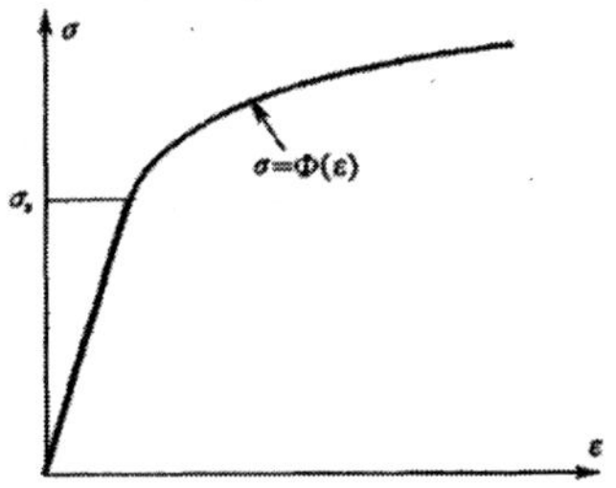

Fig. 1 Stress-strain curve of elastic band material

The elastic band not only produces plastic deformation during the extrusion process, but also is affected by the high temperature and high strain rate of the gunpowder gas. In the finite element model established in this paper, the constitutive model used for the elastic band material is the Johnson-Cook model. The Cook constitutive model is an empirical constitutive equation. Considering the effects of temperature and strain rate, the Johnson-Cook material model consists of two parts [10-11].

Part 1: The Johnson-Cook Plasticity Model

In the Johnson-Cook plasticity model, the Von Mises yield stress is a function of plastic strain, strain rate, and temperature.

$$\bar{\sigma} = (A + B\varepsilon_p^n)\left[1 + C\ln\dot{\varepsilon}^*\right]\left(1 - T^{*m}\right) \tag{2}$$

$$T^* = \frac{T - T_r}{T_m - T} \tag{3}$$

$$\dot{\varepsilon}^* = \frac{\dot{\varepsilon}_p^*}{\dot{\varepsilon}_0^*} \tag{4}$$

Where ε_p is the equivalent plastic strain; $\dot{\varepsilon}_p$ is the equivalent plastic strain rate; $\dot{\varepsilon}_0$ is the reference strain rate; T is the temperature; T_r is the reference temperature, generally taking room temperature; T_m is the melting temperature; A 、 B 、 C 、 m and n are constants.

The temperature change caused by the adiabatic process is:

$$\Delta T = \int_0^{\varepsilon_p} \chi \frac{\bar{\sigma} d\varepsilon_p}{\rho c_p} \tag{5}$$

In the formula, ρ is the material density; c_p is the constant volume specific heat; χ is the Taylor-Quinney coefficient, which represents the ratio of plastic work into heat, generally 0.85 to 0.95.

Part II: Johnson-Cook Fracture Failure Model

The Johnson-Cook fracture failure model defines damage in terms of the equivalent plastic failure strain [12]:

$$\varepsilon_f^{JC} = \left[D_1 + D_2\exp\left(D_3\sigma^*\right)\right]\left[1 + D_4\ln\dot{\varepsilon}^*\right]\left(1 + D_5T^*\right) \tag{6}$$

$D_1 \sim D_5$ is the material fracture failure parameter; σ^* is the stress triaxiality, that is, the ratio of the hydrostatic stress σ_m to the Von Mises equivalent stress σ_{eq} :

$$\sigma^* = \frac{\sigma_m}{\sigma_{eq}} \tag{7}$$

3.2 Explicit kinetic theory

The explicit dynamic analysis method is to use the central difference method to solve the dynamic finite element equations. The basic idea is that the acceleration is given by the first-order central difference of the velocity, and the velocity is given by the first-order central difference of the displacement. [13]:

$$A(t_n)=\left[V\left(t_{n+\frac{1}{2}}\right)-V\left(t_{n-\frac{1}{2}}\right)\right]/\frac{\Delta t_{n-1}+\Delta t_n}{2} \tag{8}$$

$$V\left(t_{n+\frac{1}{2}}\right)=\frac{U(t_{n+1})-U(t_n)}{\Delta t_n} \tag{9}$$

The step size of the time step and the time start and end points are defined by the following formulas.

$$\Delta t_{n-1}=t_n-t_{n-1},\quad \Delta t_n=t_{n+1}-t_n \tag{10}$$

$$t_{n-\frac{1}{2}}=\frac{t_n+t_{n-1}}{2},\quad t_{n+\frac{1}{2}}=\frac{t_{n+1}+t_n}{2} \tag{11}$$

The central difference method is a conditionally stable algorithm that requires relatively small time steps to maintain a stable state. Theoretically, the critical time step to maintain numerical stability should satisfy Equation [14]:

$$\Delta t_n \leq \frac{2}{\omega} \tag{12}$$

where ω is the highest order natural vibration frequency of the smallest element in the system.

4. Research on the extrusion process of a large-calibaer artillery shell belt

4.1 Geometric model establishment

The extrusion process of the belt is shown in Figure 2. The projectile starts from the position of the caliper, and is pushed by the gunpowder gas acting on the bottom of the bullet to overcome the extrusion resistance and advance along the axial direction of the slope until the rear end of the elastic belt is fully squeezed into the entire belt. The rifling is deep.

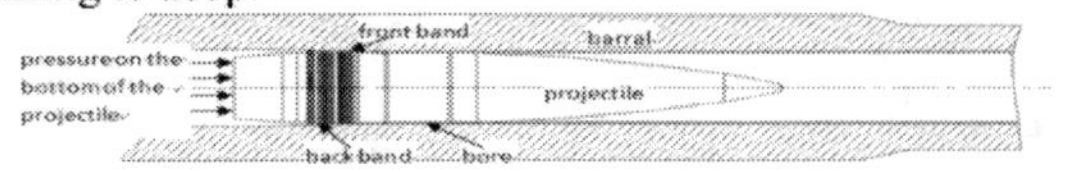

Figure 2 Schematic diagram of the extrusion process of the belt

The projectile studied in this paper consists of two elastic belts and a closed gas ring. Each of the two elastic belts has a transverse groove to accommodate the deformed material of the elastic belt. The rear elastic belt has a flange, and the rear elastic belt flange is removed from the elastic belt. The radius is large, and the rest part has a small amount of interference with the yin line of the barrel, which is conducive to the positioning of the projectile caliper. The two-dimensional cross-sectional view of the elastic belt and the closed gas ring is shown in Figure 3.

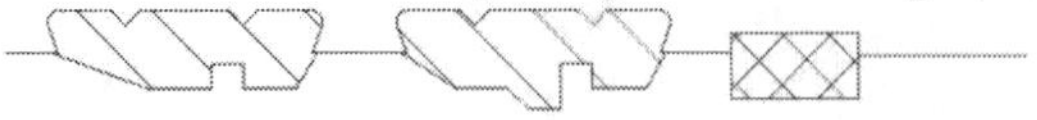

Figure 3 Two-dimensional cross-sectional view of the belt

In NX10, the geometric entity of the barrel is obtained by rotation, and the geometric entity has different tapers; in order to simplify the calculation, the rifling angle is ignored, and the straight rifling is modeled. A section of the barrel of the bore is used as the barrel model for this calculation; the two-dimensional outline of the elastic belt is rotated circularly along the axis of the barrel to form a three-dimensional solid model of the elastic belt and the closed gas ring; Obtained by axial rotation of the axis. In this paper, a large-caliber bottom concave projectile is taken as the research object. According to the actual positional relationship between the projectile body, the elastic belt and the closed gas ring, the assembly model of the projectile body and the elastic belt is established as shown in Figure 4.

Figure 4 Assembly drawing of the belt of the projectile body

4.2 Finite Element Discrete

Save the established body tube geometry as an STP format file, and import the STP format file in the Geometry module of the WORKBENCH platform. The rifling section is meshed by using the two-dimensional meshing technique, and the two-dimensional mesh of the barrel section is swept by the three-dimensional mesh SWEEP division function, so as to obtain the three-dimensional mesh of the barrel; the contact between the elastic belt and the inner wall of the groove is obtained. Algorithm-based, secondary contact algorithm, the mesh of the secondary surface needs to be subdivided. In the contact, the surface composed of the elastic belt is selected as the secondary surface, and the elastic belt element is subdivided. The finite element size of the two elastic belts and the closed air ring is selected as 0.5mm. The element is divided, and the SWEEP method is used to sweep and divide; the finite element mesh of the projectile is also divided according to the above steps, but the projectile is a secondary consideration during the extrusion process, and the relative deformation is small, so the projectile's The finite element mesh size is large, the maximum element size is 10mm, and the final divided finite element mesh is shown in Figure 5.

(a) The finite element model of the barrel

(b) The finite element model of the belt

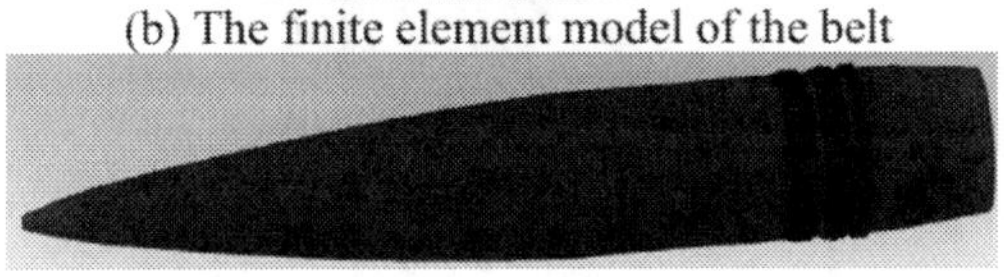

(c) Finite element assembly model of missile body and belt

Fig. 5 The finite element model of the extrusion process

4.3 Material properties

Since the material of the elastic band is softer than the barrel and projectile, the material of the barrel and the projectile is considered to be alloy steel. In this simulation, the elastic-plastic constitutive model is used, and the material parameters are shown in Table 1.

Table 1 Material parameters of projectile and barrel

	density $\rho(kg/n$	elastic Modulus $E(GP$	poisson's ratio μ	initial yield limit $\sigma_0(GP$	strength criterion
projectile	7820	2.03	0.3	0.980	Von-Mises
barrel	7820	2.03	0.3	1.030	Von-Mises

The elastic band material is red copper, which is widely used because of its good elasticity and plasticity and meets the requirements of the elastic band material, and the closed gas ring material is nylon material. Determining the parameters of the Johnson-Cook model of the elastic band material is very important to accurately simulate the extrusion process [6, 10]. The parameters of the elastic material constitutive model used in this paper are shown in Table 2.

Table 2 Parameters of Johnson-Cook constitutive model of copper material

Physical quantity	Parameter value	Physical quantity	Parameter value
density $\rho(kg/m^3)$	8960	shear Modulus $G(GPa)$	47.7
bulk modulus $K(GPa)$	143	specific heat CP	399
static yield stress $A(GPa)$	0.12	melting temperature $T_m(K)$	1356
hardening parameters $B(GPa)$	1.43	reference temperature $T_r(K)$	293
strain rate parameter C	0	hardening Index n	0.5
hardening Index $\dot{\varepsilon}_0$	1	maximum failure plastic strain	0.5

4.4 Boundary conditions

Ignoring the self-weight of the barrel and the projectile, the six degrees of freedom of the bottom end face of the barrel are restrained, and the bottom pressure curve shown in Figure 6 is applied to the bottom of the bullet.

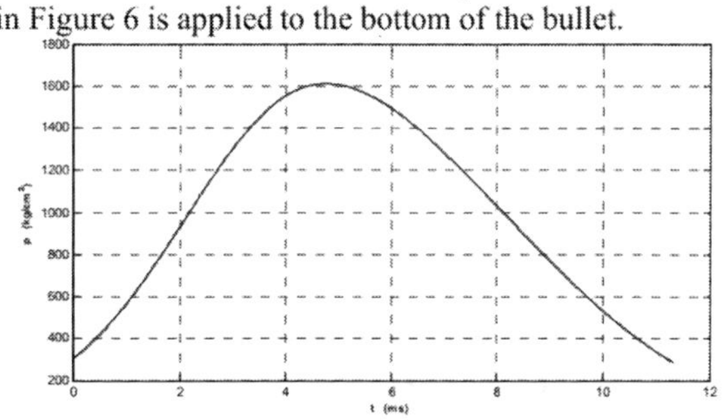

Fig. 6 Bottom pressure curve

Since the elastic belt and the projectile body belong to an interference fit, in the Explicit Dynamics module, the contact surface of the two is set to BOUNDED, the setting of the closed gas ring is consistent with the elastic belt, and the contact between the barrel or the rifling and the elastic belt adopts dynamic penalty. function method. In this paper, the relevant numerical simulation is carried out on the basis of dynamic modeling. Since the elastic-plastic constitutive, large deformation and contact problems are involved at the same time, the calculation consumes a lot of computer time. The solver used in this paper is AUTODYN, the total simulation time is 20ms, the minimum integration step is 0.01ms, the maximum integration step is 0.5ms, and the calculation results of the squeeze are saved every 1ms.

5. RESULT ANALYSIS

Figure 7 shows the resistance curve of the elastic band with the extrusion distance. It can be seen that the extrusion resistance increases continuously with the extrusion process. The maximum extrusion resistance is 83346N. At the beginning of the extrusion process, the extrusion resistance increases. Very quickly, and then it grows slowly. At the end of the squeeze, the squeeze resistance almost no longer increases, indicating that the squeeze process is completed, and the elastic bands have been embedded in the barrel rifling of the barrel. Figure 8 is the cloud diagram of the stress and deformation of the elastic belt at the beginning of the extrusion process, when the elastic belt just begins to contact the rifling. It can be seen from the figure that at this time, the maximum stress of the elastic belt is 72.5MPa, and the material of the elastic belt is still in the elastic deformation stage. The rifling stress traces can be seen from the edge, and due to the extrusion of the elastic belt by the smooth groove, the deformation of the elastic belt is mainly concentrated on the front end of the elastic belt, and the maximum deformation is 2.726mm.

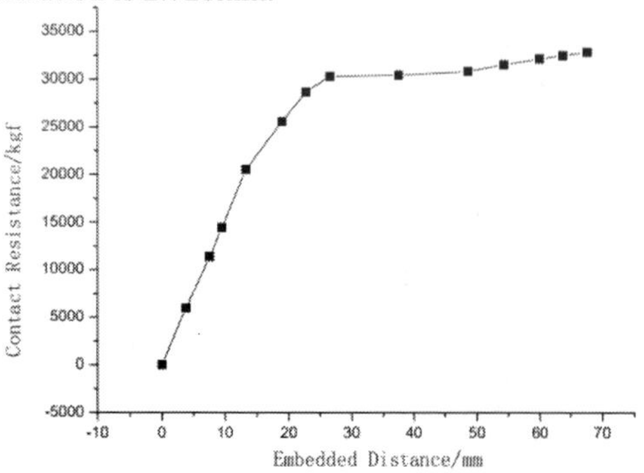

Fig.7 Variation curve of extrusion resistance with extrusion distance

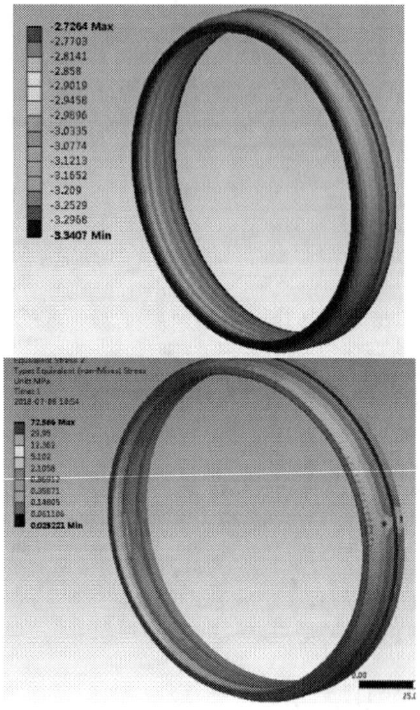

Fig.8 Cloud diagram of initial stress and deformation of front belt extrusion

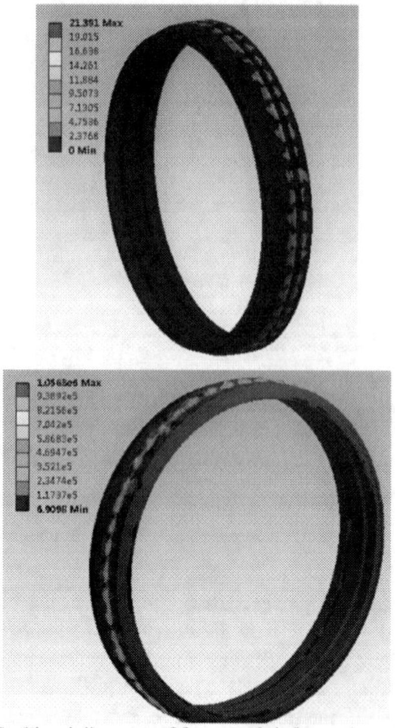

Figure 9. Cloud diagram of force and deformation after the front belt is squeezed in

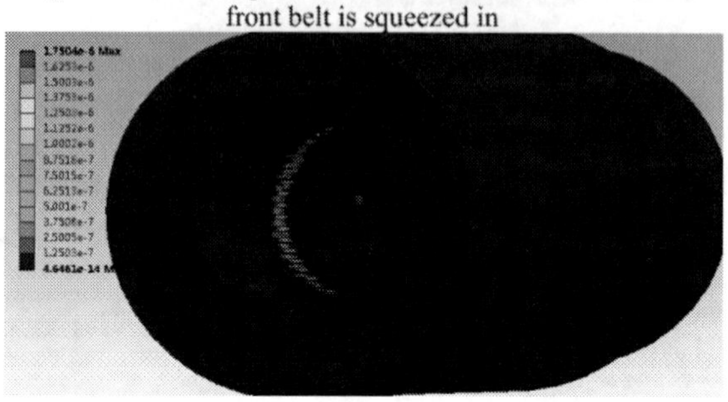

Figure 10 Force cloud diagram of the barrel during the extrusion process

Figure 9 shows the cloud diagram of the force and deformation of the elastic belt after the front elastic belt is squeezed in. It can be seen from the figure that when the front elastic belt is completely squeezed into the full rifling depth, the maximum stress of the elastic belt is 10565MPa, which exceeds the yield limit of the elastic belt material. , the elastic belt produces plastic deformation, and material flow and failure occurs at the part in contact with the rifling, resulting in a sawtooth-shaped permanent deformation similar to the shape of the rifling. It can be seen from Figure 8 that after the front elastic band is squeezed in, the maximum flow distance of the elastic band material is 21.39mm.

Figure 10 shows the force cloud diagram of the barrel after the elastic belt is completely squeezed into the rifling. It can be seen from the figure that the torus surface in contact with the elastic belt is subjected to greater force, and as the elastic belt is squeezed into the process, the force is greatly reduced. The larger force of the torus gradually moves towards the muzzle.

In order to verify the accuracy of the model, based on the test data of a simulation engraving test device to a certain caliber howitzer, the established simplified chip model was used to calculate the engraving resistance. The calculation results and test results are shown in Table 1 and Figure 2.

6. Conclusion

(1) The extrusion resistance increases continuously with the extrusion process. The maximum extrusion resistance is 83346N. In the early stage of the extrusion process, the extrusion resistance increases rapidly, and then increases slowly. At the later stage of extrusion, the extrusion resistance is almost constant. If it increases again, it shows that the entire elastic belt has completely entered the rifling of the cylindrical part of the barrel, which is in line with the physical law of the projectile extrusion process;
(2) When the elastic belt is fully squeezed into the full rifling depth, the maximum stress of the elastic belt exceeds the yield limit of the material, the elastic belt undergoes plastic deformation, and material flow and failure occur at the part in contact with the rifling, resulting in a similar shape to the rifling. The zigzag permanent deformation of the elastic band, when the extrusion is completed, the maximum flow distance of the elastic band material is about 21mm;
(3)Using the explicit dynamics method, the Johnson-Cook model is used for the elastic band material, which can not only obtain the resistance along the extrusion process, but also describe the deformation of the elastic band and the flow of the material during the extrusion process.

References

1. Jin Zhiming, Zeng Simin. Calculation and research of projectile extrusion process [J] . Journal of Ordnance Engineering, 1991.2. No.1: 7-13.
2. Xu Shaojie, Chen Xiong, Feng Xiaobing. Numerical simulation and analysis of a certain type of artillery shell belt extrusion [J] . Sichuan Ordnance Science, 2011.32. No.8: 14-15.
3. Wang Peng, Yang Guolai, Ge Jianjian, etc. Numerical simulation of elastic belt extrusion process based on Johnson-Cook constitutive model [J] . Chinese Journal of Ballistics, 2015.27. No.2: 55-56.
4. Sun Quanzhao, Yang Guolai, Wang Peng, etc. Numerical simulation research on the extrusion process of a large-caliber howitzer shell belt [J] . Journal of Ordnance Engineering, 2015.36. No.2: 206-207.
5. Li Miao, Qian Linfang, Chen Longmiao. Research on the internal ballistic characteristics of the elastic belt extrusion process [J] . Vibration and Shock, 2016, 35(23): 78-79.
6. Liu Chengzhu. Study on the structural parameters of large-caliber artillery that affect the initial disturbance of the projectile[D]. Nanjing: Nanjing University of Science and Technology, 2016: 8-11.
7. Jin Zhiming, Zeng Simin. Calculation and research of projectile extrusion process[J] . Journal of Ordnance Engineering, 1991 (1): 7-13.
8. Chen, Peter C. Analysis of Engraving and Wear in a Projectiler Rotating Band[R]. 1999.
9. South Joseph, Yioumas Aristedes, A Study of the Engraving of the M855 5.56mm Projectile [R]. 2008.
10. Lin Gui. Design of the simulation experiment device for the high-speed extrusion of the elastic belt into the rifling process [D]. Nanjing: Nanjing University of Science and Technology, 2014: 31-35.
11. Gu Wei. Numerical simulation research on belt extrusion and gun bore damage [D] . Nanjing: Nanjing University of Aeronautics and Astronautics, 2014: 27-32.
12. Jiang Zeyi. Research on projectile extrusion process [D] . Taiyuan: North Central University, 2014: 28-29.
13. Zhang Zhihong. Research on wear behavior of barrel coating based on ABAQUS [D] . Lanzhou: Lanzhou University of Technology, 2016: 7-10.
14. Zhang Hao, Zhou Yanhuang. Research on the extrusion process of submerged artillery [J]. Chinese Journal of Ballistics, 2006, 18(1): 76-77.

The influence of road surface grooving on the bearing capacity of airport road surface is analyzed

Yan Gao

Aviation Engineering Department, COMAC Flight Test Center, Shanghai, 201323, China

Abstract: Abstract: this paper is mainly to analyze splash water test pavement slot effect on airport pavement bearing capacity, according to the FAA requirements, computing pool water storage requirements, and put forward the pool building, this paper adopts the simulator ChangDao surface modeling, analysis of structures, pools on the pavement when the slot effect on airport pavement bearing capacity, explore the pool to build new methods to provide theoretical support for the follow-up.

Key words: Airport pavement; grooving; The bearing capacity

1. Introduction

In order to verify the impact of water splashing on aircraft engines during taxiing, taking off and landing on water-logged runways, water splashing tests were carried out on the newly developed aircraft. The splash test needs to build a pool on the airport pavement to meet the test requirements. This paper focuses on the technical analysis of the impact of the installation and removal of water tank test facilities on the runway during the splashing test of aircraft, and demonstrates the impact of grooving on the runway, providing theoretical support for the subsequent splashing test of new aircraft to build a pool.

2. Significance and Background

According to the FAA splash test requirements, the new aircraft should be verified before obtaining evidence of its indicators, in line with the requirements of airworthiness certification, before obtaining airworthiness certificate. This thesis mainly embarks from the splash water test, the test mainly through a complete under the specified conditions of water taking off and landing test to simulate the flight test, or in a series of large enough to not interfere with the spray splatters to simulate water demonstration area, the simulation test can achieve more stable results, also can identify and analyze the response of the engine. The main purpose of the splashing test is to verify that the splashing will not damage the aircraft structure and affect the engine, atmospheric data and other related systems in the process of taxiing, taking off and landing on the water-logged runway. The second is to obtain the aircraft performance data on the polluted runway as the input of the manual performance data.

It is known that the pool used in splash test is mainly built in two ways at home and abroad. One is the way of slotting and plugging rubber dam commonly used in the airport road surface abroad, but there is no precedent of this method in China. The other is the adhesive method, which has been used in a domestic project. The first channel surface slotting method can make the rubber dam of the pool more firmly nested into the groove. At present, this method is more commonly used abroad, but it needs to cut and slotted the airport channel surface layer, which will damage the channel surface. The second adhesive method has been used in the splashing test of a domestic project. This method is nondestructive to the airport pavement, but in the process of taxiing, it may be easy to grind out the rubber dam in the groove due to the large power, which is prone to FOD and other safety hazards.
In order to deeply analyze and explore the method of building a pool on the airport pavement, this paper takes a 4D airport as the basis and simulates the slotting of the airport pavement through the airport modeling method, analyzes and demonstrates the influence of the pavement slotting on the airport pavement, and provides theoretical support for splashing test.

3. Test plan

3.1 Brief introduction to airport road surface

A 4D airport, the road surface is cement concrete road surface， The length and width of the runway is 3600*60m, and the thickness of the pavement surface is 34cm.The upper base (cement-stabilized macadam base) is 20cm, the lower base (graded macadam base) is 20cm, the size of track panel is 5*5m, the thermal expansion groove is filled with silicone, and the PCN value of the

 https://doi.org/10.1051/matecconf/202236301030

runway is 63.There is a 6mm deep gutter along the heading distribution on the pavement. There are two longitudinal slopes of the runway. The transverse slope of the runway is double-sided and the slope is 1%. Taxiway length and width is 3600*23m.

3.2 Water tank construction scheme

3.2.1 Pool size

Water splashing test needs to set up a pool on the airport runway. Since the runway has a horizontal and longitudinal slope, in order to control the water depth on the slope of the airport runway, the water tank must be separated.In order to reduce grooving on the track surface, combined with the track panel size, a pool of 100*30m was built in the middle section of the runway (within the range of 1300m to 1500m from the south section of the runway) in this test. The channel width of the front wheel was 3m, the channel width of the main wheel was 8.5m, and the dam body of the pool was 10m apart along the course, among which the pool was located at 1.5m on both sides of the center line of the runway. The installation position of the pool is shown in Figure 1. Schematic diagram of slotting is shown in Figure 2.

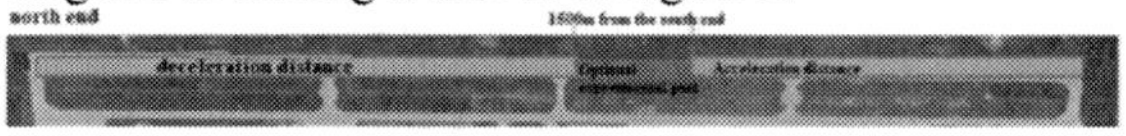

FIG. 1 Pool installation position

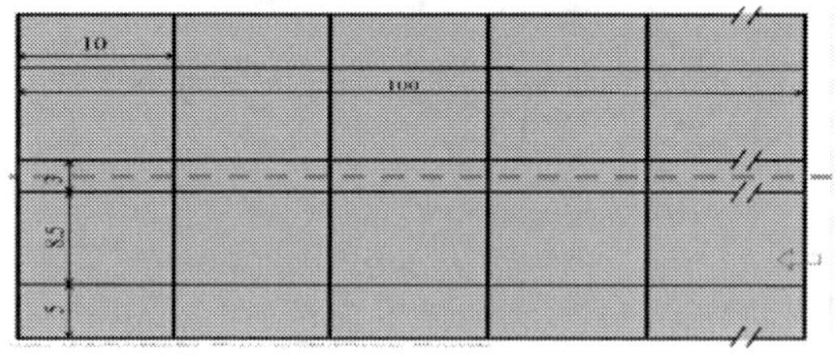

FIG. 2 Schematic diagram of slotting

3.2.2 Pool construction

In order to reduce the cutting of the airport pavement, the existing thermal expansion tank should be used as much as possible, that is, the existing thermal expansion tank should be dug out (remove the silicone filler) and inserted into the rubber belt to build the test tank. However, considering the geometrical size of the aircraft and the transverse slope of the runway, it is still necessary to cut two 100m water tanks along the heading (longitudinal cutting along the runway) at positions of 1.5m each on the center line of the airport runway.
These newly cut grooves can be designed to be identical to the existing thermal expansion grooves on the runway by using the same process used to excavate the existing thermal expansion grooves with the same road cutting machine and the same technology. The grooves required to construct the tank are 8mm wide and 3cm deep.
The pool is constructed with rubber belts, that is, rubber belts are inserted into the grooves. The water tank can be connected with the rubber belt at simple room temperature (no adhesive or other fixing device is required). The rubber belt can be cut in advance according to the length of the design of the test water tank. The rubber belt should match the size of the groove.
After consulting a lot of information, foreign splash test mostly adopt rubber belt scheme, about this groove - rubber belt scheme, you can expand to read NASA report"Problem areas associated with the construction and operation of the landing research runway at NASA Wallops Station" [RD-7].And this scheme has been widely used since the 1960s in splash tests, where rubber bands are extremely difficult to break even in water tank tests on large multi-wheel landing gear aircraft, whose tyres are large and under high pressure. At present, there are no adverse reactions in the operation of slotted airports in foreign countries. Please refer to Annex 1 for the operation of slotted airports for splash test in foreign countries.

3.2.3 Pool water injection

The section installed in the test tank will be filled with water, so it is necessary to ensure that the water level in the tank meets the test requirements during the whole test phase, so there are higher requirements for the sealing of the tank.

3.2.4 Pool to dismantle

A splash test is carried out on the waterlogged runway. After the test is completed, the rubber strip can be removed from the trench to remove all FOD on the runway. The new trench and the existing thermal expansion groove can be treated in the same way as the airport thermal expansion groove and backfilled with silicone filler.

4. Technology demonstration

4.1 The theoretical analysis

4.1.1 Runway strength analysis

a)The surface layer thickness
Under normal circumstances, the strength of the runway surface at both ends of the airport is high, and its design strength is considered according to the design full takeoff weight or maximum full takeoff weight, and the thickness of the runway surface in the middle section can be appropriately reduced compared with that at both ends. According to "Civil airport cement concrete pavement design code", the thickness of cement concrete slab, flight zone index ii for A, B should not be less than 200mm, flight zone index for C, D, E, F should not be less than 240mm.
The whole runway of the target airport is constructed according to a design index (34cm thick), with grooves in the middle of the runway and a length of only 100m. According to the design specification, "when the runway width is not less than 45m and parallel taxiways are set,

https://doi.org/10.1051/matecconf/202236301030

Concrete slabs in the middle of the runway (at least 800m away from the entrances at both ends of the runway) can be reduced to 0.9 times the thickness of concrete slabs at the end of the runway."It can be reduced by 3.4cm, that is, after the runway is slotted, the surface thickness of the middle section of the airport is still within the design specification, and meets the mandatory standard of the minimum thickness of the runway panel.

b)PCN calculate

Through qualitative analysis of PCN, that is, PCN calculation of slotting (3cm deep) section, that is, assuming that the overall thickness of the middle section of the target airport pavement changes from 34cm to 31cm (in the most unfavorable case, the actual pavement thickness is still mostly 34cm), and assuming that the 90-day design bending and tensile strength of concrete is 5.25mpa, The PCN is 52 when the plane thickness is 31cm, but the actual PCN is higher than 52. If grooving is carried out on the plane, simulation modeling is carried out for the airport.

c) Length-width ratio analysis

According to the requirements of the civil airport cement concrete pavement design code, the length to width ratio of pavement blocks should be less than 1.25. The runway thermal expansion groove is constructed before the cement is dried and cut into 3cm depth and 4mm width. After the cement is dried, the expansion groove can expand to 6-8mm and form natural cracks at the bottom. The new grooves are located at 1.5m on both sides of the center line of the runway, and are cut after the track surface is completely dry. The 3cm installation groove is a false seam, and the 5*5m plate cannot be cut. In other words, after grooving, it can still be a complete plate, without forming a narrow plate of 1.5*5m, which does not conflict with the design specification of the track surface.

4.2 Finite element modeling simulation demonstration

4.2.1 Finite element modeling

In this finite element modeling, FLAC3D software was used to establish a 5*5m channel panel. By loading the finite element model, the bearing capacity of the complete channel panel and the slotted rear panel was analyzed. The slot position is 1.5m at the middle line of the runway, the geometric dimension is 8mm wide and 3cm deep,The analysis object is mainly the surface layer of pavement structure, without considering the plastic deformation of soil foundation, so the surface layer, base layer, cushion layer and soil foundation are all set as elastic constitutive model. The model parameters of pavement structure are shown in Table 1. Modeling part of the code, see Figure 3; The geometric model is modeled as shown in Figure 4.

Table 1 Model parameters of pavement structure

The structure layer	The thickness of the pavement	Modulus of elasticity (MPa)	Poisson's ratio	The density of the pavement (kg/m^3)
The surface layer	0.34	36000	0.15	2400
At the grass-roots level	0.20	1500	0.25	2000
Cushion layer	0.30	200	0.30	1500
Soil base	4	80	0.35	1800

FIG.3 models some of the code

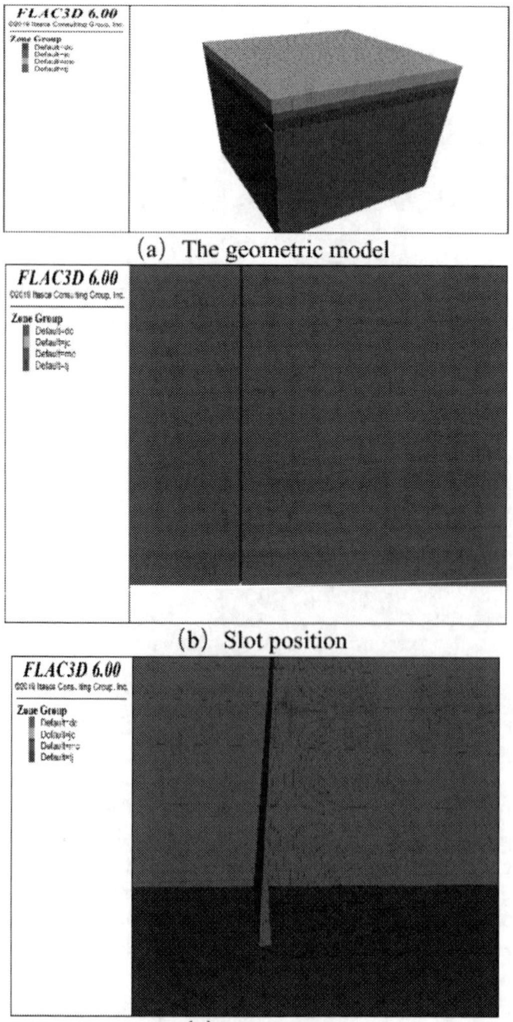

(a) The geometric model

(b) Slot position

(c) Slot details

FIG.4 Slot details

4.2.2 Aircraft wheel seal confirmed

It is assumed that the contact stress between the airplane tire and the track surface is evenly distributed within the range of wheel printing, and the size is equal to the tire pressure. In view of the different wheel load, tire pressure and tire type will affect the shape of aircraft single wheel imprint, currently in many countries (such as China, the United States and Canada, etc.) airport runway design, the wheel imprint is assumed to be a combination of rectangle and two semicircles (i.e., combined wheel imprint).See Figure 5 (a). Since semicircle is not beneficial to the establishment of finite element model in the finite element modeling process, the wheel printing area can be converted into rectangle for modeling calculation through the area equivalence principle, as shown in FIG. 5 (b).

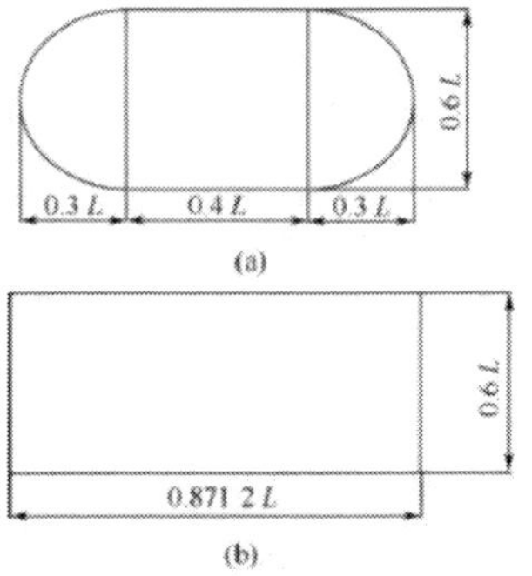

FIG.5 Wheel print area (L is wheel print length)

The calculation of wheel print area, length and width are shown in Formula (1), (2) and (3) respectively. In this loading, 737 was selected as the wheel load size of the reference model. According to the calculation, the length of wheel mark of this loading is 0.5m, the direction is along the middle line of the runway, the width is 0.3m, and the load size is 1.4mpa. The inner spacing of wheel printing is 0.3m, and the distance between loading position and center line is 3.8m.

$$A = \frac{P}{1000q} \quad (1)$$

$$L = \sqrt{\frac{A}{0.5227}} \quad (2)$$

$$W = 0.6L \quad (3)$$

Type in the: A ——Aircraft single wheel wheel printing area, unit m^2;

q——Single wheel contact pressure on main landing gear of aircraft, Take tire pressure (take1.4MPa) ;

W——Round printing width

4.2.3 Aircraft load

a)Stress and strain analysis under load at normal take-off and landing position

Aircraft load was carried out on normal pavement and slotted pavement panel respectively, and bending settlement value of pavement layer was analyzed under the action of aircraft load. This loading was under static load loading condition, and wheel load was loaded (static load), As shown in Figure 6.

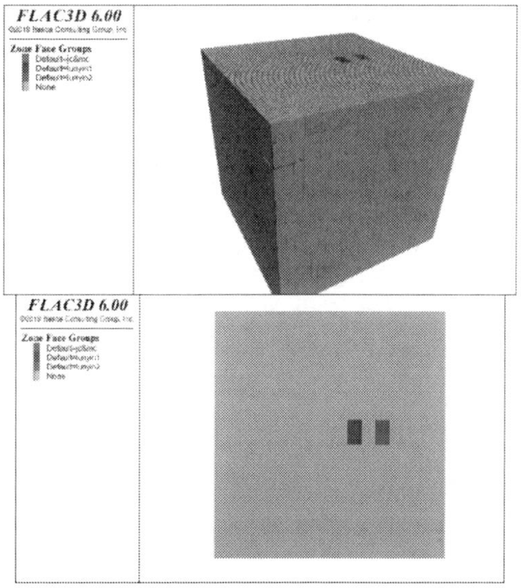

FIG.6 Wheel load loading (static load)

When the wheel load is static, the maximum deflection value at different positions of normal track surface and slotted track surface is calculated within the range of 0-5m, as shown in Table 2 and Figure 7.

Table2 Bending and settling values of normal and slotted surfaces

x/m	0	0.5	1	1.5	2	2.5	3	3.5	4	4.5	5	Max deflection
normal	4.537	4.591	4.810	5.202	5.791	6.501	7.252	7.405	7.002	6.509	6.322	7.415
grooving	4.532	4.585	4.799	5.167	5.785	6.502	7.255	7.409	7.006	6.512	6.324	7.419

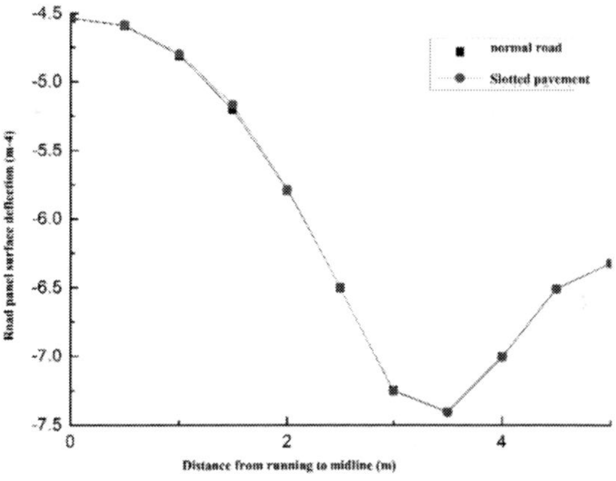

FIG.7 Comparison of deflection and subsidence of road surface

Combined with Table 2 and Figure 7, it can be seen that the curve of deflection of normal track surface and slotted track surface is almost the same under the same aircraft load at the same position. The maximum bending weight of the slotted surface is 7.419×10^{-4} m, Maximum deflection of normal track surface is 7.415×10^{-4}m, It indicates that the stiffness (resistance to deformation) of the track panel at the same position is slightly reduced after the grooving:

$$(7.419\times10^{-4}-7.415\times10^{-4})\div7.415\times10^{-4}=\mathbf{0.54\%}$$

By comparison, the left side of the slotted track is slightly lower than that of the normal track (both less than 0.01m). The normal track is shown in Fig.8 (a), and the surface of the slotted track is shown in Fig.8 (b). This value indicates that the groove has a slight influence on the load transfer capacity of the pavement.

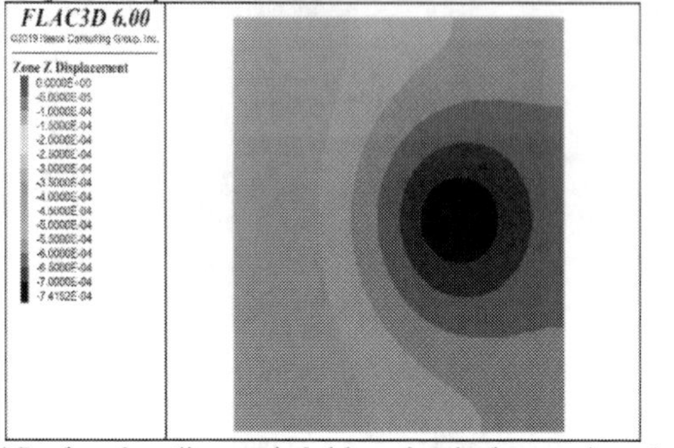

(a)Surface bending and sinking cloud of normal track

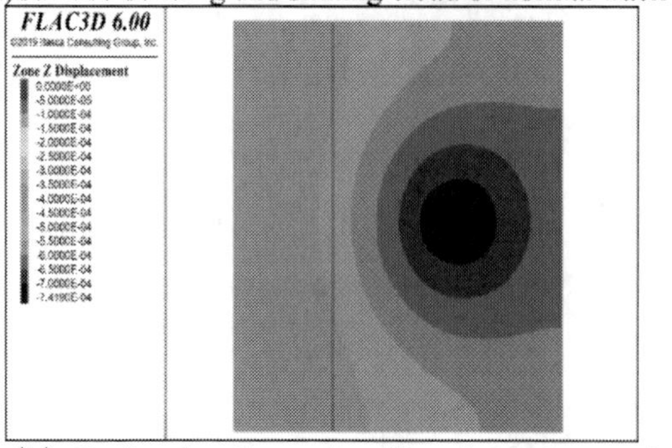

(b) Surface bending and settling cloud picture of slotted surface

FIG.8 Surface bending and settling cloud picture of slotted surface

The design strength of pavement cement concrete shall adopt the bending tensile strength of 28 days. The airfield index ii is A, B, the design strength of cement concrete is not less than 4.5mpa;The design strength of cement concrete should not be less than 5.0mpa for airports with index ii of C, D, E and F. The compressive strength of cement concrete materials is usually much higher than the flexural strength. The design strength index of cement concrete pavement panel refers to the flexural strength.Therefore, by observing the bending and tensile strength at the bottom of the runway panel under the load of the aircraft wheel, we can judge whether it exceeds the allowable stress range. In the normal takeoff and landing state (that is, the aircraft track is 3.8m away from the middle line of the runway), the stress calculation cloud diagram of the normal and slotted tracks is shown in FIG. 9. The stress state of each part of track panel is shown in Table 3.

It can be concluded that the maximum normal stress and maximum shear stress at all places on the normal and slotted surfaces are less than the designed bending and tensile strength, and the maximum stress at the bottom of the plate under load is 2.22mpa,far less than the design strength of 5.0mpa, and the range of change is very small, the groove has a small impact on the overall surface, the maximum difference is not more than 5%, that is, after the groove panel in the normal rise and fall of the aircraft will not cause bending and tensile damage risk to the plate bottom, does not affect the normal use.

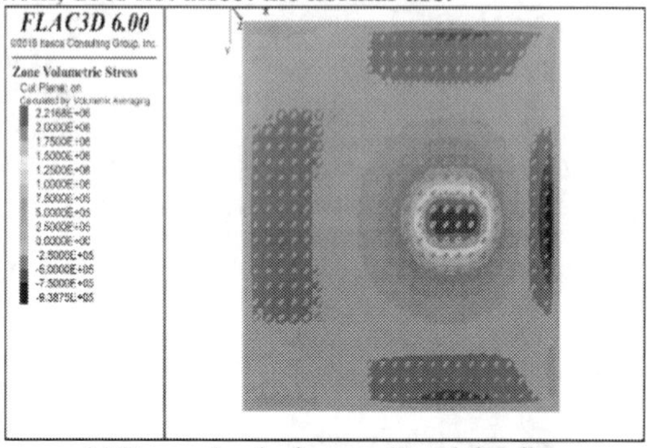

(a)Stress cloud of normal track loading

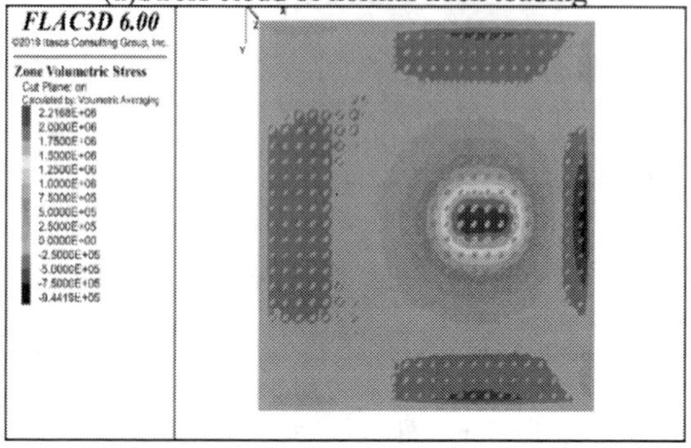

(b)Stress cloud of slotted surface loading

FIG.9 Stress cloud of slotted surface loading

Table3 Stress state of each part of track face plate(MPa)

location		Place of loading		Edge in the middle		voids	
		The surface of the plate	The bottom of the plate	The surface of the plate	The bottom of the plate	The surface of the plate	The bottom of the plate
Normal pavement	σ_{Max}	4.16	**2.22**	1.05	1.01	0.01	0.06
	τ_{Max}	0.26	0.81	0.56	0.63	0.03	0.05
Slotted surface	σ_{Max}	4.17	**2.22**	1.04	1.01	0.01	0.06
	τ_{Max}	0.26	0.81	0.54	0.63	0.03	0.06

Considering crosswind and other conditions, the main landing gear tire acted on the slot, namely, the position of the most unfavorable load, and the bending of the normal track surface and the slotted track surface, as shown in FIG. 10, and the stress and strain cloud diagram, as shown in FIG. 11. At this time, the maximum bending value increases compared with the normal take-off and landing of the aircraft position : $7.526 \times 10^{-4} - 7.419 \times 10^{-4} = 0.107 \times 10^{-4}$m.

The stiffness of main landing gear tire on slot opening is lower than that of normal main landing gear wheel track distribution**1.4%**.

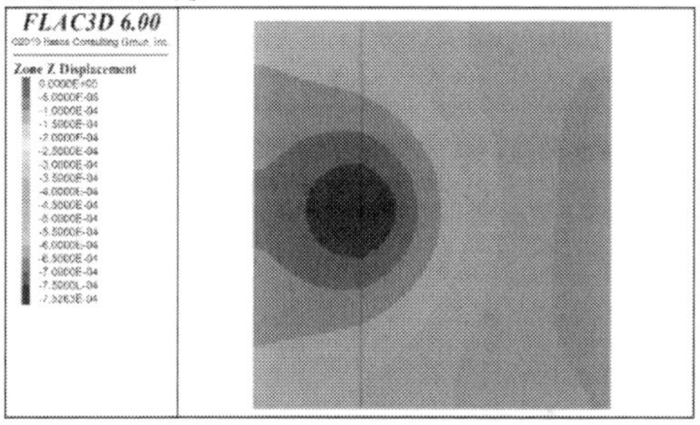

FIG.10 Cloud picture of road surface bending and settling under wheel load on groove

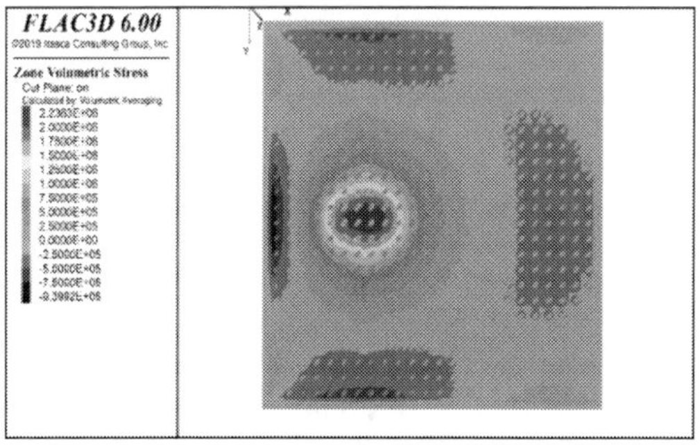

FIG.11 Stress cloud diagram at the most unfavorable load on slotted surface

By comparing the main damage index of the front and rear grooving panels -- the variation of the tensile stress at the bottom of the plate, the tensile stress at the bottom of the plate under the action of the wheel load of the main landing gear on the groove is 2.24mpa, which increases by 0.02mpa compared with the normal rise and fall, and the ratio is 0.9%. All of them did not reach the design bending tensile strength of the panel, so there would be no bending and tensile damage risk to the bottom of the plate. To sum up, slotting at 1.5m of runway middle line has little influence on bending settlement and stress strain of cement concrete pavement.

5. Conclusion

In summary, the theoretical analysis and finite element modeling analysis above, combined with the operation status of the target airport, can be concluded as follows:

1) Slotting has little influence on the runway surface. According to the finite element model simulation calculation, the grooving has little influence on vertical bending and stress of airport pavement, and has little influence on bearing capacity of pavement;

2) Slotting damages the appearance of the pavement, adding two 100m*8mm*3cm cracks and increasing the maintenance cost in the later period. However, the probability of cracks in the pavement is almost the same as that in the normal airport pavement. FOD can be avoided by strengthening the pre-flight patrol and follow-up maintenance detection;

3) For the filling of thermal expansion and cold contraction joints, after the splashing test, the rubber belt insertion part is filled with aggregate, and the same material can be used to fill the thermal expansion and cold contraction joints of the original airport surface.

References

1. Liu Deyi. Introduction to civil aviation (revision) .Civil Aviation Press of China, 2008.
2. MH/T5004-2010.Code for design of cement concrete pavement of civil airports.
3. NASA report"Problem areas associated with the construction and operation of the landing research runway at NASA Wallops Station".

The Protection of Data Privacy of College Students

Zhiheng Tao

107 S Indiana Ave, Bloomington, IN 47405, US

Abstract. Privacy data belongs to big data, and it is personal information from our lives. The common problem is privacy leaking and cyber attacks. The current solution uses a complex algorithm to encrypt data, but it is expensive and inefficient. In the paper, we combine students' information at university to make four layers based on a reliable framework named FMM. The proposed framework FMM, according to different privacy levels, will choose other encryption methods to protect these data and increase efficiency to keep trade-offs.

Keywords: Privacy, Big data analytics, Security, Encryption, Protection.

1. Introduction

With the development of the Internet, data privacy protection is taken more seriously for people. Many people always store their personal information on the Internet, and it is now prime and common data globally. Privacy protection has many layers to filter people's information and distribute it in the correct layers for encryption. Through internet development, many places support privacy protection methods and combine them with their Internet of Things (IoT) infrastructure, such as medical facilities, stores, transit, etc. In the IoT scenario, users' information will be transmitted through the IoT infrastructure to the database. These processes will have many dangerous factors to influence (cyberattacks). Since this personal information will become plaintext, and there are many encryption methods (symmetric, asymmetric, and hashing) to encrypt, it leads to different encryption methods with different speeds and security strengths.

Currently, people use three standard techniques to protect their private data: Anonymous [5], Encryption [5], and Access Control [5]. Indeed, each technique has its advantages and methods, such as anonymity, which is suitable for hiding users' names, encryption can use a different type of encryption to encrypt, and Access Control can check each access user's condition and identity at the first time. Even successful privacy protection methods still need to be built to meet users' requirements: applicable requirements to them. However, all the conditions must contain confidentiality, integrity, and availability. Data also needs to make more highly efficient encryption, to match what type of data is more suitable and which type of encryption to encrypt.

In contrast to other privacy protection models, our FMM advantages are flexible and innovative; FMM can, according to different layer features, use different encryption to encrypt each layer. To improve efficiency, we screen suitable encryption methods for each layer. In addition, our FMM can change students' requirements, and if students only want to filter their records information, they can only use our Record Layers from our FMM to filter their required information. In the paper, we will focus more on the privacy model and propose a new four-layer framework-the Face Mask Model (FMM)-at universities. To analyze how to flexibly and deeply protect students' privacy data at universities and solve how to store data from universities more securely. It is essential that the FMM reduces the risk of data privacy leaks and increases the encryption speed and flexibility to satisfy different types of students. There are three contributions to the paper.

1. We propose a new privacy framework (FMM) to help students protect their university information and provide different layers using different encryption algorithms to increase effectiveness and safety.

2. The privacy framework (FMM) contains four interaction layers: Sensitive, Bias, Record, and Base. To give us a big data scenario, we will fulfill the need to use different encryption methods to achieve high reliability and efficiency trade-offs according to that security level.

3. Flexible to match the student's requirements. This (FMM)model can satisfy different students' requirements by providing other protection solutions. FMM can further protect our data security level, thus reducing the risk of private data leaks.

2. RELATE WORK

2.1 Data Analytics

Nowadays, Zainuddin [2] proposes different potential problems regarding privacy in the Internet of Things environment. It analyzes many intelligent environments, like homes, meters, medical centers, cities, and clouds. These innovative areas [2] work remotely. It will have potential problems like people can steal information on the transmission, collect other users' data to predict their private information, directly attack physical devices: make them invalid, and also use tools to collect unencrypted user information from intelligent homes devices. These attacks are hazardous to the Internet of Things; they can use these methods to collect users' big data quickly. Also, Alshboul's model method [7] combines big data security lifecycle and security attacks, which improves data protection; it also prevents data mining influence because it is a widespread attack method and causes significant damage to the database. Then, Mehmood analyzes many new model ideas [6] that can explore different areas or types of data at the same time. People don't have to use multiple models to protect or analyze this data; it still has some areas that need strengthening in different periods, like decentralized storage and data anonymization.

2.2 Data Privacy

Schaub [8] proposes using many different policies and rules, such as transparency, security, organizational safeguards, etc., to protect students' privacy. It also protects students' confidentiality and prevents unfair analysis of students. Reidenberg's new policies suggest [8] that people care more about students' privacy and data. They notice students' privacy rights. Educational data mining and learning analytics [8] will positively support the study process and negatively hurt users' privacy. Also, El Ouazzani can block people who want to attack or leak, even if they successfully attack this data, but they still don't know which data matches which owner: which means nothing. However, the weakness of this method is that hackers can still use keywords to match this privacy. Additionally, the innovation of this method is it can block people who want to attack or leak data. Even if they successfully attack this data, but they still don't know which data matches which owner: which means nothing. Also, the weakness of this method is that hackers can still use keywords to match this privacy. And people can use this method to attack back, like the utilization of t-closeness [9] against some similar attack styles in treating categorical sensitive attributes.

2.3 Internet of Things

Internet of Things (loT) is when the user uses physical equipment connected to the Internet and has services from there, it often happens that people change the equipment without the users' authorization: get data from that method and steal data in the data transmission part. Recently, Kabalci [3] proposed how the energy Internet connects with the Internet of Things and how they combine. It [3] can provide two advantages in the energy Internet area, the first advantage is it can provide specific communication and complex network structures for different communication scenarios, and the other is that it is used to reduce power and cost to make the devices more efficient. So we can see it is an excellent idea to combine them and reduce the Internet's original weakness, making it more flexible. In early research, Medaglia [4] proposed how current technological and technical trends influence our security and privacy and analyzed the future impact. It [4] explores the IoT infrastructure between users' connections and remotely managed relationships will increase security and privacy problems because it needs data transmission: between infrastructure and users. Whereas IoT is not very safe in their infrastructure because people overlook that and invest time to upgrade, loT security design [4] tries to open pervasive and interoperable infrastructure to follow this trend.

2.4 Application Scenarios

Ziegeldorf [1] proposes a basic new model and law. It is based on the International Telecommunications Union and the IoT European Research Council to make users quickly and safely connect any devices and services on the Internet and needs a complete privacy law to support it. Primarily, they still think [1] that profiling is very dangerous to their model because people can, though, collect data to increase their attack privacy possibility. Sun [5] proposed that medical loT needs to strengthen security and privacy because medical loT will remotely watch and collect patient medical data: heart rate, etc. Medical loT [5] usually has three-layer: perception, network, and application; each layer will store important privacy data in there. It needs to promise its integrity, validity, and authenticity because this private data is only accessible to people who have authorization from the owner. The problem for the future in medical loT [5] is an insecure network, lightweight protocols for devices, and data sharing; these areas are challenging to manage and have low-security protection methods, and they need more focus on their devices and network security techniques on this area to prevent these factors from influencing it.

3. Methodology

Based on different encrypting features, different data styles need suitable encryption methods to protect them. There are four other encryption methods below.

Figure 1. DES algorithm process

According to FIGURE 1 shows, the DES encryption method is a symmetrical encryption, it uses the same key to encrypt and decrypt. It has a simple encryption algorithm: short key, and manages well to extensive data. Contrast the AES method, the AES has overprotection to these types of data.

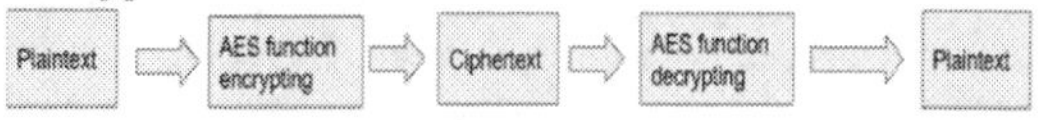

Figure 2. AES algorithm process

We describe the AES details in FIGURE 2, and the AES encryption is a symmetrical-based method. It has a high-speed and a new algorithm for encryption. Contrast the DSA method, and the DSA is not a very high-speed encryption method because DSA has high-level verification by users.

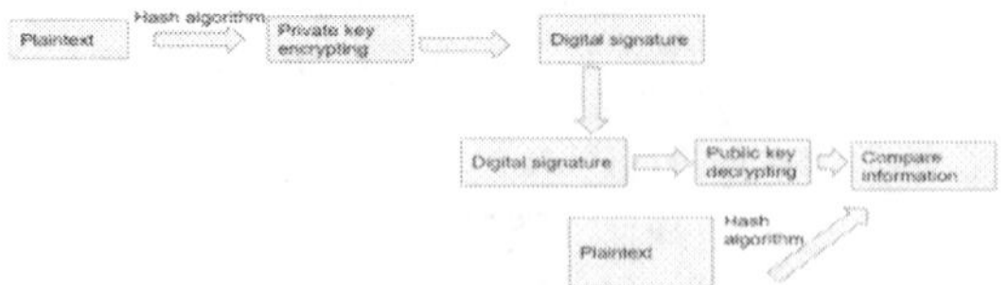

Figure 3. DSA algorithm process

The FIGURE 3 illustrates the DSA encryption method, and it is asymmetric, this method uses the signature to verify the key. It has high-level security protection, and needs a digital signature for decryption and encryption, to improve the data transmission process security. Contrast the RSA method, the RSA has low-speed decryption.

Figure 4. RSA algorithm process

As process content in the FIGURE 4, the RSA encryption method is asymmetric. The public key is that everyone can view it and the sender to encrypt, but the private key is only for the receiver to decrypt. Then, it will form different encrypt and decrypt keys to exchange. It has the highest security level protection, this method needs a public and private key exchange because the public and private decryption keys are different, it makes each key individual. Besides, the RSA key is longer than other encryption methods, so it will be difficult to break.

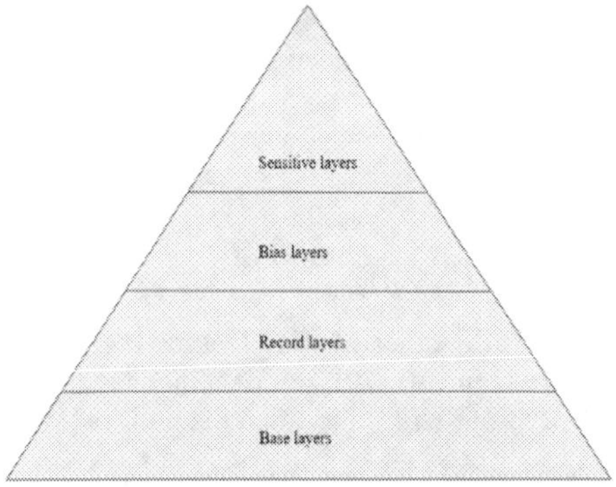

Figure 5. FMM Model Filter

Based on these four encryption algorithms, we propose a new model in FIGURE 5, to better protect and flexibly fill our private data. It has a sensitive layer, bias layer, record layer, and base layer. The base layer uses the DES method because this layer has large data and these data are not important, we only need standard protection. Second, the record layer uses the AES method because this layer also has large data, but these records data are important to users, we need safe and fast speed to protect them. Third, the bias layer uses the DSA encryption because this data will influence the user's living quality. We need signatures to verify by users. Fourth, the sensitive layers use the RSA method because this layer of data is very important to users' social credits, and some data is from the government, we need to use high-level encryption to protect them.

Furthermore, TABLE 1 illustrates each layer's details. For instance, the Base layers include full name and phone number information, the full name is Jack Duke Li and his phone number is 812-453-2433, these informations are searchable by the internet, we only need standard encryption to encrypt. The Record layers have an expense record and bus record. The expense record information is the student spends 10 dollars in the dining hall and he takes the route 9 bus back to home because he swiped his student card before entering the bus, most records information will be from that reader maschine. Then, the Bias layers include transcript and sex, if the student has 2.4 GPA and he is a male, these inforamtion will influence by jobs market: need agree by user permission. Last, the Sensitive layers cover SSN and IP address, the student's SSN is 154-75-3541 and his IP address is 11.18.238.174, these information are very important to this student's living, so we need the highest security level to protect them.

Table 1. Details of Each Layers

Layers	Function	Definition
Base Layers	Full name	A person's legal name
	Phone number	Student's phone number and emergency contact
	Major	Major and minor information
	Enrollment date	Enrollment date information for the academic team
	School name	Currently university name
	Student ID number	Student ID number infromation
	School email	Student's school email information
	Tuition	Student yearly tuition
	Class	Taken and in-progress classes
	Degree time	How long have you been stidying for your degree at univerisity
	Transfer	School transfer history information
Record Layers	Spending record	Student spending on campus, like dining, vending machines, stores, etc.
	Campus bus record	Campus bus record information
	Attendance record	Attendance record information from your classes
	Dorm record	Dorm card reader information
Bias Layers	Transcript	Transcript information
	Addresses	Local address and Home address
	Nationality	Nationality information
	Race	Race information
	Sex	Sex information
	High school privacy	High school privacy information
Sensitive Layers	Bank account	Bank account information
	IP address	Student's devices IP address information
	Medical history	Medical history information
	Criminal history	Criminal history information
	SSN	SSN information
	Gender	Gender information

3.1 DES Definition(DES Symmetric-key algorithm)

This layer is the basic information layer, there is no need for complex encrypting, but it needs to be moderate speed and efficient, and DES meets the moderate speed and large areas of data encryption requirements.

3.2 DES Example

Jake is a university student, and his legal name is Jake Duke Li, his phone number is 812-453-2433 and his email is jakeli@iu.edu, he is taking computer science, a major, at Indiana University Bloomington. His student ID is 20003598041 and his enrollment date was August 20th in 2018, he paid 56000 dollars for his second-year tuition. And he is taking 15 credits, in-person classes, for this semester and he already finished 30 credits last year. Also, he has no transfer history from another school, and he has been studying for his degree for one year at his university: currently.

3.3 AES Definition (AES Symmetric-key algorithm)

The information at this layer basically comes from records, it needs some security protection. AES has certain security, high speed, and high efficiency.

3.4 AES Example

Jake is taking route 9 of the campus bus every day morning for his first class, and his instructor will record his attendance every time. After his classes are down, he will go to school dining for his lunch: spend 10 dollars on fast food, because he needs to go back to his class at 1 p.m. to prepare for his afternoon classes. When his 5 p.m. class is down, he will take the route 9 bus again to go back to his dorm, and swipe his student ID to enter his dorm: back to his room.

3.5 DSA Definition (DSA asymmetric-key algorithm)

This layer of information is actually searchable, such as transcript and address, and information leakage will cause psychological burden on users. Also, DSA can avoid these problems because each decryption needs user approval. It lets users know how secure their information is. The following is the flow chart of encryption and decryption.

3.6 DSA Example

Jake is a male international student, his nationality is Spain and he is an Asian. He currently: local address, live in 300 Pelo Varde Dr, Bloomington, IN 47401 and his home address is B. Diego de Velázquez, 1, 35449 Pozuelo de Alarcón, Madrid, Spain. He goes to the office to see his transcript: overall GPA and each semester credits condition, etc. Also, he is checking his high school privacy, like behaviors record and GPA, etc.

3.7 RSA Definition (RSA asymmetric-key algorithm)

This layer contains sensitive information, such as bank accounts and SSN. Information leakage can lead to users losing money and credit because others use their accounts and SSN for illegal transactions. However, RSA can avoid this situation because RSA has a high security level, and only needs the consent of both parties to decrypt and encrypt. Therefore, information security has effective protection.

3.8 RSA Example

Jason's gender is male and he has a heart attack history in his medical history, he needs to make an appointment to the hospital to check every year. Before going to the hospital, he wants to see how much money he has right now, so he goes to log in to his bank account: 5247-5874-3574-7164. But, his computer warned him, he is using VPN to hide his IP address: 11.18.238.174. He has to immediately close his VPN because he has a criminal history: he used VPN to hide his location and DDOS attacked other people's computers. After this process, he fills the medical form and gives the hospital his SSN number: 154-75-3541.

4. Conclusion

FMM resolves students' privacy security potential problems from their school. Students spend lots of time on campus to stay, and the school grasps students' mass personal information. A potential problem is a data leak. Our innovation in our proposal utilizes different features of encryption methods to combine a model, and this model will, according to the encryption data type, automatically adjust the security level. Additionally, this model has four independent layers to filter the different types of data because students leave many different types of data records at school. These four independent layers can work together or independently to satisfy user preferences. Compared to other models, our model has DES, AES, DSA, and RSA advantages to protect our different types of data: flexibility. This model also has high efficiency in dealing with these data and cooperates with students' data features. In the future, FMM will develop in the school area to deeply improve the security speed and accuracy. It also could merge in the education area or sociality because these areas' data is diverse. Significantly, IoT will apply in our society in the future, and it will collect multiple data to protect from the database. This model could add an artificial intelligence element to more automatically give to different types of data or area solutions.

References

1. Ziegeldorf, J. H., Morchon, O. G., & Wehrle, K. (2014). Privacy in the Internet of Things: threats and challenges. Security and Communication Networks, 7(12), 2728-2742.
2. Zainuddin, N., Daud, M., Ahmad, S., Maslizan, M., & Abdullah, S. A. L. (2021, January). A study on privacy issues in internet of things (IoT). In 2021 IEEE 5th International Conference on Cryptography, Security and Privacy (CSP) (pp. 96-100). IEEE.
3. Kabalci, Y., Kabalci, E., Padmanaban, S., Holm-Nielsen, J. B., & Blaabjerg, F. (2019). Internet of things applications as energy internet in smart grids and smart environments. Electronics, 8(9), 972.
4. Medaglia, C. M., & Serbanati, A. (2010). An overview of privacy and security issues in the internet of things. The internet of things, 389-395.
5. Sun, W., Cai, Z., Li, Y., Liu, F., Fang, S., & Wang, G. (2018). Security and privacy in the medical internet of things: a review. Security and Communication Networks, 2018.
6. Mehmood, A., Natgunanathan, I., Xiang, Y., Hua, G., & Guo, S. (2016). Protection of big data privacy. IEEE access, 4, 1821-1834.

7. Alshboul, Y., Nepali, R. K., & Wang, Y. (2015, August). Big Data LifeCycle: Threats and Security Model. In AMCIS.
8. Reidenberg, J. R., & Schaub, F. (2018). Achieving big data privacy in education. Theory and Research in Education, 16(3), 263-279.
9. El Ouazzani, Z., & El Bakkali, H. (2020). A classification of non-cryptographic anonymization techniques ensuring privacy in big data. International Journal of Communication Networks and Information Security, 12(1), 142-152.

Advances in biomimetic mineralization of tooth enamel based on cell-free strategies

Zhuoyue Shi

Department of Stomatology, Soochow University, Suzhou 215031, China

Abstract. Tooth enamel is a highly-mineralized hard tissue covering the outermost layer of the dental crown, and amelogenesis is inseparable from the participation of necessary components such as ameloblasts, organic matrix proteins, and mineral ions, such as Ca2+ and PO43-. However, mature enamel is an acellular tissue and it is difficult to self-repair once damaged. The current treatment methods for enamel damage are filling or repairing with alloys, ceramics, or composite resins. However, the mechanical properties of these materials are quite different from the natural enamel and they can't ensure a completely closed interface with the remaining enamel surface, which usually causes a series of post-repair problems. At present, the biomimetic mineralization of tooth enamel is a research hotspot in the field of prosthodontics, and has great clinical application needs and prospects, especially the researches on cell-free strategies have made significant accomplishment. Here, based on the cell-free strategies, we review the recent knowledge from ex situ and in situ two dimensions in the remineralization of tooth ename.

Keywords: tooth enamel; enamel-like materials; in situ enamel remineralization; review

1. Introduction

As the hardest tissue in the human body, tooth enamel consists of ~96wt% mineral phase, ~2-3wt% organic phase and ~1wt% water. Based on enamel rods, the highly oriented apatite crystals and the matrix filling in between form a unique and complex hierarchical structure[1], which endows the enamel with near-perfect mechanical properties. Therefore, enamel acts as a natural barrier to protect dentin and pulp tissues from stimuli of external physical and chemical factors.

Enamel is the only tissue in the body which is secreted and then mineralized by epithelial cells. Amelogenesis is dominated by ameloblasts, whose activities can be mainly divided into two stages: secretory stage and maturation stage. In the early stage of secretion, the polarity of ameloblasts changes and the number of organelles increases in preparation for the secretion of organic matrix. Then ameloblasts secrete different types of enamel matrix proteins (EMPs), composed of 80-90% amelogenin, 10% non-amelogenin and proteases in order to create and maintain a unique extracellular matrix environment for the deposition of minerals, following the nucleation and growth of apatite crystals. In the maturation stage, ameloblasts regulate further mineralization of tooth enamel[2]. The initial mineralization product formed by enzymatic reaction is amorphous calcium carbonate (ACC), which is then transformed to more thermodynamically stable amorphous calcium phosphate (ACP), followed by hydroxyapatite (HAp) crystals[3, 4]. Generally, OH^- are easily replaced by CO_3^{2-}, so carbonate hydroxyapatite crystals are the most commonplace forms existing in the enamel, which is more easily to be dissolved by acid and demineralized[4]. Of note, the ameloblasts will gradually transform to a layer of reduced dental epithelium covering the surface of the tooth when the enamel matures, and then turn to junctional epitheliums around the dental neck with the eruption of the tooth[2]. That means mature tooth enamel is acellular tissue and cannot repair itself once it was damaged.

Under physiological conditions, enamel has a series of physicochemical changes in the normal oral cavity even without cells, including demineralization and remineralization, which are an equilibrium system, especially dental caries, the most common oral diseases, which begin to invade from enamel first. The following acid produced by oral cariogenic bacteria break the balance, and the enamel continues to demineralize and even form caries[5]. The existing filling and restoration materials are mainly alloys, composite resins, ceramics etc., whose mechanical properties are different from those of tooth enamel, and an ideal interface cannot be formed between them, which may result in the stress concentration, leading to a series of problems such as secondary caries. Therefore, a new method of tooth enamel restoration is urged.

The biomimetic mineralization of tooth enamel is a hot spot of the current researches, which can be mainly divided into two categories at present. One is based on ex-situ remineralization, imitating the basic structure of

* Corresponding author: shizhuoyue1102@163.com

enamel to artificially prepare enamel-like materials whose properties are as close to those of natural enamel as possible. The other is in-situ remineralization on the surface of the deficient enamel directly, using EMPs as templates and mimicking the microenvironment of amelogenesis, or building a biomineralization frontier to guide the regrowth of HAp crystals. Taking cell-free strategies as the main line, we review the recent advances in biomimetic mineralization of tooth enamel in the past five years ex-situ and in-situ remineralization of enamel in this article.

2. Ex-situ remineralization of tooth enamel

Bionics is the development trend of oral restorative materials. At present, various technologies and methods have been used for the preparation of enamel-like materials in vitro, which mimic natural enamel in terms of biosafety, biocompatibility, function, morphology, internal structure, etc.

2.1 Layer-by-layer (LBL) deposition techinique

The LBL deposition technique can modify the surface or prepare nanocomposites with specific structures and functions by using polyanionic electrolytes and polycationic electrolytes to alternately form matrix films. These allow the incorporation of a large number of materials (such as polymers, polysaccharides, etc.) into multilayer films and is now widely used in fields such as optics, energy, and biomedicine[6]. Yeom et al.[7] successfully prepared enamel-like columnar nanocomposites (ZnO/LBL)n in vitro by hydrothermally generating ZnO nanowires on silicon substrates, and then depositing polyallylamine (PAAm) and polyacrylic acid (PAA) around them layer by layer, where n represents the number of cycles of nanowire growth combined with layer-by-layer deposition of the PAAm/PAA matrix. Although the preparation process is in the reverse order of occurrence of enamel, it reproduces the main structural features of tooth enamel more accurately. And the relatively high organic phase content not only does no harm to the hardness, but endowed it with viscoelastic figures of merit (VFOM) similar to or higher than that of natural tooth enamel, these will reduce the low-impact damage and increase the durability. In addition, graphene oxide (GO) can also be introduced into the LBL matrix membrane, which gives it unique antibacterial adhesion properties without impairing its mechanical properties as well, especially the adhesion of Streptococcus mutans[8].

2.2 Laser sintering

Anastasiou et al.[9] are the first to demonstrate that calcium phosphate minerals can be directly laser sintered and attached to enamel without any significant thermal damage to the underlying tissue, due to the use of femto-second laser pulses and the doping of light-absorbing materials in calcium phosphate minerals. A layer of Fe^{3+}-containing fluoroapatite powder is coated on the surface of acid-etched bovine tooth enamel and irradiated with a femto-second pulsed laser, among which Fe^{3+} absorbs light and transmits heat to adjacent mineral phases, triggering selective sintering and densification of surrounding calcium phosphate crystals, thus forming a dense sintered layer of about 20 μm thickness and bonding with the surface of natural tooth enamel. It was shown that the mechanical properties of the layer are superior to existing restorative materials. However, we need more researches to optimize the properties of the sintered layer, such as hardness, stiffness, and aesthetics, for clinical translation.

2.3 Rotary evaporation

Using silk fibroin (SF) as a template, Wang et al.[10] successfully regenerated enamel-like HAp crystals on different substrates such as enamel and titanium by rotary evaporation. This material is easy to operate and cost-effective because the thickness of the regenerated crystal increases by 50 μm/day without affecting the structure of crystals, 25-50 times faster than that without rotary evaporation. Meanwhile, the growth thickness can be controlled artificially. The crystal size, morphology, arrangement and mechanical properties of the new HAp layers are highly similar to those of natural tooth enamel, but its mechanical properties still remain unsatisfactory.

2.4 New biomimetic ceramic-polymer composites

In recent decades, organic-inorganic composite materials have been widely applied in oral prosthodontics. New biomimetic ceramic-polymer composite materials consist of a ceramic skeleton and an infiltrating resin, which are closer to the natural tooth enamel, now including two main preparation processes.

2.4.1 Polymer Infiltrated Ceramic Network (PICN) Composites

Traditional dental CAD&CAM materials include metal, ceramic or resin-based materials. Because of their different mechanical properties from enamel, various problems often occur after restoration due to stress concentration. The ideal restorative material needs to mimic the mechanical properties of human tooth enamel, including hardness and elastic modulus, so the concept of PICN has been proposed[11-13]. This structure is usually prepared by infiltrating resin monomers into the continuous pores of pre-sintered ceramics, followed by polymerization, which can be prepared down to the nanometer scale at present. The SiO_2-PMMA [poly (methyl methacrylate)] composite prepared by Hiroshi Ikeda et al.[14] has hardness compatibility with human enamel, but its elastic modulus is closer to that of dentin. Kawajiri[15] et al. changed the PICN composites by mechanical properties especially elastic modulus, closer to those of enamel, and had good bonding properties with resins. Despite the above advantages, PICN materials still experience mechanical failure during use[16]. The reason behind may be that the microstructure of almost all

CAD&CAM materials is isotropic, which contradicts the anisotropic microstructure of natural tooth tissue[17].

2.4.2 Biomimetic nacre-like ceramic-polymer composites

The nacre in the inner surface of seashells is composed of 95vol% aragonite calcium carbonate plates and 5vol% protein sheets, in which the plates are 0.5 μm thick, similar to bricks, and the sheets are 20-30 nm thick, similar to mortar. It is the "brick and mortar" structure that creates a strong and tough material that provides inspiration for enamel biomimetic restorations. Tan et al.[18] used ceramic suspension as raw material to form interconnected "mineral bricks" through bidirectional freezing casting and uniaxial pressing. After surface treatment, resin monomers infiltrated into gaps between bricks and solidified to obtain a new type of bionic ceramic-polymer material, whose mechanical properties are proved closer to enamel than any of ceramic or resin materials, reducing the wear of antagonistic teeth. In addition, the nacre-like ceramic-polymer composites show not only extraordinary fracture toughness, but capacity of energy dissipation under cyclic loading and good machinability. On this basis, Algharaibeh et al.[17] arranged the "mineral bricks" into an anisotropic microstructure, which enhanced both crack growth resistance, flexural strength, elastic modulus and hardness. It is also shown that the 70% ceramic volume fraction is the best combination of strength, elasticity, hardness and toughness. The nacre-like material presents a brilliant future in CAD/CAM repair.

2.5 Multiscale engineered artifical tooth enamel

In order to mimic the hierarchical structure more precisely, Zhao et al.[19] put forward an ingenious design of the microstructure of the enamel-like material. Amorphous intergranular phase (AIP)-coated hydroxyapatite nanowires (HAp@A-ZrO2) were intertwined with polyvinyl alcohol (PVA) to construct a novel enamel-like material from nanoscale to macroscale. The existence of AIP, the strong chemical interfacial bonding force between AIP and HAp nanowires through the coordination of Zr4+ and OH-, PO43- and the confinement of PVA are the reasons for the combination of perfect but seemingly contradictory mechanical properties, such as the high stiffness, strength, hardness, viscoelasticity, toughness, etc., superior to previously reported enamel-like composites and even better than the enamel itself. Besides, the material also overcomes the tricky problem of insufficient thickness of remineralization layer, and can be mass-produced. Therefore, it is most likely the multiscale engineered material become the next-generation tooth enamel restoration material.

3. In-situ remineralization of tooth enamel

The ex-situ mineralization of tooth enamel is realized by the above-mentioned enamel biomimetic materials, although the mechanical properties of tooth enamel can be well simulated, and the ideal thickness can be achieved by design. However, like traditional filling and repairing materials, its function must rely on a good surface active system and bonding system. Therefore, there are a lot of studies by simulating the formation process of enamel,and then directly regenerate enamel-like HAp crystals on the surface of damaged enamel.

3.1 Inorganic strategies

Fluoride[20] is a gold standard for inhibiting demineralization of enamel, and the application of fluoride as a remineralization strategy has been widely used in clinical practice and achieved certain results. However, the restoration of enamel by F- is limited in children under 6 years old, due to the concern that they may swallow the paste by mistake.

3.1.1 Biomimetic hydroxyapatite

The initiation of dental caries is often related to the dissolution and destruction of enamel HAp crystals, the most ideal way is to directly supplement hydroxyapatite. Therefore, a strategy based on fluorine-free replenishment of hydroxyapatite was proposed. Using toothpaste as a carrier and using exogenous zinc carbonate hydroxyapatite nanoparticle formulation[21, 22] regularly can trigger the remineralization reaction and successfully form the deposition of hydroxyapatite nanocrystal coating on the enamel surface. It is not based on physical and chemical changes, but on the coating effect of nano-hydroxyapatite particles on damaged enamel, so as to realize the epitaxial growth of residual enamel rods and smooth the rough enamel surface. Although the newly formed hydroxyapatite continues with the natural enamel, the binding force is far from enough.

3.1.2 Polyphosphate particles

Supplementation of mineral raw materials such as phosphate and calcium can inhibit demineralization and promote remineralization. The donor of phosphate in vivo is polyphosphate (polyP) mostly produced by platelet degradation, which consists of 40-100 phosphate units connected by high-energy phosphoric anhydride bonds. Although polyP is not a constituent of HAp crystals, the studies have shown that polyP released from platelets contributes to the growth of HAp crystals, and adding amorphous calcium polyphosphate microparticles to toothpaste become a potential remineralization method[23, 24]. A protective layer of Ca-polyP deposits will be firstly formed on the surface of damaged tooth enamel, and then polyP will be hydrolyzed by alkaline phosphatase (ALP) rich in saliva, producing a large amount of orthophosphate and energy, also lowing the local ambient pH at the same time. Under the effect of activated carbonic anhydrase, the deposition of calcium phosphorus accelerates, and the phase transition of ACP to HAp crystals expedites[3]. Studies have shown that the Ca/P ratio in enamel deposits is significantly higher than that in toothpaste, indicating that the particles can fuse with enamel and promote the remineralization of demineralized enamel surfaces. Although the transformation of ACP to HAp takes a longer time, compared with direct supplementation of HAp, the regenerated layer is more tightly bound to the enamel surface, able to resistant to short ultrasound by a high-power (320W) ultrasound machine[4]. Even more attractive, the layer can seal dentinal tubules, repair dentin and enamel at the same time, which reveal a brilliant future for polyphosphate particles as an in-situ remineralization material.

3.1.3 Amorphous calcium phosphate(ACP)

Amorphous calcium phosphate may be a precursor of hydroxyapatite crystals, which is quite unstable when it exists alone and it is easy to aggregate and nucleate causing remineralization failure. Studies have shown that casein phosphopeptides (CPP)[25, 26] and some non-collagenous protein analogues (NCPs), such as polyacrylic acid (PAA)[27] and poly(amidoamine) (PAMAM)[28], can stabilize ACP nanoparticles. They use liquids, pastes, gels, etc. as carriers to form an immediate protective layer on the surface of damaged tooth enamel, and then gradually transform into a layer of hydroxyapatite crystals through the principle of dissolution-precipitation or solid-solid phase transformation. However, the remineralized layer is not the same as the natural tooth enamel, and the reason may be that ACP particles are adsorbed on the surface of HAp, causing a discontinuous particle-crystal interface, leading to the polycrystalline growth of HAp[29]. Shao et al.[30] proposed an improved method to construct a biomineralization frontier on the surface of damaged enamel through triethylamine (TEA)-stabilized calcium phosphate ion clusters (CPICs)—residual hydroxyapatite crystals are completely wrapped by amorphous precursors ACP to form a structurally continuous HAp-ACP interface and realize the perfect epitaxial growth. The hierarchical structure of tooth enamel was successfully replicated in 48 hours, and the thickness was up to 2.7μm. It is surprisingly completely consistent with the natural tooth enamel, and integrated with the remaining enamel without any gap. These realize the complete regeneration of enamel, which can be regarded as a permanent restoration. The use of TEA, a volatile organic compound, can stabilize ACP before use, and be completely volatilized without residue at room temperature, which ensures the safety of use and the mechanical strength of the mineral phase. Regarding the thickness of the repair layer, it has shown that the ideal thickness can be achieved by cyclic processing.

3.2 Strategies based on functions of organic matrix

Amelogenesis is a biomineralization process mediated by enamel matrix proteins (EMPs), which are composed of 90%-95% amelogenin and 5%-10% non-ammelogenin, synergistically controlling the initiation, organization, and orientation of enamel crystal mineralization, and are gradually degraded and cleared by proteases during enamel maturation.

3.3 Peptide strategies based on secondary structure of amelogenin

Amelogenin is the main component of organic matrix in the process of enamel biomineralization. When amelogenin is secreted from ameloblasts, it can self-assemble into nanospheres with a diameter of 10–20nm after proteolytic cleavage, playing a key role in the mineralization and organization of tooth enamel due to highly conserved N-terminal domains and C-terminal domains of amelogenin (N-Ame and C-Ame). N-Ame folds through β-sheet stacking and promotes the self-assembly of amelogenin into functional amyloid aggregates[31], serving as a template to guide the growth of HAp crystals, while C-Ame mediates the oriented growth and parallel alignment of HAp crystals. With the aggregation of ACP nanoparticles and the further phase transition of ACP to HAp, a sophisticated hierarchical structure is formed. Synthesis of full-length amelogenin can guide the orderly growth of enamel-like HAp crystals in bundles, but it demands a lot. and the protein is prone to contamination, denaturation, and difficulty in storage. Therefore, it is crucial to completely mimic N-Ame and C-Ame in a flexible and low-cost manner without using the full sequence of amelogenin for the successful application of amelogenin-inspired biomimetic strategies.

A leucine-rich amelogenin peptide (LRAP), composed of 56 non-phosphorylated amino acids, retaining only N-Ame and C-Ame splicing products, is shown to have a regulatory effect similar to full-length amelogenin. Kwak et al.[32] combined non-phosphorylated LRAP and CaP formation inhibitor PPi for biomimetic mineralization of enamel by exposing acid-etched enamel to a PPi-stabilized supersaturated CaP solution containing 0.04 mg/ml LRAP, seamless epitaxial growth of tooth enamel was successfully achieved. When PPi came into contact with HAp crystals, its original inhibitory effect on mineralization was reversed, and enamel regeneration was initiated preferentially on the acid-etched enamel surface. While LRAP preferentially interacted with the enamel crystal plane parallel to the c-axis and prevented the crystal from forming on the ab plane growth to promote directional elongation of the crystal. To simulate the physical and chemical environment of enamel formation, Fang et al.[33] took the interaction of non-amelogenin and amelogenin into the consideration of the construction of the EMPs model. First, an N-terminal modified LRAP (mLRAP) and a non-amelogenin analog (NAA) were artificially synthesized, then their interaction was realized through specific-designed domains. Due to the strong Ca2+ binding ability of NAA, when NAA and mLRAP assemble together, the aspect ratio of CaP crystals increases significantly, the co-assembled structure increases the stability of transient ACP, and the glycosylation and phosphorylation sites present in NAA can interact with HAp, which promote crystal nucleation. Although the hierarchical structure of the regenerated HAp crystals differs from that of natural tooth enamel, it is also meaningful to try to make different models for biomimetic mineralization.

An amelogenin-mimicking product PTL/C-AMG matrix film based on mimicking N-Ame and C-Ame with specific domains, with good biocompatibility, also successfully achieved epitaxial growth of HAp crystals[34]. The phasetransited lysozyme (PTL), a biomimetic of N-Ame, which is formed by lysozyme folding into amyloid aggregates in vitro based on protein phase transitions. As an interfacial anchor, it can form a layer of strong adhesive on various substrate surfaces, free from the limitation of residual HA crystals required by traditional strategies. The synthetic peptide C-AMG with the C-terminal functional domain of amelogenin added to this coating promoted the directional alignment of ACP and transformed it into ordered enamel-like HAp crystals. The interface between the regenerated HAp crystals and the remaining enamel has good bonding stability, and the hardness reaches or even exceeds that of natural tooth enamel. The only disadvantage is that its regeneration efficiency is low, and the thickness can only be limited to the micron level.

3.3.1 Hydrogel systems based on natural polysaccharide polymers

The hydrogel system can be prepared by physical or chemical cross-linking. Compared with the solution system, the physical and chemical properties of the gel-like environment are closer to the enamel matrix in the secretory stage and maturation stage, and it is easier to operate clinically and stay on the tooth surface for a long time with effective concentration. Natural polysaccharide polymers with good biosafety, compatibility and degradability are often used in the biomimetic mineralization of tooth enamel in the form of hydrogels. It can both fix the nanoscale building units of HAp crystals well, and serve as non-collagen biomaterials templates to guide the growth of HAp crystals.

Chitosan (CS) is a natural cationic polysaccharide that can form stable hydrogel networks through chemical cross-linking, whose chemical structure is similar to the main component of extracellular matrix, with antibacterial activity and strong adhesion to both soft and hard surfaces. Ruan et al.[35] established the CS-AMEL-CaP hydrogel system and introduced amelogenin into the chitosan hydrogel system. The new HA layer grows directly on the original enamel and fully adheres to the acid-etched enamel, which can reduce secondary caries caused by poor interfaces. And the hardness and elastic modulus were increased by 9 times and 4 times, respectively compared with acid-etched enamel. Under natural conditions, amelogenin cannot function without a protease, matrix metalloproteinase-20 (MMP-20), which plays a role in hydrolyzing full-length amelogenin. On the one hand, it can ensure the good growth law of HAp crystals, and on the other hand, it can prevent amelogenin from being trapped in the growing HAp crystals and prohibit protein occlusions in the enamel crystals. Ruan et al.[36] added MMP-20 to the above hydrogel system, which improved the morphology and structure of the regenerated HAp crystals, and the elastic modulus and hardness were significantly increased by 1.8 times and 2.4 times, respectively.

Agarose (A), a family member of natural polysaccharides extracted from red algae, can promote the growth of HAp crystals by forming a thermal reversible gel. A hydrogel system of chitosan (CS) and agarose (A) simultaneously working to achieve biomimetic mineralization of dental enamel was first reported recently. The CS-A hydrogel was remineralized in artificial saliva for 7 days, resulting in a calcium-to-phosphorus ratio (1.64) similar to that of natural tooth enamel, and the microhardness recovery of the enamel-like layer was 77.4%[37].

3.4 Other strategies

Mineralization repair of tooth enamel can also be achieved by rationally-designed small peptides. The self-assembled peptide P11-4 monomers and its β-sheet aggregates can diffuse below the early enamel carious lesions, triggering self-assembly through specific environments, and forming a three-dimensional network with surface features of the enamel matrix. These trigger the nucleation of HAp crystals and growth, and remineralize the surface of early carious lesions. CT showed that the remineralized specimen in the presence of P11-4 increased the mineralization rate by 68% within two weeks, which was statistically significant compared to the placebo control group[38].

Chang et al.[39] achieved efficient biomimetic mineralization for the first time using low-complexity protein segments (LCPSs). LCPSs containing phosphate or phosphonate groups serve as a novel biocompatible peptide scaffold with high water solubility, structural flexibility, and ability to form network morphologies due to weak multivalent interactions. Integrating biomineralization-related components such as matrix proteins, regulatory additives and osteogenic stimulators, LCPCs synergistically promoted the epitaxial growth of HAp crystals on the surface of acid-etched enamel, Besides the remineralized layer can reduce the adhesion of Streptococcus mutans and prevent secondary caries.

4. Summary

Natural tooth enamel has a complex and well-organized structure, so successfully regenerating tooth enamel is an extremely challenging task. For the ex-situ remineralization of enamel, namely the preparation of enamel-like materials, more attention should be paid to the design of the microstructure to achieve a finer bionic from the nanometer to the macroscopic scale. At the same time, the bonding technology should be developed accordingly to form a well-sealed interface between the restoration material and the remaining enamel surface.

For the in-situ remineralization of tooth enamel, it is necessary to fully explore the self-assembly principle of biological macromolecules such as proteins, polypeptides, DNA[40], so as to provide a more accurate template for the nucleation and growth of HAp crystals. And it is still necessary to continue to study the process and mechanism of enamel occurrence in order to realize the precise construction of the enamel mineralization microenvironment in vitro.

In terms of clinical translation, enamel-like materials are faced with the difficulty of mass production while insufficient thickness for in-situ remineralization of enamel. The diseases that lead to enamel defects, including caries, trauma, abrasion, etc., often invade enamel and dentin at the same time, but the mechanical properties of dentin and enamel differs. Thus, how to achieve the simultaneous repair of both in one method requires further research. Materials specialists and physicians working side by side in a complementary manner will certainly shed further light on the ideal remineralization materials and on their translation from the bench to the bedside.

References

1. Elsharkawy, S. and A. Mata, Hierarchical Biomineralization: from Nature's Designs to Synthetic Materials for Regenerative Medicine and Dentistry. Adv Healthc Mater, 2018. 7(18): p. e1800178.
2. Lacruz, R.S., et al., DENTAL ENAMEL FORMATION AND IMPLICATIONS FOR ORAL HEALTH AND DISEASE. Physiol Rev, 2017. 97(3): p. 939-993.
3. Ackermann, M., et al., Biomimetic transformation of polyphosphate microparticles during restoration of damaged teeth. Dent Mater, 2019. 35(2): p. 244-256.
4. Muller, W., et al., A Novel Biomimetic Approach to Repair Enamel Cracks/Carious Damages and to Reseal Dentinal Tubules by Amorphous Polyphosphate. Polymers (Basel), 2017. 9(4).
5. Selwitz, R.H., A.I. Ismail and N.B. Pitts, Dental caries. Lancet, 2007. 369(9555): p. 51-9.
6. Borges, J. and J.F. Mano, Molecular Interactions Driving the Layer-by-Layer Assembly of Multilayers. Chemical reviews, 2014. 114(18): p. 8883-8942.
7. Yeom, B., et al., Abiotic tooth enamel. Nature, 2017. 543(7643): p. 95-98.
8. Wong, H.M., Y.Y. Zhang and Q.L. Li, An enamel-inspired bioactive material with multiscale structure and antibacterial adhesion property. Bioactive Materials, 2022. 7: p. 491-503.
9. Anastasiou, A.D., et al., Exogenous mineralization of hard tissues using photo-absorptive minerals and femto-second lasers; the case of dental enamel. Acta Biomaterialia, 2018. 71: p. 86-95.
10. Wang, S., et al., Rapid regeneration of enamel-like-oriented inorganic crystals by using rotary evaporation. Mater Sci Eng C Mater Biol Appl, 2020. 115: p. 111141.
11. He, L.H. and M. Swain, A novel polymer infiltrated ceramic dental material. Dent Mater, 2011. 27(6): p. 527-34.
12. Della, B.A., P.H. Corazza and Y. Zhang, Characterization of a polymer-infiltrated ceramic-network material. Dent Mater, 2014. 30(5): p. 564-9.
13. El, Z.H., et al., Polymer infiltrated ceramic network structures for resistance to fatigue fracture and wear. Dent Mater, 2016. 32(11): p. 1352-1361.
14. Ikeda, H., Y. Nagamatsu and H. Shimizu, Preparation of silica–poly(methyl methacrylate) composite with a nanoscale dual-network structure and hardness comparable to human enamel. Dental Materials, 2019. 35(6): p. 893-899.
15. Kawajiri, Y., et al., PICN Nanocomposite as Dental CAD/CAM Block Comparable to Human Tooth in Terms of Hardness and Flexural Modulus. Materials, 2021. 14(5): p. 1182.
16. Santos, F., et al., Comparative study of the wear of the pair human teeth/Vita Enamic(R) vs commonly used dental ceramics through chewing simulation. J Mech Behav Biomed Mater, 2018. 88: p. 251-260.
17. Algharaibeh, S., et al., Fabrication and mechanical properties of biomimetic nacre-like ceramic/polymer composites for chairside CAD/CAM dental restorations. Dental Materials, 2022. 38(1): p. 121-132.
18. Tan, G., et al., Nature - Inspired Nacre - Like Composites Combining Human Tooth - Matching Elasticity and Hardness with Exceptional Damage Tolerance. Advanced Materials, 2019. 31(52): p. 1904603.
19. Zhao, H., et al., Multiscale engineered artificial tooth enamel. Science, 2022. 375(6580): p. 551-556.
20. Degli, E.L., et al., Characterization of a Toothpaste Containing Bioactive Hydroxyapatites and In Vitro Evaluation of Its Efficacy to Remineralize Enamel and to Occlude Dentinal Tubules. Materials (Basel), 2020. 13(13).
21. Bossu, M., et al., Enamel remineralization and repair results of Biomimetic Hydroxyapatite toothpaste on deciduous teeth: an effective option to fluoride toothpaste. J Nanobiotechnology, 2019. 17(1): p. 17.
22. Lelli, M., et al., Remineralization and repair of enamel surface by biomimetic Zn-carbonate hydroxyapatite containing toothpaste: a comparative in vivo study. Front Physiol, 2014. 5: p. 333.
23. Muller, W.E., et al., A biomimetic approach to ameliorate dental hypersensitivity by amorphous polyphosphate microparticles. Dent Mater, 2016. 32(6): p. 775-83.
24. Muller, W., et al., Bifunctional dentifrice: Amorphous polyphosphate a regeneratively active sealant with potent anti-Streptococcus mutans activity. Dent Mater, 2017. 33(7): p. 753-764.

25. Zhou, C., et al., Casein phosphopeptide-amorphous calcium phosphate remineralization of primary teeth early enamel lesions. J Dent, 2014. 42(1): p. 21-9.
26. Huq, N.L., et al., The Interactions of CPP-ACP with Saliva. Int J Mol Sci, 2016. 17(6).
27. Wang, Z., et al., A novel fluorescent adhesive-assisted biomimetic mineralization. Nanoscale, 2018. 10(40): p. 18980-18987.
28. Gao, Y., et al., Enamel remineralization via poly(amido amine) and adhesive resin containing calcium phosphate nanoparticles. J Dent, 2020. 92: p. 103262.
29. De Yoreo, J.J., et al., CRYSTAL GROWTH. Crystallization by particle attachment in synthetic, biogenic, and geologic environments. Science, 2015. 349(6247): p. aaa6760.
30. Shao, C., et al., Repair of tooth enamel by a biomimetic mineralization frontier ensuring epitaxial growth. Sci Adv, 2019. 5(8): p. eaaw9569.
31. Zhang, J., et al., Hydroxyapatite Formation Coexists with Amyloid-like Self-Assembly of Human Amelogenin. Int J Mol Sci, 2020. 21(8).
32. Kwak, S.Y., et al., Biomimetic Enamel Regeneration Mediated by Leucine-Rich Amelogenin Peptide. J Dent Res, 2017. 96(5): p. 524-530.
33. Fang, Z., et al., Enamel-like tissue regeneration by using biomimetic enamel matrix proteins. Int J Biol Macromol, 2021. 183: p. 2131-2141.
34. Wang, D., et al., Controlling Enamel Remineralization by Amyloid-Like Amelogenin Mimics. Adv Mater, 2020. 32(31): p. e2002080.
35. Ruan, Q., et al., An amelogenin-chitosan matrix promotes assembly of an enamel-like layer with a dense interface. Acta Biomater, 2013. 9(7): p. 7289-97.
36. Prajapati, S., et al., The Presence of MMP-20 Reinforces Biomimetic Enamel Regrowth. J Dent Res, 2018. 97(1): p. 84-90.
37. Musat, V., et al., A Chitosan-Agarose Polysaccharide-Based Hydrogel for Biomimetic Remineralization of Dental Enamel. Biomolecules, 2021. 11(8).
38. Kind, L., et al., Biomimetic Remineralization of Carious Lesions by Self-Assembling Peptide. J Dent Res, 2017. 96(7): p. 790-797.
39. Chang, R., et al., Phosphorylated and Phosphonated Low-Complexity Protein Segments for Biomimetic Mineralization and Repair of Tooth Enamel. Adv Sci (Weinh), 2022. 9(6): p. e2103829.
40. Kim, F., et al., Functionalized DNA nanostructures as scaffolds for guided mineralization. Chem Sci, 2019. 10(45): p. 10537-10542.

AUTHOR INDEX

AUTHOR INDEX